METHODS IN MOLECULAR BIOLOGY™

Series Editor
John M. Walker
School of Life Sciences
University of Hertfordshire
Hatfield, Hertfordshire, AL10 9AB, UK

For other titles published in this series, go to
www.springer.com/series/7651

RNA

Methods and Protocols

Edited by

Henrik Nielsen

University of Copenhagen, Denmark

Humana Press

Editor
Henrik Nielsen, Ph.D.
Department of Cellular and Molecular
Medicine
The Panum Institute
University of Copenhagen
Copenhagen DK-2200N, Denmark
hamra@sund.ku.dk

Additional material to this book can be downloaded from http://extras.springer.com.

ISSN 1064-3745 e-ISSN 1940-6029
ISBN 978-1-4939-5643-2 ISBN 978-1-59745-248-9 (eBook)
DOI 10.1007/978-1-59745-248-9
Springer New York Dordrecht Heidelberg London

Printed on acid-free paper

Humana Press is part of Springer Science+Business Media (www.springer.com)

Preface

This is a book about the procedures and methods that are used to describe the structure of the messenger RNAs and non-coding RNAs that are transcribed as the immediate gene products by RNA polymerase II in mammalian cells. It is intended for researchers working on a biological problem that involves characterization of the expression of "gene X." The book is focused on the structure of the RNA products of gene X and mapping of the proteins associated with these RNAs. The book is mainly intended for the non-specialist in RNA biology.

Recent insight into the transcripts generated from the mammalian genome (i.e., the transcriptome) has revealed that transcription is a far more complex phenomenon than previously thought. In a sense, the present situation is comparable to the mid-1970s when the exon–intron organization of genes was discovered. Prior to that, it was generally believed that the mature mRNA was co-linear with the gene from which it was transcribed. This view was challenged by the extraordinary size of the genomes and the puzzling observation of very long nuclear RNA molecules that were capped and polyadenylated similar to the mature mRNAs. The introduction of recombinant DNA technology allowed for a direct comparison of the genes and their RNA products and led to the surprising conclusion that almost all of the genes in vertebrates are a mosaic of mRNA encoding exons interrupted by, on the average, six introns that are subsequently removed by RNA splicing. These intronic sequences comprise 30% of the mammalian genome. In hindsight, it is interesting to note that such a manifest phenomenon could be overlooked for many years.

It now appears that we are confronted with a similar dramatic change in the view of our genes and their products. New methods designed to give a complete and unbiased view on transcription have shown that the transcriptional landscape of the genome is far more complex than previously believed. Most of the genome is transcribed, and, within a given locus, the typical picture is that of multiple, overlapping transcripts generated from both strands of the DNA. Furthermore, characterization of the mature transcripts shows that half of the capped, spliced, and polyadenylated transcripts do not encode a protein. This class of non-coding RNAs are essentially in search of a function, but their characterization is included as part of the scope of this book because their biosynthesis is parallel to that of the mRNAs and because they may belong to a parallel regulatory universe to that of their protein encoding cousins, the mRNAs.

Organization of the Volume

The volume is organized into two parts. The first part deals with preparation and analysis of RNA. The second part is about the proteins and miRNAs that bind to RNA to regulate its function. The final chapter does not follow this outline. It deals with the problems of outsourcing experimental work for high-throughput services. Each part has a conceptual chapter that introduces the new concepts in the field. Bioinformatics and experimental chapters are mixed to emphasize that bioinformatics should become an integral part of the experimental work, although this may be a bit optimistic at present. The volume contains both very basic and advanced chapters. The reason for the former is to have the basics at hand while embarking on the new and more advanced techniques.

Part I: RNA Methods

The first chapter of the volume introduces the new view on the transcriptional landscape characterized by multiple, overlapping transcripts from both strands of DNA. The complexity of transcripts derived form virtually any genomic region suggest that the operational unit in description of gene expression should be the transcript rather than the DNA from which it was transcribed.

Chapter 2 is about the basics of working with RNA. The chemical nature of RNA is briefly introduced followed by a description of how to create a working environment for RNA work in particular, with respect to maintaining the integrity of the RNA. This is followed by introductions to all of the basic procedures, including extraction, precipitation, quantitation, and storage. Recommendations for preparation of standard reagents and short protocols are also included. Another basic procedure, synthesis of RNA by in vitro transcription is described in **Chapter 3**. Beckert and Masquida provide the protocols for template preparation, synthesis of RNA, and purification of transcripts. They also discuss the synthesis of transcripts that are modified at the 5′ end or internally for specialized purposes as well as the use of ribozymes to create populations of transcripts with homogenous ends for NMR or X-ray crystallography. Continuing with a classic and very basic technique, Josefsen and Nielsen in **Chapter 4** present variations of northern blotting and hybridization analysis. Recent developments have made northern blotting analysis almost as sensitive as nuclease protection analysis and to many it remains the most convincing method for analysis of the size and quantity of an RNA transcript.

The present volume is focused on RNA polymerase II transcripts that with few exceptions are polyadenylated. These RNAs constitute 1–4% of cellular RNA and have to be purified from other RNAs in many protocols. In **Chapter 5**, Jacobsen and colleagues describe a variation of the classical oligo(dT) chromatography for purification of poly(A)$^+$ RNA using Locked Nucleic Acid (LNA) oligo(T) capture of the poly(A)$^+$. This is a very efficient method, and the chapter also serves to introduce LNA which has proven to be a particularly useful tool in many hybridization-based applications in RNA biology, including in situ hybridization and microarray analysis. The poly(A) tail of mRNA has several functions including stability and translational control which both depend on the length of the tail. Unfortunately, the tail length is quite difficult to assess. Meijer and de Moor provide a simple method for fractionation of mRNA according to tail length in **Chapter 6**. The method is based on differential elution from oligo(dT) and can be used for preparation of samples for microarray analysis. In **Chapter 7** by Yeku and Frohman, both ends of the RNA molecule are addressed. The chapter presents improvements to the Rapid Amplification of cDNA ends (RACE) technique. The method provides easy access to full-length cDNA which is of particular significance because an important aspect of diversity in gene expression involves the use of alternative 5′ and 3′ ends.

The sequencing of the human genome was a milestone in biology, and the public access to genome data organized in genome browsers is a beautiful testimony to the openness of scientific endeavors. In **Chapter 8**, Torarinsson provides a primer to two such browsers (UCSC and Ensembl) with short exercises. The following chapter, **Chapter 9**, by George and Tenenbaum, is aimed at the much more experienced researcher. Here, a comprehensive list of web-based resources for the identification and study of RNA structural motifs is presented. The list comprises databases as well as analytical tools, each with a link, a brief description and a primary literature reference. These motifs are of particular importance for understanding protein binding and regulatory functions associated with the RNA molecules. RNA motifs are also amenable to experimental analysis of their structure, and two chapters in the electronic supplementary materials present such methods. First, in ESM1, Regulski and Breaker describe the use of in-line probing in the characterization of riboswitches in the bacterial world. Riboswitches are found in mammalian systems, but the technique is applicable to all RNA structures. This chapter was originally published as Chapter 4 in *Methods in Molecular Biology*, Vol. 419, *Post-Transcriptional Gene Regulation*, edited by Jeffrey Wilusz. Then, in ESM2, Wakeman and Winkler, in addition to providing a protocol on in-line probing, present structure probing of RNA by SHAPE (Selective 2′-Hydroxyl Acylation Analyzed by Primer Extension). This is a very useful technique that has been used in structure probing of large molecules such as the HIV-1 genome. SHAPE can also be used to study the folding of RNA molecules provided that a fast-reacting acylation reagent is used. This chapter was originally published as Chapter 4 in *Methods in Molecular Biology*, Vol. 540, *Riboswitches: Methods and Protocols*, edited by Alexander Serganov.

The next two chapters deal with the most powerful of post-transcriptional modification processes: alternative splicing. This process is a major contributor to the diversity of gene products derived from the relatively few genes in the human genome. Furthermore, an increasing number of errors in gene expression leading to diseases are found to involve splicing errors. In **Chapter 10**, Zhang and Stamm provide an overview along with a description of bioinformatics tools to predict the influence of a mutation on alternative pre-mRNA splicing and the experimental testing of these predictions. Then, in **Chapter 11**, Lützelsberger and Kjems show how the classical S1-nuclease protection method can be used to quantitate alternatively spliced mRNA isoforms. The method requires no specialized equipment and allows detection of as few as a couple of hundred femtograms of a specific RNA.

RNA interference (RNAi) is the method of choice for inactivation of cellular RNA molecules. In **Chapter 12**, Sioud provides a broad review of the use of RNAi as a research tool and in therapy. After an introduction to the RNAi pathway, the rules for design of siRNA are presented. This is followed by a thorough discussion of the detection of exogenous RNA by the immune system. Particular attention is given to separation of the effects of gene silencing from unwanted effects that have led to many erroneous conclusions in the literature. **Chapter 13** by Henriksen and Einvik describes one of the ways of introducing siRNA into cells. The procedure involves construction of vectors expressing short-hairpin RNA (shRNA) that are processed into siRNA by the cellular RNAi machinery. Detailed descriptions of target site selection, shRNA construction, shRNA transfection, and target knockdown validation are provided. The most obvious method for validation of target knockdown is quantitative RT-PCR, also known as real-time PCR. Josefsen and

Lee (**Chapter 14**) describe the application of a very general method for quantitation of RNA in a sample. The chapter includes other general protocols, e.g., on RNA isolation and cDNA synthesis.

Northern blotting, nuclease protection, and qRT-PCR are used to analyze the steady-state level of RNA. Chromatin immunoprecipitation (ChIP) using RNA polymerase II antibodies is a technique that in combination with measurements of mRNA levels can be used to measure transcription rates as an alternative to the cumbersome nuclear run-on method. Nelson and colleagues have developed a fast version of ChIP outlined in **Chapter 15**. ChIP is a general method that can be used with antibodies raised against other components of chromatin to provide a detailed description of the chromatin state of individual genes.

Part II: RNP Methods

The second part opens with an introduction to the post-transcriptional operon by Tenenbaum and colleagues. The mRNAs, and probably also the non-coding RNAs, are associated with protein factors throughout their lifetime. Some remain stably bound to the RNA while others are exchanged. The proteins are involved in coupling the various steps in the processing of genetic information. Transcription factors influence the pattern of splicing, and splicing factors influence translation. Ultimately, the associated proteins dictate the cytoplasmic fate of the mRNAs. Thus, a description of the structure of mRNAs and non-coding RNAs is very incomplete without a description of their protein partners. The post-transcriptional operon is a set of monocistronic mRNAs encoding functionally related proteins that are co-regulated by a group of RNA-binding proteins. The model is used to describe data from an assortment of methods (e.g., RIP-Chip, CLIP-Chip, miRNA profiling, ribosome profiling) that globally address the functionality of mRNA. Thus, the conceptual **Chapter 16** is followed by **Chapter 17**, by Jain and colleagues from the Tenenbaum lab, describing RIP-Chip analysis in which an antibody directed toward an RNA-binding protein is used to pull-down a collection of mRNAs that are subsequently identified by microarray analysis.

A different approach to the same problem is taken by Jønsson and colleagues in **Chapter 18**. Here, a tag (FLAG-tag) is attached to the RNA-binding protein that is expressed at endogenous levels under tetracycline control. The tag is used as a handle for immunoprecipitation of RNP granules that are visualized by atomic force microscopy. Like in Chapter 17, the RNA can be recovered from the granules, and the RNA content is subjected to microarray or deep sequencing analysis. Further characterization of RNPs as well as the detailed characterization of binding of individual proteins to RNA frequently involves analysis by electrophoretic mobility shift assay (EMSA), a technique that is also known for the characterization of DNA-binding proteins. Gagnon and Maxwell have refined this technique for protein-RNA complexes and demonstrate its usefulness in **Chapter 19**.

The steady-state levels of mRNA and protein are poorly correlated for a large fraction of genes. Polysome profile analysis is a method that can be used to study the translation status of cells and to isolate and characterize mRNAs actively engaged in translation. In **Chapter 20**, Masek and colleagues introduce translational control and present methods for sucrose-gradient-based analysis of polysomes followed by extraction of RNA suitable for a wide-range of downstream applications, including microarray and qRT-PCR. miRNAs are mostly, but not exclusively, involved in translational repression. In the context of the post-transcriptional operon model, they can be considered formally equivalent to RNA-binding proteins. Many research projects involve miRNA profiling with the aim of identifying particular miRNAs that are up- or downregulated followed by a search for the targets of those identified miRNAs. Target-finding has proven to be one of the major challenges in bioinformatics. In **Chapter 21**, Lindow gives guidelines on how to use the existing tools for target-finding.

Many of the protocols described in this volume end with a sample for subsequent analysis by high-throughput technologies, such as deep sequencing or microarray analysis. In many research institutions, the options are to have this analysis done in a core facility or as a commercial service. **Chapter 22** provides some hints to the non-specialist with respect to choice of analytical tool and sample preparation for the outsourcing of experiments.

One of the surprises of the human genome project was the small number of genes (ca. 25,000) identified compared to that of, say, fruit flies (14,000) and nematodes (19,000). The new insights have challenged the concept of the gene and shown that a simple counting of the number of genes completely misses the point in understanding the complexity of an organism. The new view on the transcriptional landscape and the appreciation of the role that proteins play in the processing and interpretation of genetic information can account for many more products and much more sophisticated regulatory networks than the traditional DNA view. It is our hope that this volume will help researchers to reveal many new examples of this.

Finally, I would like to thank the authors for their contributions and for their patience during the preparation of this volume. Special thanks go to the editors at MiMB, who have been very supportive. One of the characteristics of the contributions to MiMB is the solidarity among scientists that is expressed in the willingness by the authors to share protocols and the very direct advice that is given in the extensive notes sections. It is my sincere hope that this volume lives up to the tradition.

Henrik Nielsen

Contents

Contributors

TIMOTHY E. BARONI • *Department of Biomedical Sciences, School of Public Health, Gen* NY* Sis Center for Excellence in Cancer Genomics, University at Albany-SUNY, Rensselaer, NY, USA*

BERTRAND BECKERT • *Department of Cellular and Molecular Medicine, The Panum Institute, University of Copenhagen, Copenhagen, Denmark*

KAROL BOMSZTYK • *Department of Pharmacology, School of Medicine, University of Washington, Seattle, WA, USA*

JOCHEN BUECHNER • *Department of Pediatrics, University Hospital of North-Norway, Tromsø, Norway*

SRIDAR V. CHITTUR • *Department of Biomedical Sciences, School of Public Health, Gen* NY* Sis Center for Excellence in Cancer Genomics, University at Albany-SUNY, Rensselaer, NY, USA*

JAN CHRISTIANSEN • *Department of Biology, University of Copenhagen, Copenhagen, Denmark*

CORNELIA H. DE MOOR • *RNA Biology Group, School of Pharmacy, Centre for Biomolecular Sciences, University of Nottingham, University Park, Nottingham, NG7 2RD, UK*

OLEG DENISENKO • *University of Washington Medicine at Lake Union, University of Washington, Seattle, WA, USA*

TIFFANY DEVINE • *Department of Biomedical Sciences, School of Public Health, Gen* NY* Sis Center for Excellence in Cancer Genomics, University at Albany-SUNY, Rensselaer, NY, USA*

CHRISTER EINVIK • *Department of Pediatrics, University Hospital of North-Norway, Tromsø, Norway*

JENS ERIKSEN • *Laboratory of Oncology, Herlev University Hospital, Herlev, Denmark*

TROND FLÆGSTAD • *Department of Pediatrics, University Hospital of North-Norway, Tromsø, Norway; Department of Pediatrics, Institute of Clinical Medicine, University of Tromsø, Tromsø, Norway*

MICHAEL A. FROHMAN • *Department of Pharmacology, Center for Developmental Genetics, Stony Brook University, Stony Brook, NY; Center for Molecular Medicine, Stony Brook University, Stony Brook, NY, USA*

KEITH T. GAGNON • *Department of Pharmacology, University of Texas Southwestern Medical Center, Dallas, TX, USA*

AJISH D. GEORGE • *Department of Biomedical Sciences, School of Public Health, Gen* NY* Sis Center for Excellence in Cancer Genomics, University at Albany-SUNY, Rensselaer, NY, USA*

JØRN REMI HENRIKSEN • *Department of Pediatrics, University Hospital of North-Norway, Tromsø, Norway*

NANA JACOBSEN • *Exiqon, Vedbaek, Denmark*

RITU JAIN • *Department of Biomedical Sciences, School of Public Health, Gen* NY* Sis Center for Excellence in Cancer Genomics, University at Albany-SUNY, Rensselaer, NY, USA*

LARS JØNSON • *Department of Clinical Biochemistry, Copenhagen University Hospital, Copenhagen, Denmark*
KNUD JOSEFSEN • *The Bartholin Institute, Copenhagen University Hospital, Copenhagen, Denmark*
JØRGEN KJEMS • *Department of Molecular Biology, University of Århus, Århus C, Denmark*
YING C. LEE • *Cellular and Metabolic Research Section, Biomedical Institute, University of Copenhagen, Copenhagen, Denmark*
MORTEN LINDOW • *Santaris Pharma A/S, Hørsholm, Denmark*
CECILIE LØKKE • *Department of Pediatrics, Institute of Clinical Medicine, University of Tromsø, Tromsø, Norway*
MARTIN LÜTZELBERGER • *Institute of Genetics, Technical University of Braunschweig, Braunschweig, Germany*
TOMÁŠ MAŠEK • *Department of Genetics and Microbiology, Charles University in Prague, Prague, Czech Republic*
BENOÎT MASQUIDA • *Architecture et Réactivité de l'ARN, Université de Strasbourg, CNRS, IBMC, Strasbourg, France*
E. STUART MAXWELL • *Department of Molecular and Structural Biochemistry, North Carolina State University, Raleigh, NC, USA*
HEDDA A. MEIJER • *Toxicology unit, Medical Research Council, Hodgkin Building, University of Leicester, Lancaster Road, Leicester LE, 9HN, UK*
JOEL NELSON • *Molecular and Cellular Biology Program, University of Washington, Seattle, WA, USA*
FINN CILIUS NIELSEN • *Department of Clinical Biochemistry, Copenhagen University Hospital, Copenhagen, Denmark*
HENRIK NIELSEN • *Department of Cellular and Molecular Medicine, The Panum Institute, University of Copenhagen, Copenhagen, Denmark*
PETER STEIN NIELSEN • *Exiqon, Vedbaek, Denmark*
LUIZ O. PENALVA • *Department of Cellular and Structural Biology, Children's Cancer Research Institute, San Antonio, TX, USA*
MARTIN POSPÍŠEK • *Department of Genetics and Microbiology, Faculty of Science, Charles University in Prague, Prague, Czech Republic*
MOULDY SIOUD • *Molecular Medicine Group, Department of Immunology, Institute for Cancer Research, The Norwegian Radium Hospital, Montebello, Oslo, Norway*
STEFAN STAMM • *Department of Molecular and Cellular Biochemistry, Biomedical Biological Sciences Research Building, College of Medicine, University of Kentucky, Lexington, KY, USA*
SCOTT A. TENENBAUM • *College of Nanoscale Science and Engineering, Nanoscale Constellation, University at Albany-SUNY, Rensselaer, NY, USA*
ELFAR TORARINSSON • *Division of Genetics and Bioinformatics, Department of Basic Animal and Veterinary Science, University of Copenhagen, Frederiksberg C, Denmark; Department of Natural Sciences, Faculty of Life Sciences, University of Copenhagen, Frederiksberg C, Denmark*
LEOŠ VALÁŠEK • *Laboratory of Regulation of Gene Expression, Institute of Microbiology, Academy of Sciences of the Czech Republic, Prague, Czech Republic*

Oladapo Yeku • *Department of Pharmacology, Center for Developmental Genetics, Stony Brook University, Stony Brook, NY, USA*
Zhaiyi Zhang • *Department of Molecular and Cellular Biochemistry, Biomedical Biological Sciences Research Building, College of Medicine, University of Kentucky, Lexington, KY, USA*

Part I

RNA Methods

Chapter 1

The Transcriptional Landscape

Henrik Nielsen

Abstract

The application of new and less biased methods to study the transcriptional output from genomes, such as tiling arrays and deep sequencing, has revealed that most of the genome is transcribed and that there is substantial overlap of transcripts derived from the two strands of DNA. In protein coding regions, the map of transcripts is very complex due to small transcripts from the flanking ends of the transcription unit, the use of multiple start and stop sites for the main transcript, production of multiple functional RNA molecules from the same primary transcript, and RNA molecules made by independent transcription from within the unit. In genomic regions separating those that encode proteins or highly abundant RNA molecules with known function, transcripts are generally of low abundance and short-lived. In most of these cases, it is unclear to what extent a function is related to transcription *per se* or to the RNA products.

Key words: Pervasive transcription, promoter, ncRNA, antisense.

1. Introduction

Genetic information is stored in DNA and is expressed through copying into RNA molecules by the action of RNA polymerase in a process known as transcription. The RNA molecules (transcripts) can either be the final product themselves (e.g., ribosomal RNA and transfer RNA) or be messenger RNA used to instruct the synthesis of protein molecules that in these cases are the ultimate products of gene expression. Together the RNA molecules transcribed from a genome are referred to as the transcriptome. In the traditional view of the transcriptional landscape, i.e., the map of transcripts derived from the genome, the transcripts are made from well-defined transcription units with discrete boundaries scattered over the genome. In some organisms (typical of

H. Nielsen (ed.), *RNA*, Methods in Molecular Biology 703,
DOI 10.1007/978-1-59745-248-9_1, © Springer Science+Business Media, LLC 2011

prokaryotes), the functional transcripts are co-linear with the genome. In other organisms (typical of eukaryotes), transcripts are made as larger precursors that are subsequently processed into smaller mature transcripts. The understanding of the complexity of the transcriptome and the view on the transcriptional landscape is rapidly changing these years. The central observation is that most, if not all, of the genome is transcribed such that multiple, overlapping transcripts derived from both strands of DNA are being produced throughout the genome. This realization is driven by application of new experimental methods and large-scale bioinformatics analysis of gene expression. The new view calls for a revision of central concepts in molecular biology, such as the gene, the promoter (driving transcription initiation), and the transcription unit. An introduction to the new discoveries is provided in the following.

In the traditional approach to studying gene expression, a complex RNA sample was fractionated by gel electrophoresis, blotted onto a membrane, and analyzed by hybridization analysis using labeled probes representing individual genes. This yielded relatively little information and was much improved with the introduction of array analysis that can be thought of as an "inversion" of the experimental strategy. Here, large collections of probes fixed to a membrane or a chip are hybridized with labeled, complex RNA. This type of array analysis is biased in the sense that the analysis is limited by the selection of probes on the membrane or chip. These would typically be chosen to represent already known transcripts. To overcome this, an unbiased way of interrogating the transcriptional complexity was introduced with the tiling array analysis in which overlapping oligonucleotides covering an entire region of the genome is bound on the chip. Another unbiased method that was introduced more recently is deep sequencing in which an RNA sample is copied into DNA molecules that are sequenced in parallel. New sequencing technologies allows for the analysis of millions of parallel sequencing reads. The breakthrough in understanding of the transcriptional complexity came first from application of tiling arrays to the mouse and human genome (1, 2). A rough calculation of the transcriptional complexity prior to this would be that 10–20% of the genome (one DNA strand only) was transcribed into discrete RNA molecules. This was based on an estimated 21–23,000 protein coding genes that require transcription of 1.2% of the genome for specifying the amino acids and considering that mRNAs initially are made as intron-containing precursors that are processed into mature mRNAs. However, the tiling array studies showed that >90% of both strands of the genome are transcribed into multiple, overlapping RNA molecules. Furthermore, this was based on analysis of RNA from selected cells and did not take into account the complexity of transcripts when all types of cells at all stages in development were considered. The majority of the transcriptional output is non-polyadenylated transcripts derived from what was previously considered intergenic regions. The nuclear complexity exceeds that of the cytosolic compartment but the new transcripts are not confined to the nucleus and both polyadenylated and non-polyadenylated unannotated transcripts are found in both compartments (3, 4). The most updated annotation of the human genome is probably provided through the ENCODE project (*Encyclopedia of DNA elements*) (**http://www.genome.gov/10005107**).

Much of the newly discovered transcriptional complexity is low abundance transcripts with steady-state levels of less than 1–10 copies per cell (5). Transcripts at the lower end of this range could in principle be due to transcriptional "noise" resulting from stochastic transcription. However, care should be taken not to confuse steady-state levels with transcriptional activity and disregard low abundance transcripts. Within cells, RNA surveillance systems rapidly remove some transcripts. In cells that have been crippled in their RNA surveillance systems (e.g., by elimination of nuclear exosome activity), many transcripts have increased steady-state levels that are comparable to those of known functional RNAs. In fact there are examples of regulatory systems that are dependent on rapid turnover of regulatory RNAs (6, 7). In support of the idea that many RNAs are transcribed at high frequency but rapidly turned over, analysis of localization of RNA polymerase II by chromatin precipitation experiments show similar occupancy at these sites and those of RNAs with high steady-state levels. One other concern is that most of the transcriptional complexity is poorly conserved in sequence when human and mouse is compared (2, 8, 9). A classical criterion for functional importance is conservation at the sequence or RNA structure level. However, there is an increasing emphasis on lineage-specific traits, not least as part of the search of what defines a human.

It is now widely accepted that the genome is pervasively transcribed. This raises questions on the functions of the transcripts that are produced, most of which are not understood at present. First, transcription *per se* rather than the RNA being produced could be functional in many cases. Transcription impacts the structure of chromatin and could be required to keep the chromatin accessible to regulatory factors. Second, many new functional RNAs have indeed been identified. The current estimate of RNA genes is in the range 4–5,000 compared to 21–23,000 protein coding genes and is rapidly increasing. New families of functional RNA species and RNA structural domains are annotated in databases such as Rfam (**http://www.sanger.ac.uk/Software/Rfam/**) (10) and described through the RNA WikiProject (11). An interim term "ncRNA" is frequently used to classify these RNAs simply to state that they are non-protein coding RNAs. The term is not very useful when applied both to small RNAs with distinct functions such as the modification guide RNAs or the regulatory miRNAs and to long RNA molecules that are transcribed by RNA polymerase II, capped, polyadenylated, and spliced but not encoding a protein. Incidentally, the discovery of the latter group of RNAs is one of the most important new discoveries. In the days of describing gene expression by construction of cDNA libraries of expressed sequence tags (EST libraries), many cDNAs were found not to have an open reading frame for protein coding. The cDNAs were made by priming at the polyA tail with oligodT, and it was believed that these cDNAs were copied from mRNA with very long 3′ UTRs (untranslated regions). When methods became available for synthesis of full length cDNAs, it became clear that these RNAs were devoid of open reading frames. Surprisingly, this class of RNAs may constitute half of the RNA polymerase II output (1, 12, 13). Only few of these have been thoroughly analyzed for function. As an example, NRON is an alternatively spliced RNA that is found as transcripts ranging in size from 0.8 to 3.7 kb. It is bound to 11 different proteins and the complex functions as a specific regulator of transcription factor NFAT nuclear trafficking (14). There are other examples of ncRNAs with similar functions and it is quite possible, given the number of such RNAs, that this is a general and as yet unanticipated layer of gene regulation. A third type of function is related to the abundance of sense–antisense pairs of transcripts. Depending on the method applied, the occurrence of antisense transcripts in mammalian genomes have been estimated from a few percent to 72% of the corresponding sense transcripts (12, 15). Some early studies based on cDNA probes may have suffered from the propensity of reverse transcriptase to synthesize a second strand cDNA using the first strand as primer and template by a loop-back mechanism. However, this mechanism can be inhibited by actinomycin D (16) and some data sets have been corrected accordingly. The extent of antisense regulation and the mechanisms (e.g., transcriptional interference or production of double-stranded RNA) remains to be explored but there are several well-documented cases of functional antisense RNAs (17–19). Finally, a fourth type of unanticipated function of transcripts is related to the recent discovery of new promoter-associated transcripts. Their description requires a brief introduction to the eukaryotic RNA polymerase II promoter.

2. The Promoter

The traditional view of transcription is that it is driven from a promoter that recruits the RNA polymerase directly or indirectly through the action of transcription factors and furthermore

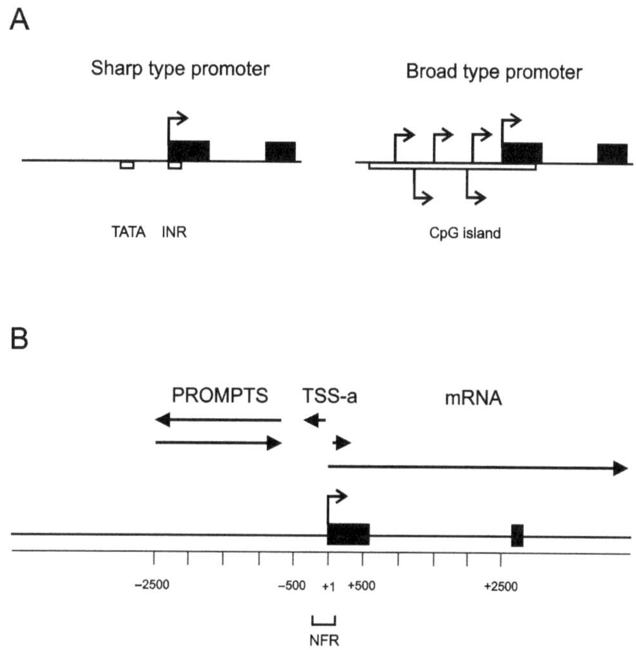

Fig. 1.1. **a** Two main types of promoters in the human genome. The sharp type of promoters (10–20%) initiates transcription at a well-defined site and typically has TATA-boxes and initiator (INR) elements. The broad type of promoters (>50%) initiates over a few hundred base pairs and are typically characterized by the presence of CpG-islands. **b** A close-up of the promoter region showing short bi-directional transcripts (TSS-a) originating from the nucleosome-free region (NFR) comprising the transcription start site (*broken arrow*). Unstable promoter proximal transcripts (PROMPTS) in both orientations and peaking in the region –500 to –2,500 are revealed in cells that are deficient in RNA turnover.

determines the direction of transcription. The prototype of promoters is one that harbors a TATA-box located approximately 30 base pairs upstream of the start site and initiates transcription at a well-defined site (sharp type of promoters; *see* **Fig. 1.1a**). This turns out not to be representative of promoters in the human genome – even in protein coding genes. First of all, only 10–20% of these genes have a TATA-box. The remainder of the genes recruits the polymerase in a TATA-box independent manner, primarily (>50%) in regions rich in 5′ CpG sequences (20). Second, transcription initiation does not take place at a specific site in most genes but is scattered over several hundred base pairs. This phenomenon was revealed by genome-wide cap analysis of gene expression (CAGE) (21). In this method, RNAs carrying an m^7G-cap, the hallmark of a 5′ end representing transcription initiation, are affinity purified and used as templates for making small DNA fragments that are sequenced and mapped on the genome. From this type of analysis it appeared that both TATA-box containing and TATA-less genes used both types of initiation but that the scattered initiation (broad type of promoters) was predominantly associated with the TATA-less promoters. An even bigger surprise was recently revealed from several studies of promoter associated transcripts (*see* **Fig. 1.1b**). One study based on deep sequencing of human RNAs smaller than 200 nt disclosed classes of RNA mapping to the 5′ (*p*romoter-*a*ssociated *s*mall RNAs; PASRs) and 3′ ends (*t*erminator-*a*ssociated *s*mall *R*NAs; TASRs) of transcription units (22, 23). The sites of origin coincide with regions that frequently are nucleosome free and the PASR align with CAGE tags indicating that they are capped. Another sequencing study of murine small RNAs reported a class of uncapped 20–90 nt TSS-a RNAs (*t*ranscription *s*tart *s*ite-*a*ssociated *RNA*s) (24). These RNAs flank the transcription start site with peaks of antisense at −250, and sense at +50, respectively. These transcripts are like PASR and TASR sufficiently abundant to be detected by northern blotting. Transcripts in both directions co-localized with chromatin markers of transcription initiation (RNA polymerase II and H3K4-trimethylated histones) but a marker of transcription elongation (H3K79-dimethylated histones) was only found downstream of the transcription start site, i.e. in association with synthesis of sense transcripts. Finally, a study of nascent transcripts in human cells found promoter proximal transcripts (*n*uclear *r*un-*o*n *RNA*s; NRO-RNAs) flanking the transcription start site in 30% of the genes (25). The peaks of antisense and sense transcripts mapped to the same positions as in the above mentioned mouse study. The three studies together shows that transcription initiation at the eukaryotic RNA polymerase II promoter is bi-directional. The initiation mechanism and the functional implications of this phenomenon are not known. It is also not known why synthesis of long transcripts eventually takes place in the sense direction only and whether this phenomenon is related to conversion of polymerases from pausing to processive mode as seen in the regulation of many genes. A completely unanticipated class of transcripts was recently found by tiling array analysis of exosome-depleted human cells, which are cells that are deficient in RNA turnover (26). These promoter upstream transcripts (PROMPTS) are variable length, polyadenylated transcripts that are transcribed from both strands and show overlapping distribution in the upstream −500 to −2,500 region. They are correlated with active genes and their presence is dependent on the nearby promoter. Based on their instability, they may be related to cryptic unstable transcripts (CUTs) in yeast. These are predominantly found as divergent transcripts from promoter regions of *bona fide* genes and are believed to have regulatory functions (6, 7). The promoter-associated transcripts show that transcription initiation is a complex phenomenon. This may not come as a surprise given that promoter elements and transcription factor binding sequences are short sequences that occur frequently in the genome. Thus, transcription initiation can be viewed a dual task of recruitment of the RNA polymerase and suppression of initiation at illegitimate start sites, primarily by formation of inhibitory chromatin structure (27).

3. The Transcription Unit

The transcription unit has been a useful concept because methods are known for mapping the ends of a transcript. The site of polyadenylation is used to define the 3′ end of transcription units because the actual site of termination of transcription generally is unknown and considered to be of little interest. The clear definition of a transcription unit contrasts the definition of a gene that originally was introduced as a physical entity corresponding to an observable phenotype. In many genes, distant sequence elements influence the expression and thus the phenotype in subtle ways and make it virtually impossible to delineate the gene as a physical entity. In the traditional view, transcription units have distinct boundaries. Although this is still the case in a strict sense, the picture is blurred by the observation from genome-wide studies that transcripts from a genomic region generally have multiple 5′ and 3′ ends. In a specialized type of cloning procedure, short tags representing the extreme 5′ and 3′ ends of a transcript are joined as "ditags" and sequenced. Analysis of such ditags revealed that the large majority of transcriptional units have alternative transcription start sites and polyadenylation sites with an average of 1.32 5′ start/3′ end and 1.83 3′ end/5′ start (1). This result can to a large extent be explained by the existence of broad type promoters (*see* **Fig. 1.1a**). Adding to the picture of 5′ end heterogeneity is the observation from the ENCODE project that >65% of the genes were alternatively transcribed from previously unnoted upstream promoters that on the average were located more than 100 kb upstream (2, 28). These transcripts typically were spliced to incorporate upstream exons that extend the 5′ UTR and thus the regulatory potential of the mRNA but could also add protein-coding exons to the mRNA. At the 3′ end, a recent mouse study showed that the 3′ UTRs of mRNAs tend to be longer due to alternative polyadenylation at later stages in development (29). Longer 3′ UTRs increase the potential for post-transcriptional regulation and the observation emphasizes the dynamical nature of the transcription unit. Transcription termination is generally not considered important in gene expression studies. It is worth to note that a genome-wide study of nascent transcripts showed that the RNA polymerase on average travels 10 kb beyond the polyadenylation site (30). This additional RNA is normally turned over rapidly, but at least the termination-associated transcription activity has the potential to influence the expression of nearby genes by transcriptional interference.

The transcriptional activity of a hypothetical region of the human genome encoding a protein is depicted in **Fig. 1.2**. To simplify the picture, the transcription unit is drawn with a sharp type of promoter (*see* **Fig. 1.1a**). The figure is drawn to scale to represent a unit that is average in most respects. The basic primary transcript is composed of seven exons (average 200 nt) interrupted by six introns. Overall, protein coding exons and introns make up 1.2 and 30%, respectively, of the genome and this ratio was used in the figure. In the default mature mRNA, the 5′ UTR will be 150 nt and the 3′ UTR 520 nt with a post-transcriptionally added polyA tail of 150–250 residues. The resulting 2.3-kb mRNA encodes a protein of 476 amino acids. Using this basic transcription unit as the starting point several variant and additional transcripts are noted from the region:

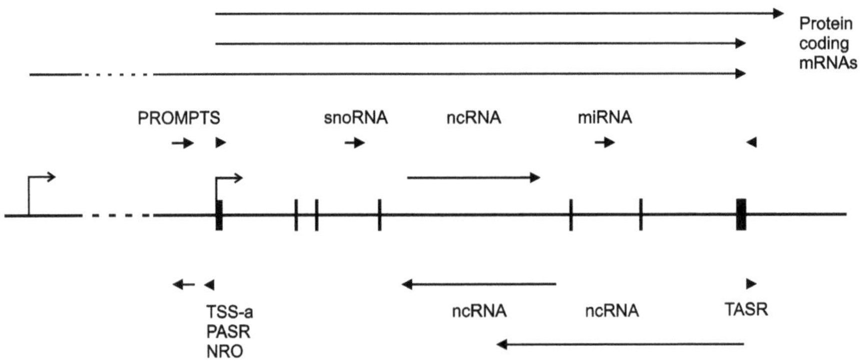

Fig. 1.2. Transcription map of a generalized region of the human genome. Regular protein coding mRNAs are transcribed from the main promoter in the region (*broken arrow*) or from a far upstream promoter belonging to a different transcriptional region. Some transcripts are 3′ extended due to suppression of the default polyadenylation site and use of alternative sites. Small transcripts are generated in both orientations from the 5′ end (TSS-a, PASR, NRO RNA) and 3′ end (TASR) and overlapping, unstable transcripts are found upstream of the main promoter (PROMPTS). Small nucleolar RNAs (snoRNAs) and microRNAs (miRNAs) are made by processing of intron RNA. Non-coding RNAs (ncRNAs) are produced in both sense and antisense orientations.

- Variant transcripts are initiated from far upstream promoters that in some cases are associated with other transcription units. These transcripts results in incorporation of additional exons derived from upstream transcription units or intergenic region into the mature mRNA.
- Other variant transcripts are extended to downstream polyadenylation sites by suppression of the default site. These transcripts have longer 3′ UTR with additional potential for post-transcriptional regulation.
- Unstable transcripts (PROMPTS) that are dependent on the main promoter in the region are generated from both

the forward and the reverse strand in the upstream region between –500 and –2,500. The function of these is currently unknown.

- Flanking the transcription start site, short transcripts are made in both directions (PASR, TSS-a, and NRO RNAs). To simplify the picture, three different types of observations described above are merged into a single class of transcripts in the figure although it is unclear whether these observations are of the same phenomenon. In any event, these transcripts are intimately associated with the mechanism of transcription initiation in a way that is currently unclear. Some transcripts in the sense orientation may be paused transcripts that are involved in a regulatory mechanism in which the polymerase is shifted from a paused to a processive mode.
- SnoRNAs (*s*mall *n*ucle*o*lar *RNA*s) of 60–300 nt are made from within introns in protein coding genes. These are stable transcripts made by processing of the intron RNA. They are chiefly of two types (box C/D methylation guides and box H/ACA pseudouridylation guides) that specify sites of modification of other RNAs, mainly ribosomal RNA. In other eukaryotes, these RNAs are predominantly made from independent genes or polycistronic transcripts.
- Many miRNAs (*mi*cro *RNA*s) are similarly made by processing of intron RNA in humans. The mature miRNAs are 21–23 nt RNAs that acts as translational repressors targeting a substantial fraction of the mRNA population (current estimates range from 30 to 90%).
- NcRNAs (*N*on-*c*oding *RNA*s) are transcribed from both the forward and the reverse strand and from different start sites. Those transcribed from the reverse strand are the main source of antisense RNAs found abundantly throughout the genome and has a potential to regulate the expression of the protein coding transcripts made from the forward strand in various ways.
- TASRs (*t*erminator-*a*ssociated *s*hort *R*NAs) are found in both orientations transcribed from a region corresponding to the polyadenylation site and also found to be relatively devoid of nucleosomes. These RNAs are preferentially associated with active genes and their function is unknown.

Figure 1.2 describes a generalized picture of transcription of a protein coding region of the human genome. One of the fascinating things of our time is that similar information relating to a gene of specific interest is publicly accessible in genome browsers such as the UCSC (**http://genome.ucsc.edu/**) and Ensembl (**http://www.ensembl.org/index.html**) browsers. Furthermore, these databases have links for each entry to many other types of biological and medical information. Thus, it is possible to get an overview of a gene of interest that may be sufficient to generate relevant hypotheses simply by browsing the databases.

Complicated as it may seem at present, the full story of the transcriptional complexity of the human genome has probably not yet been told. Interesting new information is likely to come from studies of co-transcriptional events and analyses on how transcripts are used in modifying DNA and chromatin. Due to its ability to fold into intricate conformations, RNA has the ability to bind specifically to small molecule ligands, including metabolites and to catalyze biochemical reactions. This is well known from riboswitches and ribozymes found in other systems (31). There are currently no known riboswitches described from the human genome and the ribozymes are only represented by the fundamental catalytic activities in the ribosome, the spliceosome (not definitively proven), and RNase P and a few sporadic small cleavage ribozyme of uncertain biological function (32). However, it is noteworthy that bioinformatics analyses reveals hundreds of thousands of sequences that appears to be conserved at the RNA structure level indicating that much more is to be found in relation to functional RNA molecules (33, 34). What is clear is that RNA is taking over the central stage in our efforts to understand the complex structure and expression of genetic information. This is challenging the gene as the basic operational unit in molecular biology. Just like alternative splicing makes it is impossible to decide whether a given sequence belongs to an exon or an intron at the DNA level without reference to the mature mRNA transcript, it is generally not possible to assign a DNA sequence to a single transcript and thus to a single phenotype. It has been suggested that it is time to consider the individual transcript as the operational unit in place of the gene (5). This would be in concert with the evolutionary view on genetic information. Or, in other words, the RNA World is still around!

References

1. Carninci, P., Kasukawa, T., Katayama, S., Gough, J., Frith, M. C., Maeda, N., Oyama, R., Ravasi, T., Lenhard, B., Wells, C., et al. (2005) The transcriptional landscape of the mammalian genome. *Science* 309, 1559–1563.
2. Birney, E., Stamatoyannopoulos, J. A., Dutta, A., Guigo, R., Gingeras, T. R., Margulies, E. H., Weng, Z., Snyder, M., Dermitzakis, E. T., Thurman, R. E., et al. (2007) Identification and analysis of functional elements in 1% of the human genome by the ENCODE pilot project. *Nature* 447, 799–816.
3. Cheng, J., Kapranov, P., Drenkow, J., Dike, S., Brubaker, S., Patel, S., Long, J., Stern, D., Tammana, H., Helt, G., et al. (2005) Transcriptional maps of 10 human chromosomes at 5-nucleotide resolution. *Science* 308, 1149–1154.
4. Kiyosawa, H., Yamanaka, I., Osato, N., Kondo, S. and Hayashizaki, Y. (2003) Antisense transcripts with FANTOM2 clone set and their implications

for gene regulation. *Genome Res* 13, 1324–1334.
5. Gingeras, T. R. (2007) Origin of phenotypes: genes and transcripts. *Genome Res* 17, 682–690.
6. Neil, H., Malabat, C., Ubenton-Carafa, Y., Xu, Z., Steinmetz, L. M. and Jacquier, A. (2009) Widespread bidirectional promoters are the major source of cryptic transcripts in yeast. *Nature* 457, 1038–1042.
7. Xu, Z., Wei, W., Gagneur, J., Perocchi, F., Clauder-Munster, S., Camblong, J., Guffanti, E., Stutz, F., Huber, W. and Steinmetz, L. M. (2009) Bidirectional promoters generate pervasive transcription in yeast. *Nature* 457, 1033–1037.
8. Bertone, P., Stolc, V., Royce, T. E., Rozowsky, J. S., Urban, A. E., Zhu, X., Rinn, J. L., Tongprasit, W., Samanta, M., Weissman, S., et al. (2004) Global identification of human transcribed sequences with genome tiling arrays. *Science* 306, 2242–2246.
9. Kampa, D., Cheng, J., Kapranov, P., Yamanaka, M., Brubaker, S., Cawley, S., Drenkow, J., Piccolboni, A., Bekiranov, S., Helt, G., et al. (2004) Novel RNAs identified from an in-depth analysis of the transcriptome of human chromosomes 21 and 22. *Genome Res* 14, 331–342.
10. Gardner, P. P., Daub, J., Tate, J. G., Nawrocki, E. P., Kolbe, D. L., Lindgreen, S., Wilkinson, A. C., Finn, R. D., Griffiths-Jones, S., Eddy, S. R., et al. (2009) Rfam: updates to the RNA families database. *Nucleic Acids Res* 37, D136–D140.
11. Daub, J., Gardner, P. P., Tate, J., Ramskold, D., Manske, M., Scott, W. G., Weinberg, Z., Griffiths-Jones, S. and Bateman, A. (2008) The RNA WikiProject: community annotation of RNA families. *RNA* 14, 2462–2464.
12. Katayama, S., Tomaru, Y., Kasukawa, T., Waki, K., Nakanishi, M., Nakamura, M., Nishida, H., Yap, C. C., Suzuki, M., Kawai, J., et al. (2005) Antisense transcription in the mammalian transcriptome. *Science* 309, 1564–1566.
13. Claverie, J. M. (2005) Fewer genes, more noncoding RNA. *Science* 309, 1529–1530.
14. Willingham, A. T., Orth, A. P., Batalov, S., Peters, E. C., Wen, B. G., Za-Blanc, P., Hogenesch, J. B. and Schultz, P. G. (2005) A strategy for probing the function of noncoding RNAs finds a repressor of NFAT. *Science* 309, 1570–1573.
15. He, Y., Vogelstein, B., Velculescu, V. E., Papadopoulos, N. and Kinzler, K. W. (2008) The antisense transcriptomes of human cells. *Science* 322, 1855–1857.
16. Perocchi, F., Xu, Z., Clauder-Munster, S. and Steinmetz, L. M. (2007) Antisense artifacts in transcriptome microarray experiments are resolved by actinomycin D. *Nucleic Acids Res* 35, e128.
17. Krystal, G. W., Armstrong, B. C. and Battey, J. F. (1990) N-myc mRNA forms an RNA-RNA duplex with endogenous antisense transcripts. *Mol Cell Biol* 10, 4180–4191.
18. Thrash-Bingham, C. A. and Tartof, K. D. (1999) aHIF: a natural antisense transcript overexpressed in human renal cancer and during hypoxia. *J Natl Cancer Inst* 91, 143–151.
19. Hongay, C. F., Grisafi, P. L., Galitski, T. and Fink, G. R. (2006) Antisense transcription controls cell fate in *Saccharomyces cerevisiae*. *Cell* 127, 735–745.
20. Sandelin, A., Carninci, P., Lenhard, B., Ponjavic, J., Hayashizaki, Y. and Hume, D. A. (2007) Mammalian RNA polymerase II core promoters: insights from genome-wide studies. *Nat Rev Genet* 8, 424–436.
21. Kodzius, R., Kojima, M., Nishiyori, H., Nakamura, M., Fukuda, S., Tagami, M., Sasaki, D., Imamura, K., Kai, C., Harbers, M., et al. (2006) CAGE: cap analysis of gene expression. *Nat Methods* 3, 211–222.
22. Kapranov, P., Cheng, J., Dike, S., Nix, D. A., Duttagupta, R., Willingham, A. T., Stadler, P. F., Hertel, J., Hackermuller, J., Hofacker, I. L., et al. (2007) RNA maps reveal new RNA classes and a possible function for pervasive transcription. *Science* 316, 1484–1488.
23. Affymetrix/Cold Spring Harbor Laboratory ENCODE Transcriptome Project (2009) Post-transcriptional processing generates a diversity of 5′-modified long and short RNAs. *Nature* 457, 1028–1032.
24. Seila, A. C., Calabrese, J. M., Levine, S. S., Yeo, G. W., Rahl, P. B., Flynn, R. A., Young, R. A. and Sharp, P. A. (2008) Divergent transcription from active promoters. *Science* 322, 1849–1851.
25. Core, L. J. and Lis, J. T. (2008) Transcription regulation through promoter-proximal pausing of RNA polymerase II. *Science* 319, 1791–1792.
26. Preker, P., Nielsen, J., Kammler, S., Lykke-Andersen, S., Christensen, M. S., Mapendano, C. K., Schierup, M. H. and Jensen, T. H. (2008) RNA exosome depletion reveals transcription upstream of active human promoters. *Science* 322, 1851–1854.
27. Buratowski, S. (2008) Transcription. Gene expression–where to start? *Science* 322, 1804–1805.
28. Gerstein, M. B., Bruce, C., Rozowsky, J. S., Zheng, D., Du, J., Korbel, J. O.,

Emanuelsson, O., Zhang, Z. D., Weissman, S. and Snyder, M. (2007) What is a gene, post-ENCODE? History and updated definition. *Genome Res* 17, 669–681.
29. Ji, Z., Lee, J. Y., Pan, Z., Jiang, B. and Tian, B. (2009) Progressive lengthening of 3′ untranslated regions of mRNAs by alternative polyadenylation during mouse embryonic development. *Proc Natl Acad Sci USA* 106, 7028–7033.
30. Core, L. J., Waterfall, J. J. and Lis, J. T. (2008) Nascent RNA sequencing reveals widespread pausing and divergent initiation at human promoters. *Science* 322, 1845–1848.
31. Serganov, A. and Patel, D. J. (2007) Ribozymes, riboswitches and beyond: regulation of gene expression without proteins. *Nat Rev Genet* 8, 776–790.
32. Salehi-Ashtiani, K., Luptak, A., Litovchick, A. and Szostak, J. W. (2006) A genomewide search for ribozymes reveals an HDV-like sequence in the human CPEB3 gene. *Science* 313, 1788–1792.
33. Washietl, S., Pedersen, J. S., Korbel, J. O., Stocsits, C., Gruber, A. R., Hackermuller, J., Hertel, J., Lindemeyer, M., Reiche, K., Tanzer, A., et al. (2007) Structured RNAs in the ENCODE selected regions of the human genome. *Genome Res* 17, 852–864.
34. Torarinsson, E., Sawera, M., Havgaard, J. H., Fredholm, M. and Gorodkin, J. (2006) Thousands of corresponding human and mouse genomic regions unalignable in primary sequence contain common RNA structure. *Genome Res* 16, 885–889.

Chapter 2

Working with RNA

Henrik Nielsen

Abstract

Working with RNA is not a special discipline in molecular biology. However, RNA is chemically and structurally different from DNA and a few simple work rules have to be implemented to maintain the integrity of the RNA. Alkaline pH, high temperatures, and heavy metal ions should be avoided when possible and ribonucleases kept in check. The chapter outlines the specific precautions recommended for work with RNA and describes some of the modifications to standard protocols in molecular biology that are relevant to RNA work. The methods are applicable to all types of RNA and require a minimum of specialized equipment.

Key words: RNA structure, RNA folding, RNA degradation, RNase inhibitors, RNA purification.

1. Introduction

RNA is chemically different from DNA in two aspects: (1) the base thymine in DNA is replaced by the base uracil in RNA. Uracil lacks the methyl group found at carbon atom 5 (C5) in thymine; (2) the sugars in RNA have an OH group attached to C2 and are thus ribose sugars rather than deoxyribose sugars as found in DNA. In addition to these two general differences, a large number of nucleotide modifications are known from certain RNA molecules, in particular ribosomal RNA and tRNA.

The chemical differences between RNA and DNA have profound structural consequences. The sugars in RNA almost exclusively adopt the C3′-endo conformation because the 2′OH otherwise would sterically clash with the attached base. The 2′OH group can engage in the formation of hydrogen-bonding thereby adding structural versatility to RNA. Many RNA molecules are

H. Nielsen (ed.), *RNA*, Methods in Molecular Biology 703,
DOI 10.1007/978-1-59745-248-9_2, © Springer Science+Business Media, LLC 2011

highly structured (ribosomal RNA, tRNA and most small RNA molecules) and others have structural elements separated by unstructured parts (many mRNAs). The nucleotides in RNA can form Watson–Crick base pairs similar to those found in DNA because uracil can form a base pair with adenine that is isosteric with the A-T base pair found in DNA. Uracil can also form the so-called wobble base pair with guanine. In addition, a very large number of non-Watson–Crick base pairs (1) contribute to the structural versatility of RNA. RNA helices are of the A-form and different from the B-form that is general in DNA. It is stabilized by hydrogen-bonding between the H2′ atom and O4′ of the neighboring residue. Helices are typically formed by intramolecular base pairings that make up the scaffold of the RNA. In structured RNA molecules, the remaining residues are typically involved in formation of non-Watson–Crick base pairs that in combination can form RNA motifs (2).

The presence of the 2′OH in RNA also has the practical consequence of making RNA chemically labile in many experimental situations. The 2′OH can be activated for nucleophilic attack at the neighboring phosphodiester bond by alkaline pH. Furthermore, heavy metal ions can promote the attack of the 2′OH on the phosphodiester bond. Thus, special precautions have to be taken when working with RNA compared to DNA.

2. Creating a Working Environment for RNA Work

There are four experimental circumstances that have to be avoided in RNA work: alkaline pH, high temperatures, metal ions, and the presence of RNases. The pH has to be kept at neutral or slightly acidic to avoid the activation of the ribose 2′OH for attack at the phosphodiester bond that will result in strand cleavage observed as degradation of the RNA. High temperatures will similarly promote degradation of the RNA. Thus, RNA work is carried out at 0–4°C whenever possible. The deleterious heavy metal ions can in some cases be removed from critical solutions by treatment with an ion exchange resin (e.g., Chelex 100 (Sigma)). In many cases, 0.1 mM EDTA is included in solutions for RNA work to keep certain metal ions in check. Obviously, in many types of experiments, it is necessary to compromise on the experimental circumstances to meet the requirement, e.g., protein binding to RNA, enzymatic treatment or a requirement to keep the RNA in a native structural conformation. In these cases, incubation of the RNA at non-optimal conditions should be kept as short as practically possible.

The avoidance of exposure of the RNA to RNases has two aspects – endogenous RNases and contaminating RNases in the experimental environment. Endogenous RNases are RNases originating from the same biological material as the RNA. Typically these RNases are found in distinct cellular compartments away from the RNA but are set free during cell lysis. For this reason, lysis is mostly carried out in the presence of a denaturing reagent, such as guanidinium thiocyanate (3, 4). This will suffice for most biological materials, but some tissues, in particular placenta and pancreas are notoriously difficult to handle in this respect and procedures have to be carried out very fast and at low temperatures. Denaturation with guanidinium thiocyanate will also destroy native RNA–protein interactions and RNA structure. If preservation of either of these is critical to the experiment, RNase inhibitors such as vanadyl ribonucleoside complexes, heparin, or peptide RNase inhibitors (see below) must be applied.

Contamination with RNases is frequently an issue in the RNA laboratory. Laboratory equipment can be contaminated by RNases from biological materials or from procedures that include the use of RNases. Current protocols for plasmid minipreps, UV-crosslinking experiments, RNase protection analysis, and in vitro translation, include a step for removal of RNA by RNase A digestion. RNase A is problematic because it is highly resistant to high temperatures and chemical treatment. Such contamination is best avoided by using disposable plasticware for RNA experiments and by using alternatives to the aggressive RNase A for other procedures, e.g., RNase T_1, sometimes in combination with the double-strand specific RNase V1. In case laboratory equipment has to be re-cycled and used for RNA experiments, care should be taken to eliminate contaminating RNases. Glassware is baked for 1–2 h at 200°C or, alternatively, treated with a mixture of chromic and sulphuric acids followed by a rinse with EDTA-containing diethylpyrocarbonate-treated water (*see* **Section 11**). Non-disposable plasticware can be treated with 0.1 M NaOH, 1 mM EDTA and rinsed with water. In general, solutions should be either autoclaved or sterile-filtered in order to prevent RNase contamination resulting from microbial growth. Sterile-filtration is the method of choice and should be used for smaller volumes because autoclaving can lead to release of microbial RNases into solution. A simple work-rule to prevent microbial growth is to store solutions as frozen aliquots.

Another source of contaminating RNases is the so-called "finger-RNases" from human skin. For this reason, it is important to wear disposable plastic or latex gloves. In our experience, it is quite possible to work with high molecular weight RNA and RNases simultaneously at the workbench as long as the workplace is kept tidy and well-organized. However, pipettes equipped with filter-tips should be used when pipetting concentrated RNase solutions to avoid contamination of the pipette with RNase containing aerosols. Utensils such as plastic tips and tubes are generally free of RNases if they are used directly from the original packing. Autoclaving of these utensils are unnecessary and could even lead to the introduction of RNases by handling. Teflon-coated (Sorenson) or siliconized tubes are recommended for RNA work because sample loss due adsorption to surfaces is minimal with these tubes. Critical chemicals should be set aside and reserved for RNA work to avoid contaminations.

3. Quantification

Nucleic acids are almost always quantificated by UV-spectroscopy at 260 nm. The rule of thumb is that 1 A_{260} unit equals a concentration of 40 μg/mL of single-stranded RNA. The exact value depends on the sequence of the RNA, the structure of the RNA (because of hypochromicity due to base stacking), and pH. Many factors influence the value indirectly by influencing the folding state of the RNA, e.g., ionic strength, type of ion, presence of EDTA and denaturants, and temperature. For these reasons, quantification should be made in buffered salt solutions rather than in water. For a detailed discussion of the extinction coefficient of RNA, *see* (5).

Most erroneous quantifications result from the presence of impurities in the RNA sample. Typical examples are proteins, aromatics such as phenol, and millimolar concentrations of compounds such as 2-mercaptoethanol or dithiothreitol that are included in some protocols for RNA isolation. Some of these problems are diagnosed by measuring the absorption of the sample at other wavelengths. A pure sample of RNA has an A_{260}/A_{280} ratio of 2 ± 0.05. Contamination with proteins that have an absorption maximum at 280 nm (mainly because of tyrosine residues) or phenol (absorption maximum at 270 nm) will lower this ratio. A typical problem in quantification of in vitro transcripts is insufficient separation from nucleotides in the transcription reaction. This will obviously not be revealed by the A_{260}/A_{280} ratio. Gel filtration using spin-columns is a fast and efficient way to purify the RNA in this case. Another problem is contamination with gel components after gel purification of RNA. In these and other difficult cases, RNA can be quantificated by comparison to a known standard in gel electrophoresis (*see* **Section 10**). This procedure is usually accurate within a factor of 2.

One practical problem in UV-spectroscopy is the loss of sample in the procedure. Normally, the RNA concentration should be higher than 5 μg/mL corresponding to a reading of 0.1–0.15 of A_{260} to obtain a reliable measurement. In old spectrophotometers using 1-mL cuvettes, this would take around 4 μg of RNA. The GeneQuant apparatus (GE Healthcare) can use a 7-μL cuvette and the Nanodrop spectrophotometer (Saveen Biotech) can make reliable measurements on a sample of 1 μL in the 1 pg/μL–3,000 ng/μL range.

Alternatives to UV-spectroscopy are phosphate analysis (6), or a fluorometric assay (7). The latter is based on binding of a compound, e.g., RiboGreen (Molecular Probes), to the RNA followed by fluorescence measurements (excitation at 480 nm, emission at 520 nm) and can be applied to concentrations in the 1–1,000 pg/μL range. One advantage of the RiboGreen method is that it is RNA-specific in contrast to UV-spectroscopy that also measures DNA.

4. Extraction with Organic Solvents

Extraction with phenol is used to purify RNA from proteins in biological samples or following incubation of RNA with enzymes (8). Most proteins are denatured in aqueous phenol and will either be preferentially soluble in the phenol phase or become insoluble in both phases. Therefore, if a mixture of proteins and nucleic acids is extracted with aqueous phenol, the denatured proteins will partition preferentially into the phenol phase or appear as an insoluble interphase, after separation of the two phases by centrifugation. The RNA is recovered by precipitation with ethanol from the upper, aqueous phase. Phenol is often used in combination with chloroform (1:1) because this mixture is more strongly denaturing than phenol alone and because poly(A)$^+$ mRNA may be soluble in phenol under some conditions (see below). Often a small amount of isoamyl alcohol is added in order to increase the surface tension, which facilitates the separation of the two phases. This mixture is referred to as "PCI" (phenol:chloroform:isoamylalcohol) and is commercially available (e.g., Invitrogen). Oxidation of phenol results in a reddish color and solutions that turn reddish should not be used. Addition of 8-hydroxyquinoline to the aqueous phenol, which then turns yellow, prevents the oxidation. Considerable amounts of nucleic acids may be trapped by denatured proteins in the interphase, which should therefore be re-extracted with buffer, followed by extraction of the combined water phases. How many times one should extract a sample in order to get a maximal recovery and purity of the nucleic acids depends on the type of material and the concentrations of nucleic acids and proteins. A single extraction is usually sufficient after enzymatic reactions, while biological material normally requires two or three extractions. Phenol in the water phase can be removed by extraction with chloroform.

In practice, an efficient extraction of the proteins often requires prior removal of Mg^{2+} and Ca^{2+} with EDTA and denaturation of the proteins with ionic detergents (e.g., sodium dodecyl sulfate (SDS)) or chaotropic salts (e.g., guanidinium thiocyanate). Phenol extraction of whole cells or cell nuclei normally requires a preceding proteolytic digestion. This is usually done by preincubation with proteinase K, which is a serine protease that becomes activated by denaturation (e.g., by SDS).

Salt concentration, pH, and temperature also have to be considered in phenol extraction procedures. If the salt concentration is sufficiently high, the RNA is completely displaced to the water phase. Extraction can be performed in 0.3 M sodium acetate or in a standard DNA extraction buffer such as STE, preferably with a slightly lower pH than used for DNA extraction (0.1 M NaCl, 1 mM EDTA, 10 mM Tris-HCl, pH 7.0). These buffers allow for subsequent precipitation of the RNA from the aqueous phase without adjustment of the salt. Extraction of DNA must occur at pH > 7 because of its tendency to form an interphase at pH 7, and DNA is selectively solubilized in the phenol phase at lower pH (9). In contrast to DNA, the phase distribution of bulk RNA is independent of pH, and RNA can therefore be selectively isolated by extraction at pH below 7. However, poly(A)$^+$ mRNA is partially soluble in the phenol phase below pH 7.6 and has to be extracted with a mixture of phenol and chloroform at pH 5–9, or with phenol at pH 9. Phenol extraction is often performed at ambient temperature or at 0°C in order to inhibit nucleases, but extraction of nucleic acids from different types of tissues may require higher temperatures (50°C–60°C). The two phases get turbid by even a slight drop in temperature during extraction because of the segregation of water and phenol, which become less miscible at lower temperature.

Protocol for standard deproteinization of RNA:

1. Adjust the volume of the sample to 100–200 μL (in order to be able to recover the aqueous phase with minimal loss in the pipetting step) by addition of H_2O and 3 M sodium acetate (pH 5.2) to make the sample 0.3 M in sodium acetate.
2. Add 1 volume of phenol saturated with TE: 10 mM Tris-HCl, pH 7.5, 0.1 mM EDTA and vortex.
3. Centrifuge the sample for 5,000×*g* for 4 min to separate the phases. Transfer the upper, aqueous phase to a new tube.
4. Repeat steps 2 and 3 using a 1:1 mixture of phenol:chloroform saturated with TE.
5. Repeat steps 2 and 3 using chloroform (to remove traces of phenol).
6. Precipitate the aqueous phase with 3 volumes of ice-cold 96% ethanol (*see* **Section 7** for considerations on this step).

5. Desalting and Removal of Nucleotides

Buffer change and removal of nucleotides and short primers are conveniently done by gel filtration using spin-columns. Such columns can be made from home-made suspensions of Sephacryl or Sephadex and 1 mL syringes or they can be purchased ready to use (GE Healthcare). Generally, the use of spin columns involves a prespin to remove the storage buffer from the column. Then the sample (typically 50–100 μL) is loaded and the column is given a short spin (e.g., 735×*g* for 2 min). The RNA is collected in the flow-through and the low molecular weight components (salts, nucleotides, short primers) are retarded in the gel matrix. In our experience, the commercial spin-columns are free of RNases when used directly from the packing.

Desalting can also be done by dilution followed by ethanol precipitation. Removal of nucleotides can be done by ethanol precipitation after addition of ammonium acetate to 2.0–2.5 M (*see* **Section** 7). Two consecutive precipitations at these conditions will result in removal of 99% of the nucleotides.

6. Gel Purification

RNA molecules up to a few thousand nucleotides can be purified by denaturing polyacrylamide gel electrophoresis. Since the method is based on diffusion from a gel slice into an elution buffer, the recovery depends strongly on the composition of the gel and the size of the RNA. The RNA is localized in the gel by UV-shadowing. The gel is wrapped in plastic wrap, placed over a sheet of Xerox paper or a thin-layer chromatography plate and inspected under UV-light (e.g., a UV_{254} handheld lamp). Due to absorption of the UV-light by the RNA, the RNA band will appear as a shadow on the Xerox paper or TLC-plate. A gel slice (as small as possible) is cut out using a disposable scalpel and placed in tube. At this stage, many protocols recommend crushing of the gel with a pipette tip to facilitate the elution. However, polyacrylamide fragments will make subsequent pipetting steps difficult, and we prefer to compromise on the recovery and leave the gel slice intact. Approximately 400 μL of elution buffer (0.25 M sodium acetate (pH 6.0), 1 mM EDTA) and 200 μL of buffer saturated phenol are added and the tube wrapped in parafilm to avoid leakage. Depending on the size of the RNA, elution can be performed at room temperature with continuous shaking in a few hours or over night in the cold room with or without shaking. Following elution, the liquid is transferred to a second tube and given a brief spin to separate the two phases. The aqueous phase is transferred to a new tube, extracted with chloroform to remove traces of phenol, and precipitated (*see* **Sections 4 and 7** for details on this). The recovery is usually well in excess of 50%. The gel slice can be subjected to a second round of elution to increase the recovery.

An alternative to elution by diffusion is electroelution. Here, the gel slice is placed in a dialysis bag containing electrophoresis buffer and placed in the electrophoresis chamber perpendicular to the electrical field. After 1 h of electrophoresis at 10 V/cm, the field is reversed for a few minutes, and the RNA can be recovered from the buffer in the dialysis bag. Several companies have made specialized electrophoresis units for electroelution of nucleic acids (e.g., BioRad).

7. Precipitation

RNA is recovered from aqueous solutions by precipitation with ethanol (10). RNA can be precipitated from solutions with a concentration as low as 20 ng/mL provided that the ionic strength is sufficiently high (0.2 M NaCl, 0.3 M NaAc, 0.8 M LiCl, or 2–2.5 M NH_4Ac). For recovery of RNA from more dilute solutions or for efficient recovery of low molecular weight RNAs, a co-precipitant is included (see below). The standard protocol involves adjusting the salt concentration (e.g., to 0.3 M NaAc from a 3-M stock) followed by addition of 2.5–3 volumes of 96% ice-cold ethanol. The sample is then left for 5 min in a dry ice bath or 15 min at −20°C or 4°C. The time and temperature depends on the size and the concentration of the RNA with longer times and lower temperatures required for small fragments and low concentrations. Next, the precipitate is recovered by centrifugation, typically at 12,000×*g* for 15 min at 4°C. Centrifugation time and centrifugal force are the most critical parameters and should be increased for small fragments and low RNA concentrations. The pellet is localized and the ethanol removed by aspiration. This is frequently in two steps. First, most of the ethanol is removed. The tube is then given a brief spin to collect the remainder of the ethanol that now can be efficiently removed. Two simple measures will facilitate the recovery of the pellet. The hinge of the tube is placed upward in the centrifuge such that the pellet will form in a predictable spot in the tube on the hinge side of the bottom. Teflon-coated (Sorenson) or siliconized tubes should be used to avoid sticking of the RNA to the plastic walls. Furthermore, the pellet formed in these tubes is more distinct and easier to locate. As mentioned below, a co-precipitant conjugated to a dye can by used to ease the detection of the pellet. Following the aspiration of the ethanol, remaining salt in the pellet can be removed by washing once with 70% ethanol or several times with 70% ethanol containing 0.25 M NH_4Ac, which prevents solubilization of the nucleic acids. The 70% ethanol wash is conveniently used to wash the sides of the tube to remove traces of salt. If the pellet is disturbed during the ethanol wash, a brief centrifugation is applied to re-collect the pellet. The pellet is dried rapidly (in a few minutes) and effectively in a vacuum centrifuge (traces of NH_4Ac will evaporate as NH_3 and HAc) or alternatively, left at 65°C for 5–10 min to evaporate the ethanol traces.

Different salt are used for different purposes in ethanol precipitations. Sodium acetate (0.3 M; pH 5.2) is used for routine precipitation. Sodium chloride (0.2 M) can be used if the sample contains SDS because the detergent in this case remains soluble in 70% ethanol. Lithium chloride (0.8 M) is used when large volumes of ethanol are used because it is very soluble in ethanolic solutions and does not co-precipitate. A special application is to precipitate large RNA molecules (ribosomal RNA and mRNA) without precipitation of small RNA molecules (tRNA and others). This is done by addition of LiCl to 0.8 M without addition of ethanol. After mixing and leaving on ice for at least 2 h, the sample is centrifuged for 15,000×*g* for 20 min at 0°C to recover the large RNA molecules. LiCl should be avoided for certain downstream applications. Chloride ions inhibit RNA-dependent DNA polymerases used for reverse transcription. Ammonium acetate (2.0–2.5 M) is used to reduce the co-precipitation of nucleotides. This method can not be applied if the RNA subsequently is to be used as substrate for bacteriophage T4 Polynucleotide Kinase because this enzyme is inhibited by ammonium ions.

The classical co-precipitant is tRNA (typically from *E. coli* or yeast) added from a stock solution of 10 mg/mL to a final concentration of 10–20 μg/mL. Some companies sell tRNA from RNase-deficient strains of *E. coli* (*E. coli* MRE600; Ambion). The drawback of using tRNA is that it interferes with UV-spectroscopy and several enzymatic treatments of the sample RNA (e.g., labelling with Polynucleotide Kinase). A very good alternative is glycogen isolated from muscle (Ambion). This is added from a stock solution of 5 mg/mL to a final concentration of 50–150 μg/mL. Glycogen does not interfere with UV-readings and is not a substrate for enzymes that act on RNA. Glycogen with a covalently attached dye is sold as "Glycoblue" (Ambion). This allows easy detection of the precipitate but is relatively expensive. Linear acrylamide (Ambion) can be used as a co-precipitant for selective precipitations of RNA >20 nt. It is used at 10–20 μg/mL diluted from a 5 mg/mL stock.

Protocol for standard ethanol precipitation of RNA:

1. Adjust the sample to 0.3 M NaAc by addition from a 3 M stock (pH 5.2).
2. Add 3 volumes of 96% ice-cold ethanol.
3. Leave on ice for 15 min.
4. Centrifuge at 12,000×*g* for 15 min at 4°C.
5. Remove the ethanol by aspiration. Spin briefly and remove the remainder of the ethanol.
6. Wash the sides of the tube and the pellet with 200 μL of 70% ethanol. Remove the ethanol.
7. Dry the pellet a few minutes in a vacuum centrifuge or at 65°C for 5–10 min.
8. Resuspend the RNA in double-distilled or similar quality of H_2O.

8. Storage

For short-term use (weeks), RNA is stored at a concentration of 1–10 μg/mL in double-distilled or similar quality H_2O at slightly acidic pH at –20°C. Alternatively, TE (10 mM Tris-HCl, pH 7.5, 0.1 mM EDTA) can be used as storage buffer. At these conditions, RNA structures are unfolded and for some applications, re-folding (*see* **Section 9**) is required. For longer term storage (months), storage at –80°C is preferred. For even longer storage (years), we prefer storage as ethanol precipitates at –20°C or –80°C. A temporary storage medium (RNAlater; Qiagen) is sold to preserve tissues for subsequent RNA extraction. Ethanol precipitates (as wet pellets) is a convenient way to ship RNA.

9. RNA Re-folding

The function of RNA molecules is critically dependent on their structure. During isolation of RNA, the structure is usually disrupted by unfolding due to the presence of denaturants, such as guanidinium thiocyanate or the removal of Mg^{2+} – ions by metal ion chelators (mostly EDTA). Thus, refolding or renaturation of the RNA becomes an issue. This is far from being a trivial problem. RNA molecules fold during transcription and the transcription rate as well as the co-transcriptional association of proteins affects the folding. These conditions are impossible to re-create in a test tube. A useful approximation is to use a renaturation protocol that takes into account that RNA molecules generally fold in a hierarchical fashion with secondary structure formation (helices) preceding the formation of tertiary structure (2, 11). The two steps have different requirements for Mg^{2+} – ions and this is used to separate them:

1. Heat denature the RNA at 90°C for 1 min in 20 mM Tris-HCl, pH 7.8, 140 mM KCl.
2. Transfer to 60°C and leave for 15 min.
3. Cool slowly to 30°C over a 15 min period.
4. Add $MgCl_2$ to a final concentration of 2.5 mM and leave at 30°C for 15 min.
5. Transfer to 0°C.

Ideally, renaturation should result in a population of molecules that all have the native fold. In reality, some molecules may end up in a misfolded conformation. This is a serious problem in structural analysis of RNA that requires a homogenous RNA population. In other types of experiments, non-native RNA forms may out-titrate protein factors and invalidate functional assays. Stringent controls or assessment of the folding state of the RNA are necessary in these types of experiments. It is recommended to consult the literature to design the experiments to conform to the state-of-the art for the relevant type of experiment.

One simple way to examine the conformational homogeneity of the renatured RNA is non-denaturing gel electrophoresis. There are several different types of gels that can be used. It is important to preserve the structure of the RNA by including Mg^{2+} – ions and avoiding denaturants and high temperatures. One example is to use a standard TBE electrophoresis buffer supplemented with 5 mM $MgCl_2$ and 50 mM KCl and run the gel at 4 V/cm at room temperature or in the cold room.

10. Gel Electrophoresis

RNA can be analyzed by electrophoresis in agarose- and polyacrylamide gels. Denaturing agarose gels are used for northern blotting analysis of RNAs in the size range of mRNAs and to assess the quality of whole cell RNA extracts. The most commonly used denaturant is formaldehyde. In a gel run of whole cell RNA on a 1% formaldehyde-agarose gel, three bands are normally seen. These are (from top to bottom) the large subunit ribosomal RNA (LSU rRNA), the small subunit ribosomal RNA (SSU rRNA), and a composite band including 5.8S and 5S ribosomal RNAs, tRNAs, and a multitude of other small molecular weight RNAs. The relative intensities of the two upper bands can be used for assessing the integrity of the RNA. LSU rRNA is approximately twice the size of SSU rRNA and band should therefore be twice the intensity of the SSU rRNA band if the RNA is intact.

Smaller RNA molecules (less than 1,500 nt) can be analyzed on denaturing (7 M urea) polyacrylamide gels. Polyacrylamide gels are cast by polymerization of acrylamide into long chains in the presence of N,N′-methylenbisacrylamide ("bisacrylamide") as a crosslinker. The polymerization process is initiated by ammonium persulfate and catalyzed by N,N,N′,N′-tetramethylendiamine (TEMED). The pore size of the gel depends on the chain length as well as the level of crosslinking (i.e., of the concentration of acrylamide as well as of bisacrylamide). Polyacrylamide gels have greater capacity than agarose gels and RNAs isolated from acrylamide gels (*see* **Section 6**) are exceptionally pure. When made in a sequencing format, polyacrylamide gels can separate small RNA molecules (up to 150 nt) that only differ in size by a single nucleotide.

RNA gels are stained by ethidium bromide after electrophoresis. It exhibits a weak orange fluorescence (520 nm) when irradiated with UV-light, and the fluorescence intensity increases dramatically by binding to nucleic acids, – most to double-stranded molecules (binding by intercalation), and somewhat less to single stranded molecules.

11. Diethylpyrocarbonate (DEPC) and RNase Inhibitors

Once the endogenous RNases in the biological material have been eliminated, there is little need to include RNase inhibitors in further steps of RNA manipulation provided that the general work rules are followed. Many protocols recommend treatment of H_2O and solutions with diethylpyrocarbonate (DEPC) prior to use in RNA work. It should be recalled that this reagent modifies adenosines (in fact, it is a standard reagent in chemical probing of RNA) and for this reason is potentially harmful to the RNA. The procedure is to treat the solution with 0.1% DEPC for at least 12 h at 37°C, and then heat it to 100°C for 15 min or autoclave it in order to remove unreacted DEPC. DEPC reacts with amines and solutions containing Tris can not be treated with DEPC. We do not recommend this extensive use of DEPC and suggest that its use is limited to cleaning of labware (including electrophoresis tanks) that have been exposed to RNase A.

Another popular way of dealing with RNases is to use the placental RNase inhibitor RNasin or RNases with similar properties from other sources (e.g., ANTI-RNase from Ambion). This is a potent inhibitor of neutral pancreatic RNase A type enzymes. It can be purchased as isolated from pancreatic extract or as a recombinant protein. The native form should be avoided because it contains a large amount of the RNase angiogenin that is released from RNasin by heating and in the presence of reducing agents. The recombinant form is sold under many different names (RNasin (Promega), RiboLock (Fermentas), RNAguard (GE Healthcare)) and is used at 1 U/μL. It works by binding to the RNase in a 1:1 ratio and care must be taken to avoid denaturation or oxidation of the inhibitor with resulting release of the RNase. In our experience the inclusion of RNase inhibitors is unnecessary in most protocols unless the RNA is exposed to extracts derived from biological materials.

12. Notes on a Few Standard Reagents

Water: The quality of the water used for making solutions is essential to RNA work. In earlier days, water for RNA work was typically double glass-distilled water. Now, other methods for purifying the water are common and water purified from salt and in particular heavy metals by ion exchange or reverse osmosis are suitable for RNA work. It is an advantage that the water is slightly acidic to prevent OH^- induced RNA degradation.

Buffers: The most common buffers are Tris (pK_a 8.1 at RT) used in the pH range 7.0–9.0 and HEPES (pK_a 7.5 at RT) used in the pH range 7.0–8.0. The buffers are made as 1 M stocks adjusted to the desired pH with HCl (Tris) or KOH (HEPES). It is important to keep in mind that these buffers are temperature sensitive, Tris (–0.028/°C) more so than HEPES (–0.014/°C).

3 M sodium acetate, pH 5.2: The stock solution made for routine ethanol precipitations is made by dissolving 408.1 g of sodium acetate · $3H_2O$ in 800 mL of water followed by adjustment of the pH to 5.2 with glacial acetic acid. Finally, the volume is adjusted to 1 L with water, dispensed into aliquots, and sterilized by sterile-filtration or autoclaving.

TE: 10 mM Tris-HCl, pH 7.5, 0.1 mM EDTA is mixed from stock solutions of Tris-HCl and EDTA. "TE" is used in the literature to describe a number of solutions containing 10 mM Tris-HCl titrated to different pH values and with 1 mM or 0.1 mM EDTA. In RNA work, the concentration of Mg^{2+} is critical for the folding state of RNA and the EDTA-concentration in this version of TE is kept low not to interfere with this.

Phenol, Phenol:Chloroform, Phenol:Chloroform:Isoamylalcohol: If the RNA lab has little expertise in handling hazardous chemicals, it is recommended that these organic solvents are purchased as ready-to-use solutions (e.g., Invitrogen).

References

1. Leontis, N. B., Stombaugh, J., Westhof, E. (2002) The non-Watson-Crick base pairs and their associated isostericity matrices. *Nucleic Acids Res* 30, 3497–3531.
2. Leontis, N. B., Lescoute, A., Westhof, E. (2006) The building blocks and motifs of RNA architecture. *Curr Opin Struct Biol* 16, 279–287.
3. Chirgwin, J. M., Przybyla, A. E., MacDonald, R. J., Rutter, W. J. (1979) Isolation of biologically active ribonucleic acid from sources enriched in ribonuclease. *Biochemistry* 18, 5294–5299.
4. MacDonald, R. J., Swift, G. H., Przybyla, A. E., Chirgwin, J. M. (1987) Isolation of RNA using guanidinium salts. *Methods Enzymol* 152, 219–227.
5. Hartmann, R. K., Bindereif, A., Schön, A., Westhof, E. (2005) *Handbook of RNA Biochemistry*. Wiley-VCH, Weinheim.
6. Murphy, J. H., Trapane, T. L. (1996) Concentration and extinction coefficient determination for oligonucleotides and analogs using a general phosphate analysis. *Anal Biochem* 240, 273–282.
7. Jones, L. J., Yue, S. T., Cheung, C. Y., Singer, V. L. (1998) RNA quantitation by fluorescence-based solution assay: RiboGreen reagent characterization. *Anal Biochem* 265, 368–374.
8. Wallace, D. M. (1987) Large- and small-scale phenol extractions. *Methods Enzymol* 152, 33–41.
9. Chomczynski, P., Sacchi, N. (1987) Single-step method of RNA isolation by acid guanidinium thiocyanate-phenol-chloroform extraction. *Anal Biochem* 162, 156–159.
10. Wallace, D. M. (1987) Precipitation of nucleic acids. *Methods Enzymol* 152, 41–48.
11. Kjems, J., Egebjerg, J., Christiansen,J. (1998) *Analysis of RNA-Protein Complexes In Vitro*. Elsevier, Amsterdam.

Chapter 3

Synthesis of RNA by In Vitro Transcription

Bertrand Beckert and Benoît Masquida

Abstract

In vitro transcription is a simple procedure that allows for template-directed synthesis of RNA molecules of any sequence from short oligonucleotides to those of several kilobases in μg to mg quantities. It is based on the engineering of a template that includes a bacteriophage promoter sequence (e.g. from the T7 coliphage) upstream of the sequence of interest followed by transcription using the corresponding RNA polymerase. In vitro transcripts are used in analytical techniques (e.g. hybridization analysis), structural studies (for NMR and X-ray crystallography), in biochemical and genetic studies (e.g. as antisense reagents), and as functional molecules (ribozymes and aptamers).

Key words: T7 RNA polymerase, in vitro transcription, template purification.

1. Introduction

RNA is conveniently synthesized by in vitro transcription using the components of bacteriophage systems. The RNA polymerase (RNAP) is a single subunit of about 100 kDa that is highly specific for its 23-bp promoter sequence. With these two simple components, it is possible to make transcripts ranging in size from less than 30 nt to well over 10^4 nt in scales from μg to mg amounts. The most frequently used systems are the T3, T7, and SP6 systems. Here, in vitro transcription is exemplified by the T7 system derived from the T7 phage of *E. coli* established many years ago (1). In vitro transcripts can be used as hybridization probes, in RNase protection or interference experiments, as antisense reagents, for analysis of RNA-binding proteins, to elucidate RNA structure by structure probing, NMR or X-ray crystallography, or as functional molecules (e.g. aptamers and ribozymes). The

H. Nielsen (ed.), *RNA*, Methods in Molecular Biology 703,
DOI 10.1007/978-1-59745-248-9_3, © Springer Science+Business Media, LLC 2011

emphasis in this chapter is the synthesis of transcripts in small scale for probes and simple biochemical applications. For a more comprehensive discussion of in vitro transcription, *see* Gruegelsiepe et al. (2).

The basic strategy is to place the sequence of interest downstream from the T7 promoter. The promoter covers the sequence ranging from –17 to +6 with +1 being the first nucleotide of the transcribed region (*see* **Fig. 3.1**). Thus, there is not complete freedom in the choice of the sequence at the very 5′-end of the in vitro transcript. Most T7 promoters, like class III promoters (3), have G's at +1, +2, and +3, and the first two G's are critical for transcriptional yield. The alternative class II promoters initiate with an A and have a similar preference for G's at +2 (4). The template for transcription can be (1) a plasmid that typically has the promoter for in vitro transcription immediately upstream from a polylinker for cloning the sequence to be transcribed, (2) a PCR product that has the T7 promoter as part of the 5′-oligonucleotide used in the PCR reaction, and (3) two annealed oligonucleotides that carries the T7 promoter sequence and the template to be transcribed (in this case, only the T7 promoter part of the template needs to be double-stranded) (*see* **Fig. 3.2**). Most plasmid cloning vectors have one or more promoters for in vitro transcription upstream of multiple cloning sites (MCS) (e.g. the pBluescript (Stratagene) and pGEM (Promega) series). An alternative strategy consists in cloning a DNA fragment including a T7 promoter immediately 5′ of the sequence to be transcribed in order to avoid the presence of nucleotides derived from the MCS in the transcript. In this case plasmids like pUC18 and pUC19 are

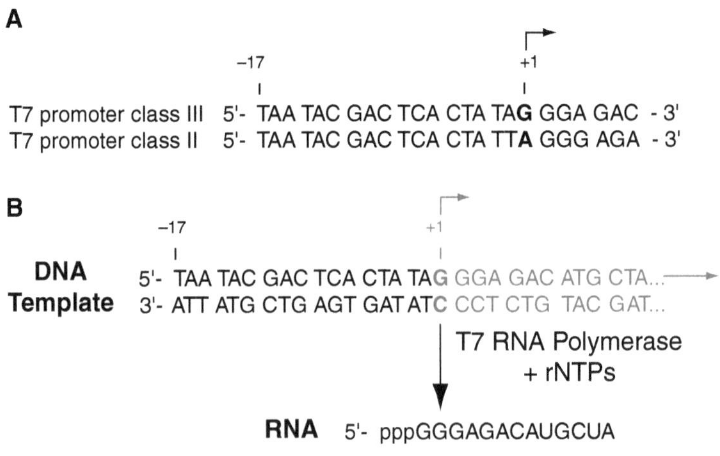

Fig. 3.1. **a** Consensus sequence of (class III and class II) T7 RNA polymerase promoter with indication of the +1 nucleotide (*bold*; corresponds to the first nucleotide in the transcript). **b** When the DNA template is incubated in the presence of T7 RNA polymerase and rNTPs, a transcript is made as indicated with a triphosphate at the 5′-end.

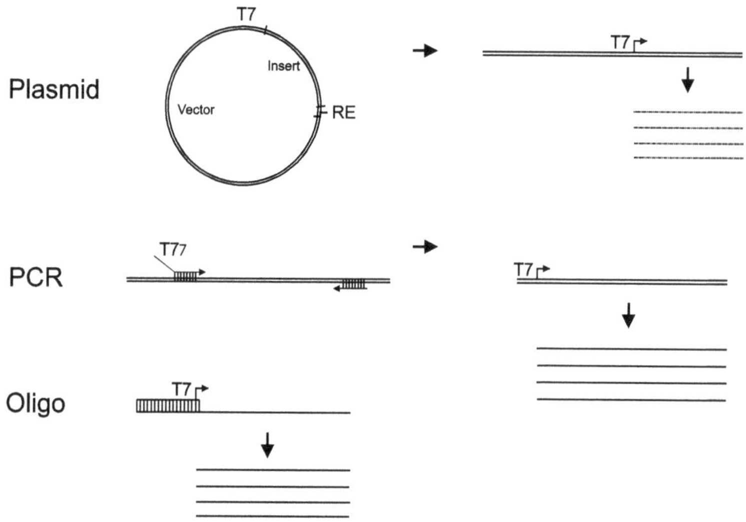

Fig. 3.2. Three different types of DNA templates for in vitro transcription. In the *upper panel*, a circular plasmid with the insert of interest cloned between a T7 promoter and a unique restriction enzyme site is linearized and transcribed from the promoter to yield multiple RNA transcripts terminated by "running-off" the template. In the *middle panel*, a DNA template (genomic DNA, cDNA, or a cloned fragment) acts as a template in PCR with a 5′-primer containing a T7 promoter (with no complementarity to the template) fused to a specific sequence complementary to the sequence of interest and a similarly specific 3′-primer. The resulting PCR-product is transcribed into RNA. In the *lower panel*, a short oligo corresponding to the T7 promoter sequence is annealed to an oligo that has the complementary sequence fused to a template sequence of interest. The partially double-stranded oligos can be transcribed into short RNAs.

preferred due to the absence of a built-in T7 promoter. Cloned templates are used for long transcripts (> 100 nt) and annealed oligo's for very short transcripts. When large amounts of RNA are needed, it is better to use a cloned template in order to generate enough template using simple and economical techniques based on bacterial culture and plasmid extraction. When small amounts are needed, PCR-products are probably the most convenient due to the flexibility in design of the template and the ease of its production.

Transcription termination in the natural setting occurs at specific terminator sites called Rho-independent terminators (5). In this mechanism, the 3′ end of the mRNA forms a hairpin structure about 7–20 base pairs in length directly followed by a U-rich stretch (6). The hairpin formation promotes pausing of the RNA polymerase and leads to the disruption of the transcription complex. However, for in vitro transcripts, termination usually intervenes by "run off," that is when the polymerase falls off at the very end of the template. With the PCR and oligo templates this is defined by the ends of the template products. With cloned templates this is achieved by linearizing the plasmid by restriction enzyme digestion downstream from the sequence of interest.

The average rate of in vitro transcription is 200–260 nt/s and the frequency error about 6 × 10^{-6} (7). In addition, the use of artificial templates for T7 transcription can result in sequence heterogeneities at the 5′ and 3′ ends of transcripts. For some applications, like in NMR or X-ray crystallography, homogeneity of the ends is crucial. Some sequences located at the 5′ end of DNA templates render the T7 RNAP inaccurate during the initiation of transcription. For example, when the template sequence starts with a stretch of 5–6 G residues, untemplated G residues can be integrated in the transcripts (8). If the 5′ end of the sequence does not start with guanine residues but with 5′C_{+1}AC/G as in the human mitochondrial lysyl and prolyl-tRNAs, transcription will occur but leads to incorporation of one additional nucleotide (preferentially a purine) or to skipping of the +1 and +2 residues (9). It is likely that other sequences could present similar transcription defects. One solution to problems like these is to fuse a cleavage ribozyme 5′ to the RNA of interest (10, 11). In this case, the natural +1 to +6 residues of the natural T7 promoter can be used regardless of the starting sequence of the RNA of interest guaranteeing efficient transcription and efficient control of the 5′ sequence content. The 3′ end of the transcript can similarly be heterogeneous. During run-off transcription T7 RNAP has a tendency to incorporate one or several non-templated nucleotides at the 3′-end, thus leaving the pool of transcripts with heterogeneous 3′-ends. This problem is addressed by incorporating a sequence that encodes a *cis*-acting cleavage ribozyme like the Hepatitis delta virus (HDV ribozyme) at the 3′-end of the template (*see* **Fig. 3.3**) (11). By using an optimized HDV ribozyme, homogenous RNA 3′ends can be easily generated even at low Mg2+ concentration (12). During transcription, the HDV ribozyme folds into an active conformation and cleaves the transcript (*see* **Fig. 3.3**). However, the competition between the folding of the RNA of interest and the folding of the HDV ribozyme could lead to reduced cleavage efficiency. This problem normally can be tackled by optimization of temperature, pH and salt conditions (13).

Another concern can be the concentration of rNTPs in the course of the transcription reaction. This problem arises when one of the nucleotides is used at limiting concentrations e.g. during synthesis of radioactive body-labelled transcripts. During the initiation process, the RNA polymerase initially produces short, abortive oligoribonucleotides of 9–12 nt in length. At some point, the polymerase switches to processive transcription leading to full-length products. If the first 9–12 nucleotides are rich in a nucleotide that is used at limiting concentrations (e.g. several U's when attempting to make a transcript labelled at high specific activity with [α-^{32}P]UTP), the switch to processive transcription is made more difficult and the ratio between full length and abortive transcripts decreases. As a consequence of this phenomenon, [α-^{32}P]GTP is frequently avoided as a label because G's are inherently rich at the 5′-end of the transcripts.

In vitro transcription protocols are easily modified to allow for synthesis of modified transcripts. T7 RNAP can initiate transcription with guanosine or GMP to obtain 5′-OH or 5′-monophosphate ends. The latter gets more easily dephosphorylated as compared to a triphosphate 5′ end for subsequent 5′ end labelling using [γ-^{32}P]ATP and T4 polynucleotide kinase. Dinucleotides (e.g. ApG) or various cap analogues, e.g. 7-methyl-guanosine (to obtain mRNA transcripts with native-like 5′-ends) can also be used for transcription initiation. The cap nucleotide protects the transcript against degradation by 5′ exonucleases present in extracts and supports translation of the transcript. T7 RNA polymerase use variety of modified nucleoside 5′ triphosphates for internal modification by incorporation. Biotinylated or digoxigenylated nucleotides can be incorporated to make non-radioactive probes for hybridization. Photoreactive nucleotides can be incorporated for synthesis of modified RNAs for various biochemical analyses. The nucleotide analog interference mapping method (NAIM, *see* Suydam and Strobel (14) for review) also relies on the property of the T7 RNA polymerase to incorporate modified nucleotides in transcripts. In this method, 5′-O-(1-thio)-nucleoside triphosphate analogs that are commercially available (GlenResearch, VA, USA) are incorporated at a 5% rate by transcription. After purification of the RNA using an activity assay specific to the studied RNA, iodine cleavage is performed so as to identify residues that are important for activity. The wild-type T7 RNA polymerase or the mutant Y639F (15) (Epicentre, WI, USA), which also allows efficient incorporation of nucleotides with a modified 2′ position, such as 2′-deoxy or 2′-fluoro can be used in this case. (*See* Gruegelsiepe (2) for a more detailed discussion of the applications of modified transcripts.)

All the protocols below describe the various procedures for in vitro transcription from plasmid- and PCR-derived templates (*see* **Fig. 3.2**). All these protocols provide simple methods to produce RNA by using a commercial T7 RNA polymerase. However, the commercial T7 RNA polymerase could be easily replaced by an in-house T7 RNA polymerase made by expression and purification of an His-tagged T7 RNA polymerase (plasmid pT7-911Q (16)). Then follow protocols for making unlabelled and ^{32}P-labelled transcripts. The protocols are for small-scale transcriptions, but they can be scaled up without problems. Similarly, the specific activity of the radioactive transcripts can be altered by adjusting the ratio between UTP and [α-^{32}P]UTP. Depending on the use of the transcript, a simple phenol:chloroform extraction directly followed by an ethanol precipitation of the transcript may be sufficient. Transcripts that are used as hybridization probes are purified by gel-filtration to get rid of the unincorporated nucleotides for reasons of radiation hazards and to allow for a simple evaluation of the probe. A protocol for gel filtration and a simple calculation of the specific activity of the probe is included. In other cases, gel purification of the transcripts is required and a simple protocol for this ends the chapter.

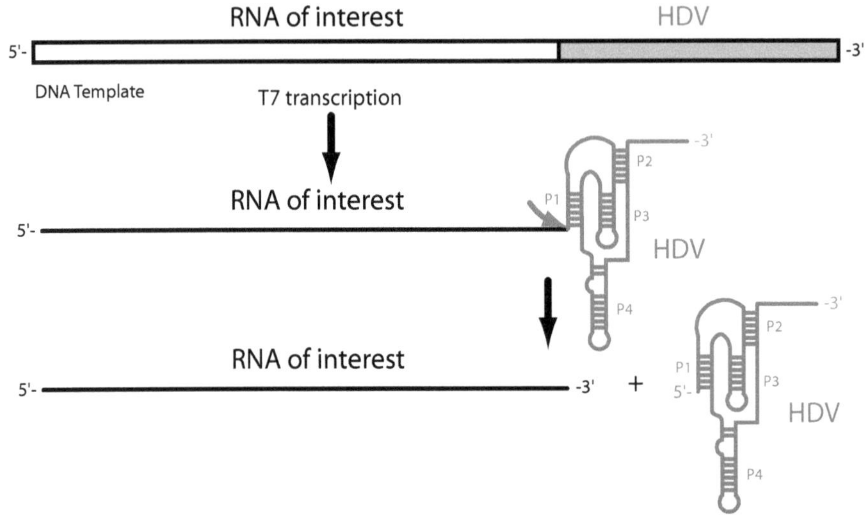

Fig. 3.3. The 3′ cassette allowing for obtaining homogeneous RNA 3′ ends. The transcribed DNA molecule (linearized plasmid, PCR product) includes an extra cassette downstream from the sequence encoding the RNA of interest. This cassette (*grey*) is transcribed into a self-splicing ribozyme (the HDV ribozyme). The cleavage activity of the HDV ribozyme leads to the release of the RNA of interest bearing a 2′,3′-cyclic phosphate group at the 3′ end.

2. Materials

2.1. Templates for In Vitro Transcription

2.1.1. Plasmid DNA Templates for In Vitro Transcription

1. Plasmid including the sequence to be transcribed downstream from a T7 promoter and upstream from a unique restriction enzyme site to be used for linearization (*see* **Note 1**).
2. Restriction enzyme and corresponding buffer.
3. Proteinase K.
4. Phenol:chloroform:isoamylalcohol (25:24:1).
5. 96% ethanol.
6. 70% ethanol.
7. TE 8.0 (10 mM Tris-HCl, pH 8.0, 0.1 mM EDTA).

2.1.2. PCR Templates for In Vitro Transcription

1. Template DNA (genomic DNA, cDNA or a cloned fragment inserted into a vector).
2. Oligonucleotides designed to amplify the sequence of interest (*see* **Note 2**).
3. Thermostable DNA polymerase with proof-reading activity such as *Pfu*I.
4. 10× polymerase buffer (usually provided by the supplier of the polymerase; *see* **Note 3**).
5. 10× dNTP-mix (2 mM of each dNTP).
6. PCR clean-up kit (e.g. GenElute™ PCR Clean-Up Kit Sigma).

2.2. In Vitro Transcription

2.2.1. In Vitro Transcription of Unlabelled Transcripts

1. Template DNA (*see* **Section 2.1.2**) at 1 μg/μL of a 3 kb linearized plasmid or 0.2 μg/μL of a 600-bp PCR-product. This will result in a final concentration of T7 promoter in the transcription of ~20 nM.
2. 10× polymerase buffer: 100 mM NaCl, 80 mM $MgCl_2$, 20 mM spermidine, 800 mM Tris-HCl, pH 8.0.
3. 100 mM DTT.
4. 10× rNTP mix: 10 mM of each rNTP.
5. T7 RNA polymerase (20 U/μL).

2.2.2. In Vitro Transcription of ^{32}P-Labelled Transcripts

1. Template DNA at 1 μg/μL of a 3 kb linearized plasmid or 0.2 μg/μL of a 600-bp PCR-product. This will result in a final concentration of T7 promoter in the transcription of ~20 nM.
2. 10× polymerase buffer: 100 mM NaCl, 80 mM $MgCl_2$, 20 mM spermidine, 800 mM Tris-HCl, pH 8.0.
3. 100 mM DTT.
4. 10× rNTP mix "low UTP" for radio-labelled transcripts: 1 mM UTP, 10 mM of each of ATP, CTP, and GTP (*see* **Note 4**).
5. T7 RNA polymerase (20 U/μL).
6. [α-^{32}P]UTP (3,000 Ci/mmol; 10 mCi/mL) (this corresponds to ~3 μM in UTP, *see* **Note 5**).

2.3. Purification

2.3.1. Purification of Transcripts by Gel Filtration

1. Sephacryl S-200 columns (GE Healthcare).

2.3.2. Gel Purification of Transcripts

1. Denaturing polyacrylamide gel.
2. TBE 10× electrophoresis buffer.
3. Ethidium bromide staining solution.
4. Elution buffer: 0.25 M sodium acetate, pH 6.0, 1 mM EDTA.
5. Phenol saturated with elution buffer.
6. Glycogen.
7. 96% ethanol.
8. TE 7.6 (10 mM Tris-HCl, pH 7.6, 0.1 mM EDTA).

3. Methods

3.1. Templates for In Vitro Transcription

3.1.1. Plasmid Templates for In Vitro Transcription

1. Digest the (RNase-free) plasmid DNA (e.g. 100 μg) with an appropriate restriction enzyme that cleaves downstream of the T7 promoter and the segment to be transcribed.
2. Add proteinase K to a final concentration of 50 μg/mL and incubate for 30 min at 37°C in order to remove the restriction enzyme from the template DNA.
3. Extract twice with one volume of phenol-chloroform (*see* **Note 6**).
4. Precipitate the template with 2.5 vols of 96% ethanol.
5. Resuspend the DNA to 1 μg/μL in TE 8.0.
6. Run an aliquot (e.g. 0.5 μg) of the DNA on an agarose gel to check the linearization of the plasmid (*see* **Note 7**).

3.1.2. PCR Templates for In Vitro Transcription

1. Design the oligos for PCR-amplification.
2. Make a standard PCR reaction.
3. Purify the PCR product using a commercial PCR clean-up kit (GenElute™ PCR Clean-Up Kit Sigma) according to the manufacturer's instructions.

3.2. In Vitro Transcription

3.2.1. In Vitro Transcription of Cold (i.e. Unlabelled) Transcripts

1. Set up the transcription reaction by adding the components in a siliconized or Teflon-coated tube in the following order at room temperature (*see* **Note 8**):

- 5 μL of 5× transcription buffer
- 4 μL of 10× rNTP mix
- 2.5 μL of 100 mM DTT
- 11.5 μL DEPC-treated dH_2O
- 1 μL of template DNA (linearized plasmid or PCR-product)
- 1 μL 10 U of the appropriate (in this case T7) RNA polymerase (*see* **Note 9**)
- Incubate for 30–60 min at 37°C.

3.2.2. In Vitro Transcription of ^{32}P-Labelled Transcripts (see ***Note 5*** for ^{32}P-Handling)

1. Set up the transcription reaction by adding at room temperature the components in a siliconized or Teflon-coated tube in the following order:
   - 5 μL of 5× transcription buffer
   - 4 μL of 10× rNTP mix "low UTP"
   - 2.5 μL of 100 mM DTT
   - 6.5 μL DEPC-treated dH_2O
   - 1 μL of template DNA (linearized plasmid or PCR-product)
   - 5 μL of 3,000 Ci/mmol, 10 mCi/ml [α-^{32}P]UTP
   - 1 μL 10 U of the appropriate (in this case T7) RNA polymerase
2. Incubate for 30–60 min at 37°C.

3.3. Purification

3.3.1. Purification of Transcripts by Gel Filtration

1. Prepare the column according to the manufacturer's recommendation (usually a brief, low-speed spin to remove storage buffer).
2. Add the transcription reaction on top of the column and spin briefly (typically 2 min) at low speed (735×*g*).
3. Collect the eluate containing the transcript. Most of the unincorporated nucleotides are retained in the column. If the transcript is radioactive, an aliquot can be removed and used for estimation of the specific activity without further purification (*see* **Note 10**).

3.3.2. Gel Purification of Transcripts

1. 1. Run the transcription mixture on a denaturing polyacrylamide gel (*see* **Note 11**; *see* **Fig. 3.4**). The type of gel depends on the size of the transcript to be purified, but in most cases, a 5% polyacrylamide gel will be appropriate.
2. Visualize the RNA by ethidium bromide staining or UV_{254}-shadowing over Xerox paper. Radioactive transcripts are detected by autoradiography using fluorescent markers to help in alignment of the gel and autoradiogram.

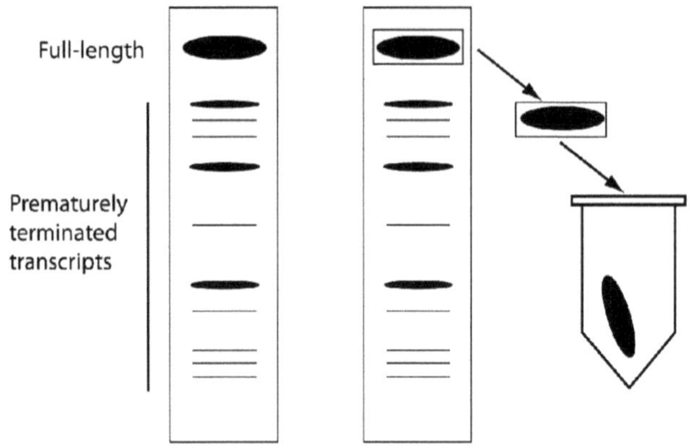

Fig. 3.4. Gelelectrophoretic separation of a transcription reaction. In addition to the full-length transcript, several prematurely terminated transcripts are seen. The full-length transcript can be excised from the gel and eluted into a buffer from which it can be recovered. Premature termination is typical when the concentration of one nucleotide is lowered to favour synthesis of radioactive transcripts of high specific activity. The presence of sequences in the template that resemble terminators or other sequences that are difficult to transcribe will similarly result in short transcripts.

3. Excise the full-length transcript using a scalpel. Avoid carrying over excessive amounts of polyacrylamide.
4. Place the gel slice in a tube containing 400 μL of elution buffer and an equal volume of phenol (*see* **Notes 12 and 13**).
5. Shake the tubes at room temperature for several hours or over night in the cold room (4°C). The time required will depend on the size of the RNA and the acrylamide gel concentration.
6. Spin and transfer all the liquid to a new tube.
7. Spin and transfer the aqueous phase to a new tube. Add 4 μL of glycogen and 1,200 μL of ethanol to precipitate the RNA.
8. Resuspend in dH_2O or TE buffer.

4. Notes

1. A restriction enzyme that leaves 5′-protruding ends is preferred in the linearization of the plasmid because T3 and T7 polymerases can initiate transcription from the ends of DNA fragments. This type of initiation is most prevalent with 3′-protruding termini followed by blunt ends and 5′-protruding termini. Non-specific initiation is suppressed in transcription buffers with increased (100 mM) NaCl

concentration. However, this will also result in a decrease of the total transcription efficiency by approximately 50%.

2. The 5′-oligo should incorporate the class III T7 promoter sequence: 5-′TAATACGACTCACTAT AGG(G) or the class II promoter sequence for ApG transcription starter: 5′-TAATACGACTCACT ATTAG (*see* **Fig. 3.1**) both of them directly followed by specific target sequence. For this and the 3′-oligo, we typically use 15- to 20-mer sequences with a T_m around 50°C as calculated adding 2°C for each A or T in the sequence and 4°C for each C or G. This simple approach for designing oligos rarely fails. However, it is also possible to use software made to optimize primer design, such as Primer3 found at **http://frodo.wi.mit.edu**.
3. The free [Mg^{2+}] must be adjusted according to the nucleotide concentration. Since each nucleotide chelates one Mg^{2+} ion, the total [Mg^{2+}] should exceed the total nucleotide concentration by approximately 5 mM.
4. Any of the four rNTPs can be used as label. The main concern is to avoid using a nucleotide that is prevalent in the first 10–12 nucleotides of the transcript and this criteria will in many cases argue against GTP because G's are required at +1 and +2 and preferred at +3 positions.
5. ^{32}P is a high energy β-emitter. Avoid exposure to the radiation and radioactive contamination. Wear disposable gloves when handling radioactive solutions. Check your gloves and pipettes frequently for radioactive contamination. Use protective laboratory equipment (protective eyeglasses, Plexiglas shields) to minimize exposure to radiation. Dispose of radioactive waste in accordance with the rules and regulations established at your institution.
6. To increase the recovery in extractions of small volumes it is sometimes advisable to increase the volume of the sample prior to extraction. For DNA samples this can be done by addition of DEPC-water.
7. Incomplete digestion can be due to suboptimal conditions or the possibility that some of the DNA was not exposed to the enzyme. As a result, subsequent transcription will lead to transcripts of the full plasmid including vector sequences. To avoid this, siliconized or Teflon-treated tubes should be used in the restriction enzyme digestion and the sample should be given a brief spin after the addition of the enzyme to collect all of the components in the bottom of the tube. One other possibility is to transfer the sample to a new tube before the next step. In this way, droplets on the side of the tube that were not exposed to the enzyme are not carried over to subsequent steps.

8. The order of assembling the reaction is to avoid spermidine precipitation of the template DNA, especially at low temperatures.
9. Alkaline pyrophosphatase can be added to the transcription reaction at 2 ng/μL. The phosphatase we use is purified from *E. coli* and commercially available at Sigma-Aldrich. This hydrolase cleaves the insoluble pyrophosphate into phosphate. Hence, the RNA pellet obtained by ethanol precipitation of the transcription reaction is free of pyrophosphate, which greatly facilitates further solubilization in an appropriate buffer. Furthermore, the hydrolysis of pyrophosphate drives the chemical equilibrium towards the formation of pyrophosphate, which means enhancing the polymerization of the RNA by the T7 RNAP and improving the transcription yield.
10. RNA labelled to a high specific activity is unstable and should be used within a couple of weeks if full-length RNA is required.
11. As an alternative to elution by diffusion, the RNA can be electro-eluted from the gel slice placed in a dialysis bag in an electrophoresis chamber (1 h at 10 V/cm in TBE) or using dedicated commercial equipment.
12. In some protocols the gel slice is crushed or freeze-thawed. In our experience this will give rise to difficulties with small pieces of polyacrylamide in downstream steps. We prefer to avoid this and have not experienced less recovery of transcript from this.
13. Break the hinge of the tube by pressing it against the table and wrap in parafilm. This will prevent leakage from the tube during shaking.

References

1. Milligan, J. F., Uhlenbeck, O. C. (1989) Synthesis of small RNAs using T7 RNA polymerase. *Methods Enzymol* 180, 51–62.
2. Gruegelsiepe, H., Schön, A., Kirsebom, L. A., Hartmann, R. K. (2005) Enzymatic RNA synthesis using bacteriophage T7 RNA polymerase, in: (Hartmann, R. K., Bindereif, A., Schön A., Westhof E., eds.), *Handbook of RNA Biochemistry*. WILEY-VCH Verlag GmbH & Co. KGaA, Germany, pp. 3–21.
3. Milligan, J. F., Groebe, D. R., Witherell, G. W., Uhlenbeck, O. C. (1987) Oligoribonucleotide synthesis using T7 RNA polymerase and synthetic DNA templates. *Nucleic Acids Res* 15, 8783–8798.
4. Huang, F., Yarus, M. (1997) 5′-RNA self-capping from guanosine diphosphate. *Biochemistry* 36, 6557–6563.
5. Jeng, S. T., Gardner, J. F., Gumport, R. I. (1990) Transcription termination by bacteriophage T7 RNA polymerase at rho-independent terminators. *J Biol Chem* 265, 3823–3830.
6. Dunn, J. J., Studier, F. W. (1983) Complete nucleotide sequence of bacteriophage T7 DNA and the locations of T7 genetic elements. *J Mol Biol* 166, 477–535.

7. Brakmann, S., Grzeszik, S. (2001) An error-prone T7 RNA polymerase mutant generated by directed evolution. *Chembiochem* 2, 212–219.
8. Pleiss, J. A., Derrick, M. L., Uhlenbeck, O. C. (1998) T7 RNA polymerase produces 5′ end heterogeneity during in vitro transcription from certain templates. *RNA* 4, 1313–1317.
9. Helm, M., Brule, H., Giege, R., Florentz, C. (1999) More mistakes by T7 RNA polymerase at the 5′ ends of in vitro-transcribed RNAs. *RNA* 5, 618–621.
10. Fechter, P., Rudinger, J., Giege, R., Theobald-Dietrich, A. (1998) Ribozyme processed tRNA transcripts with unfriendly internal promoter for T7 RNA polymerase: production and activity. *FEBS Lett* 436, 99–103.
11. Price, S. R., Ito, N., Oubridge, C., Avis, J. M., Nagai, K. (1995) Crystallization of RNA-protein complexes I. Methods for the large-scale preparation of RNA suitable for crystallographic studies. *J Mol Biol* 249, 398–408.
12. Schurer, H., Lang, K., Schuster, J., Morl, M. (2002) A universal method to produce in vitro transcripts with homogeneous 3′ ends. *Nucleic Acids Res* 30, e56.
13. Bevilacqua, P. C., Brown, T. S., Nakano, S., Yajima, R. (2004) Catalytic roles for proton transfer and protonation in ribozymes. *Biopolymers* 73, 90–109.
14. Suydam, I. T., Strobel, S. A., Daniel, H. (2009) Nucleotide analog interference mapping. *Methods Enzymol* 468, 3–30.
15. Sousa, R., Padilla, R. (1995) A mutant T7 RNA polymerase as a DNA polymerase. *EMBO J* 14, 4609–4621.
16. Ichetovkin, I. E., Abramochkin, G., Shrader, T. E. (1997) Substrate recognition by the leucyl/phenylalanyl-tRNA-protein transferase. Conservation within the enzyme family and localization to the trypsin-resistant domain. *J Biol Chem* 272, 33009–33014.

Chapter 4

Efficient Poly(A)$^+$ RNA Selection Using LNA Oligo(T) Capture

Nana Jacobsen, Jens Eriksen, and Peter Stein Nielsen

Abstract

This chapter describes a method for the isolation of intact polyadenylated mRNA using LNA oligo(T) capture. The method enables efficient isolation of poly(A)$^+$ RNA directly from guanidinium thiocyanate (GuSCN)-containing cell or tissue extract by combining the design of biotinylated LNA oligo(T) capture probes with subsequent immobilization of the captured poly(A)$^+$ RNA onto streptavidin-coated magnetic particles. In contrast to DNA oligo-dT and polyT PNA based mRNA isolation techniques, the LNA oligo(T) capture method allows poly(A) selection in the presence of 4 M GuSCN cell lysis buffer, which is needed for efficient inactivation of endogenous RNases. In addition, LNA oligo(T) facilitates highly efficient poly(A)$^+$ isolation at elevated temperatures compared to standard oligo(dT) technology. The successful use of the LNA oligo(T) capture method in recovery of mRNA from human cells and the subsequent use of the mRNA in northern blotting analysis, RT-PCR and qRT-PCR are demonstrated.

Key words: Poly(A)$^+$ RNA, mRNA, LNA, affinity purification.

1. Introduction

Efficient selection of intact polyadenylated mRNA from eukaryotic cells and tissues is an essential step for a wide selection of functional genomics applications, including full-length cDNA library construction, EST sequencing, northern and dot-blot analyses, gene expression profiling by microarrays and quantitative real time PCR. The key to successful selection of intact poly(A)$^+$ RNA is fast extraction of total RNA from cells and tissues using strong denaturing agents to disrupt the cells with the simultaneous denaturation of endogenous RNases followed by mRNA sample preparation from the extracted total RNA (1–3). Chirgwin et al. (2) improved the isolation of biolog-

H. Nielsen (ed.), *RNA*, Methods in Molecular Biology 703,
DOI 10.1007/978-1-59745-248-9_4,

ically active total RNA from tissues enriched in ribonucleases by homogenization in the chaotropic salt guanidine thiocyanate (GuSCN) and 2-mercaptoethanol followed by ethanol precipitation or by sedimentation through a cesium chloride cushion. GuSCN effectively denature secondary–tertiary protein and nucleic acid structures (4). In addition to promoting efficient cell lysis, its use in extraction buffers at a high (4 M) concentration also leads to concomitant inhibition of endogenous proteases and nucleases, including RNases (2, 5). The method was further modified by Chomczynski and Sacchi (3) to a single-step extraction of total RNA by the acid-guanidine thiocyanate-phenol-chloroform protocol. At the low pH used in this protocol, the RNA is displaced to the water phase while the DNA is selectively solubilized in the phenol phase thus eliminating the ultra-centrifugation step of the guanidinium-CsCl method (2). Yet another method has applied extraction with buffer-saturated phenol followed by proteinase K treatment to prevent RNA degradation (6).

Since most eukaryotic mRNAs contain tracts of poly(A) tails at their 3′-termini, polyadenylated mRNA can be selected by oligo(dT)-cellulose chromatography. Although peptide nucleic acid (PNA) analogues have recently been used for poly(A)$^+$ RNA isolation (7), oligo(dT) continues to be the most exploited affinity ligand in mRNA sample preparation (3). A single-step poly(A)$^+$ RNA isolation method has been described using streptavidin-coated superparamagnetic beads (8). While the direct method significantly reduces the handling and purification time, the need for high salt concentration in the stabilization of the dT-A duplexes often results in co-purification of non-polyadenylated RNAs. Moreover, the poly(A) selection is carried out directly in crude cell lysates without the presence of RNase inhibitors, thereby increasing the mRNA susceptibility to RNA degradation.

Locked nucleic acid (LNA) oligonucleotides (*see* **Fig. 4.1**) comprise a class of bicyclic RNA analogues having an exceptionally high affinity towards their complementary DNA and RNA target molecules (9, 10). We have developed a method for highly efficient isolation of intact poly(A)$^+$ RNA based on LNA-T's increased affinity to complementary poly(A) tracts (11). This allows for direct isolation of poly(A)$^+$ RNA from 4 M GuSCN-lysed cell extracts. In addition, the LNA substituted oligo(dT) probe enables efficient isolation of poly(A)$^+$ RNA from extracted total RNA samples in a low salt binding buffer. Here, we describe the protocol for isolation of poly(A)$^+$ RNA from 4-M GuSCN lysates by the combination of a biotinylated LNA oligo(T) capture probe and paramagnetic streptavidin beads.

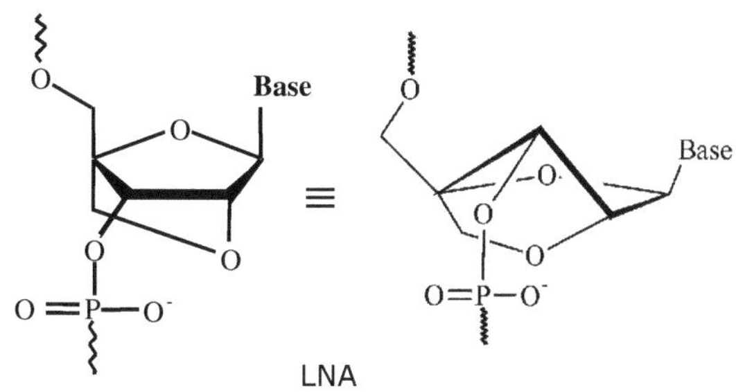

Fig. 4.1. Two representations of the chemical structure of an LNA nucleotide. The *right hand side* shows the LNA nucleotide in the 3′-endo (N-type) conformation.

2. Materials

1. 100 μM stock of biotinylated LNA oligo(T) capture probe (Exiqon) (*see* **Note 1**).
2. Lysis buffer: 4 M guanidinium thiocyanate, 25 mM Na-citrate, pH 7.0, 0.5% (w/v) sodium *N*-lauroyl sarcosinate (*see* **Note 2**).
3. Binding buffer: 20 mM Tris-HCl, pH 7.5, 0.5 M NaCl, 1 mM EDTA, pH 7.5, 0.1% (w/v) sodium *N*-lauroyl sarcosinate.
4. Washing buffer: 20 mM Tris-HCl, pH 7.5, 0.1 M NaCl, 1 mM EDTA, pH 7.5, 0.1% (w/v) sodium *N*-lauroyl sarcosinate (*see* **Note 3**).
5. TE buffer: 10 mM Tris-HCl, 1 mM EDTA, pH 7.5.
6. Quartz sand, baked at 220°C for 12 h.
7. Pestle.
8. Streptavidin-coated magnetic particles (e.g. Roche).
9. Magnetic separator (e.g. the PickPen system from BioNobile).
10. Yeast tRNA, diluted to 1 μg/μL in TE-buffer.
11. Thermomixer (Eppendorf).
12. Siliconized, RNase-free microcentrifuge tubes (e.g. from Ambion, ABI).
13. 3 M sodium acetate solution, pH 5.5.
14. Glycogen carrier, 5 mg/mL (Ambion, ABI).
15. 96% ethanol.
16. 70% ethanol.
17. RNase-free distilled/deionized water (dH_2O).

3. Methods

3.1. Sample Preparation

1. Thaw the cell or tissue sample (e.g. cells stored in RNAlater (Ambion, ABI)) (*see* **Note 4**).
2. Centrifuge at 4,000×*g* for 2 min and carefully remove the supernatant.
3. Add 200 μL of lysis buffer containing 10 mM dithiothreitol (DTT) and vortex briefly (*see* **Note 5**).
4. Add a small spatula quartz sand covering 3–5 mm of the bottom of a 1.5-mL microcentrifuge tube and disrupt the tissue/cells for 2 min on ice using a pestle in order to homogenize the sample.
5. Dilute the cell extract corresponding to 10^6 cells/50 μL in lysis buffer containing 10 mM DTT.
6. Heat the lysate at 65°C for 30 min on a thermomixer at moderate mixing avoiding the debris to precipitate.
7. Incubate for 10 min on ice and centrifuge the tube briefly (e.g. at 16,100×*g* for 1 min) and transfer the supernatant to a clean tube or directly proceed directly to the poly(A)$^+$ RNA capture (*see* pkt. 3.3).

3.2. Pre-blocking and LNA-Binding of Streptavidin-Coated Magnetic Particles

1. Pipette 60 μL of streptavidin-coated magnetic particles in suspension into a microcentrifuge tube for each sample preparation.
2. Use a magnetic separator to collect the particles on the inside of the tube wall and remove the supernatant without disturbing the particles.
3. Release the particles by removing the tube from the magnetic separator and add 100 μL of 1 μg/μL yeast tRNA in TE-buffer.
4. Keep the particles in suspension and incubate at room temperature (RT) for 5 min in order to pre-block the particles.
5. Wash the particles in 100 μL of TE-buffer using the magnetic separator to collect the particles and remove the supernatants.
6. Add to each tube 100 μL of binding buffer and add 200 pmol biotinylated LNA oligo(T) (*see* **Note 3**).
7. Incubate for 5 min at 37°C at moderate mixing to avoid sedimentation.
8. Collect the particles using the magnetic separator, remove supernatant and release the particle into 200 μL of binding buffer.

9. Repeat the washing step. Avoid the particles to dry out, completely.

3.3. Poly(A)+ RNA Isolation

1. Collect the streptavidin-coated particles using the magnetic separator and remove the supernatant.
   To the tube containing particles transfer the cell-free extract and release and resuspend the particles.
2. Incubate at 37°C for 5 min on a thermomixer at gentle mixing in order to bind the poly(A)+ RNA to the particles (*see* **Note 6**).
3. Collect the particles in the magnetic separator remove supernatant and release the particle into 200 μL of washing buffer.
4. Repeat the washing step twice.
5. Remove as much as possible of the supernatant without disturbing the particle pellet and add 50 μL of dH_2O to the tube.
6. Incubate at 65°C for 10 min in order to elute the poly(A)+ RNA from the particles (gentle mixing) and leave on ice for 5 min.
7. Collect the particles with the magnetic separator and carefully transfer the supernatant containing the poly(A)+ RNA to a clean tube.
8. Centrifuge the tube briefly (e.g. at 16,100×*g* for 1 min) and transfer the supernatant to a clean tube without transfer of remaining magnetic particles.

3.4. Ethanol Precipitation

1. Precipitate the poly(A)+ RNA by addition to the sample of 0.1× volume of 3 M sodium acetate, glycogen carrier to 150 μg/mL and 2.5 vols of 96% ethanol. Leave at –20°C overnight (*see* **Note 7**).
2. Centrifuge the tube at 16,100×*g* for 30 min at 4°C.
3. Remove the supernatant and wash the pellet with ice-cold 70% ethanol.
4. Dry the pellet at RT.
5. Dissolve the pellet in a small volume of dH_2O; centrifuge briefly to collect droplets. The poly(A)+ RNA can now be quantitated and adjusted to the appropriate concentration for subsequent use.

3.5. Analysis of Purified Poly(A)+ RNA

LNA oligo(T) capture of poly (A)+ RNA has successfully been applied to several cell types, including yeast, *C. elegans* and human cells (11). **Figure 4.2** illustrates the isolation of poly(A)+ RNA directly from 4 M GuSCN-lysed human cells and the subsequent

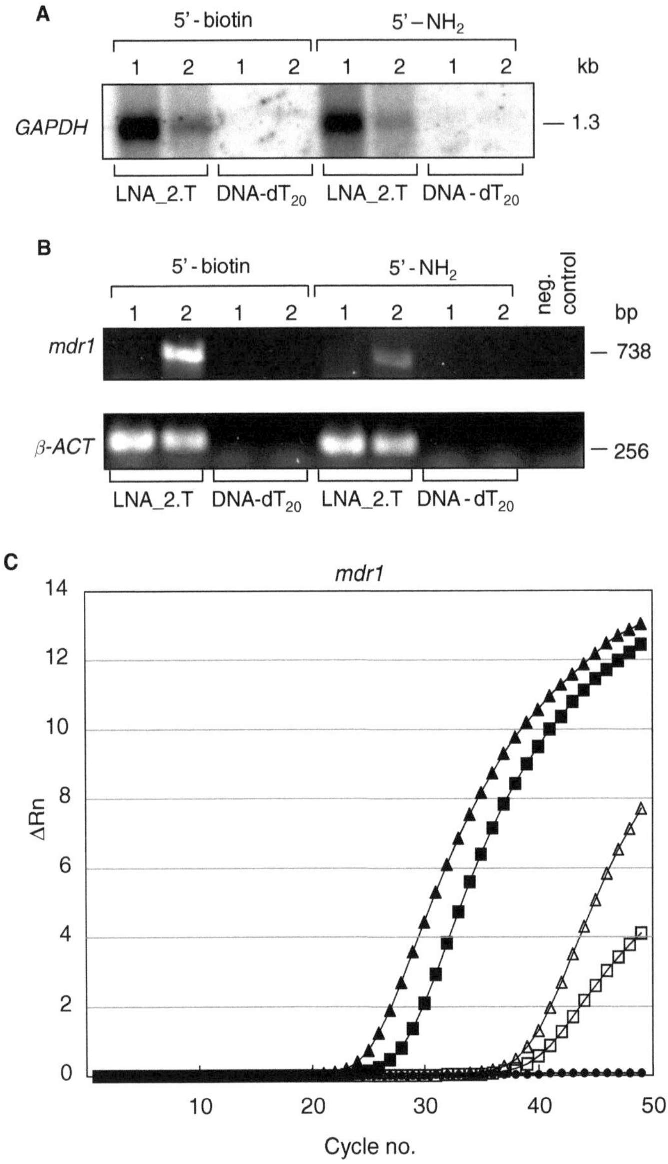

Fig. 4.2. Analysis of poly(A)$^+$ RNA isolated directly from 4 M GuSCN-lysed human K562 and K562/VCR erythroleukemia cells by LNA oligo(T) capture. **a** Northern blot analysis of the poly(A)$^+$ RNA samples selected from 4 M GuSCN-lysed human K562 (1) and K562/VCR (2) cells, respectively, using the 5′-biotinylated or 5′-NH_2-modified LNA_2.T affinity probe and the corresponding DNA oligo-dT_{20} control probes. The filter was hybridized with a ^{32}P-labelled DNA fragment for the mouse GAPDH mRNA. **b** Ca. 100 ng of poly(A)$^+$ RNA purified from the human K562 (1) and K562/VCR (2) cell lines was used as template for RT-PCR assays for human *mdr1* and *β-actin*. The amplicon sizes were 256 and 738 bp for the *β-actin* and *mdr1* mRNAs, respectively. The RT-PCR products were electrophoresed in a 1% native agarose gel and visualized by staining with Gelstar. A negative PCR control without template was performed for each assay. **c** Representative amplification plots of quantitative real-time RT-PCR assays for the human *mdr1* transcript using mRNA samples isolated from human erythroleukemia cells as template. The poly(A)$^+$ RNAs were selected either using the biotinylated LNA_2.T affinity probe from K562 cells (*solid triangle*) and K562/VCR cells (*solid square*); or by the 5′-NH_2-modifed LNA_2.T affinity probe from K562 cells (*open triangle*) and K562/VCR cells (*open square*). The plots relate the PCR cycle number to the change of detected, baseline-corrected fluorescence (ΔR_n). The small, *solid circle* depicts the fluorescence generated from the no template control reaction.

characterization by northern blotting analysis, RT-PCR and quantitative real-time PCR. The cells were a human erythroleukemia cell line derived from a chronic myeloid leukaemia patient in blast crisis (K562) and similar cells selected for resistance to the chemotherapeutic drug vincristine (K562/VCR). The yield was approximately 300 ng of poly(A)$^+$ RNA from 1 $\times$ 10^6 K562 cells with two different kinds of LNA probes (5′-biotinylated or 5′-NH_2-modified LNA_2T), whereas no mRNA could be captured with the DNA-dT_{20} control probes. Northern blot analysis of the poly(A)$^+$ RNA samples revealed a single 1.3-kb mRNA species for the human *GAPDH* gene in the K562 and K562/VCR sample preparations selected with both LNA_2.T affinity probes (*see* **Fig. 4.2a**). RT-PCR assays for the human *β-actin* mRNA revealed single cDNA fragments of the expected size in all four LNA_2.T-selected mRNA templates, whereas no PCR products were detected after 30 cycles of amplification from the DNA-dT_{20}-selected control samples (*see* **Fig. 4.2b**). In contrast, RT-PCR for the human multidrug resistance gene *mdr1* generated the 738-bp PCR amplicon in the K562/VCR cell line, but not in the drug-sensitive K562 cell line, implying that the *mdr1* gene is overexpressed in K562/VCR cells, presumably reflecting their significantly increased resistance to the chemotherapeutic drug vincristine. This result was corroborated by quantitative real-time RT-PCR that revealed an average increase of four orders of magnitude in *mdr1* expression relative to control *β-actin* mRNA in the vincristine-resistant K562/VCR cell line compared to the sensitive K562 cells (*see* **Fig. 4.2c**). It is our estimate that given the average yield of 300 ng per 1 $\times$ 10^6 cells and a fivefold dilution of the cDNA reaction made for qRT-PCR, a single LNA oligo(T) sample preparation would allow quantification of 33 different mRNAs in triplicate using real-time PCR assays. The fact that we were successful in substituting the biotinylated LNA_2.T affinity probe with the NH_2-modified LNA_2.T probe strongly suggests that the LNA oligo(T) method is amenable to automation for streamlined, high-throughput expression profiling by real-time PCR by covalently coupling the probe to solid, pre-activated surfaces, such as microtitre plate wells or magnetic particles.

4. Notes

1. The LNA described in the present work is a 20-mer oligo-dT with a 5′-biotin and every second residue substituted with an LNA-residue. This LNA is referred to as LNA_2.T: 5′-Biotin-**T**^{L}T **T**^{L}T **T**^{L}T **T**^{L}T **T**^{L}T **T**^{L}T **T**^{L}T **T**^{L}T **T**^{L}T

T^LT-3′. LNA oligo(T) capture probes can be synthesized at other lengths and with varying degrees of LNA substitution. Substitution of a DNA oligo$(dT)_{20}$ oligonucleotide with LNA-T results in significantly increased thermal duplex stabilities in all LNA oligo(T) designs measured, corresponding to an increase in melting temperature ranging from +2.8 to +6.0°C per LNA thymidine monomer. A fully substituted LNA-T_{20} has a T_M of above 95°C, indicating an exceptionally high thermal stability that would not allow efficient elution of the captured poly$(A)^+$ RNA from the affinity ligand. By comparison LNA_2.T shows a T_M of 70.8°C and an increase of 30°C compared to an all-DNA control probe. Thus, the LNA_2.T affinity probe represents an adequate compromise between increased duplex thermal stability and melting of the dA-T duplexes in elution buffer.

2. The optimal GuSCN concentration for the reference DNA oligo-dT_{20} probe was found to be 0.5 M GuSCN in accordance with previous results reported with oligo(dT) chromatography (12). In contrast RNA recovery with the LNA_2.T probe was not affected by increasing the GuSCN concentration in the binding buffer showing comparable yields of ca. 80% in the entire range from 0.5 to 4 M GuSCN.
3. A high recovery of 70–100% has been observed for both the reference DNA-dT_{20} and LNA_2.T affinity probes in the high salt concentration range of 0.2–0.5 M NaCl in the binding buffer. However, in the low salt range of 50–100 mM, a significantly decreased recovery has been observed with the reference DNA-dT_{20} probe, while the recovery is between 80 and 90% with the LNA oligo(T) affinity ligand, indicating that a low salt, high hybridization stringency window can be employed in combination with the LNA oligo(T) affinity probe without compromising the mRNA yield.
4. The protocol can also be employed using 50 μg of purified whole cell RNA as the starting material.
5. Dithiotreitol (DTT) should be added to the lysis buffer immediately before use. Stock solution of 1 M DTT in dH_2O can be stored at –20°C in aliquots.
6. We were successful in exploiting the biotin–streptavidin coupling chemistry in our mRNA isolation procedure by limiting the hybridization time to 5 min in order to prevent streptavidin from denaturation even in the presence of 4 M GuSCN. This is in accordance with previous studies reporting that streptavidin is highly resistant to denaturation by guanidine hydrochloride (13–15). Alternatively, we have

demonstrated the utility of the LNA oligo(T) sample preparation method employing a 5′ NH_2-modified LNA_2.T affinity probe coupled covalently to pre-activated magnetic particles, thus overcoming the potential problem of denaturation by GuSCN.

7. The ethanol precipitation step can be carried out in many different ways depending on the experimental situation. As an example, the precipitation time can be shortened to 15 min by placing the samples in a dry ice/ethanol bath followed by a high-speed centrifugation at 4°C.

References

1. Aviv, H., Leder, P. (1972) Purification of biologically active globin messenger RNA by chromatography on oligothymidylic acid-cellulose. *Proc Natl Acad Sci USA* 69, 1408–1412.
2. Chirgwin, J. M., Przybyla, A. E., MacDonald, R. J., Rutter, W. J. (1979) Isolation of biologically active ribonucleic acid from sources enriched in ribonuclease. *Biochemistry* 18, 5294–5299.
3. Chomczynski, P., Sacchi, N. (1987) Single-step method of RNA isolation by acid guanidinium thiocyanate-phenol-chloroform extraction. *Anal Biochem* 162, 156–159.
4. von Hippel, P. H., Wong, K. Y. (1964) Neutral salts: the generality of their effects on the stability of macromolecular conformations. *Science* 145, 577–580.
5. MacDonald, R. J., Swift, G. H., Przybyla, A. E., Chirgwin, J. M. (1987) Isolation of RNA using guanidinium salts. *Methods Enzymol* 152, 219–227.
6. Frazier, M. L., Mars, W., Florine, D. L., Montagna, R. A., Saunders, G. F. (1983) Efficient extraction of RNA from mammalian tissue. *Mol Cell Biochem* 56, 113–122.
7. Phelan, D., Hondorp, K., Choob, M., Efimov, V., Fernandez, J. (2001) Messenger RNA isolation using novel PNA analogues. *Nucleosides Nucleotides Nucleic Acids* 20, 1107–1111.
8. Hornes, E., Korsnes, L. (1990) Magnetic DNA hybridization properties of oligonucleotide probes attached to superparamagnetic beads and their use in the isolation of poly(A) mRNA from eukaryotic cells. *Genet Anal Tech Appl* 7, 145–150.
9. Koshkin, A. A., Singh, S. K., Nielsen, P., Rajwanshi, V. K., Kumar, R., Meldgaard, M., Olsen, C. E., Wengel, J. (1998) LNA (Locked Nucleic Acid): synthesis of the adenine, cytosine, guanine, 5-methylcytosine, thymine and uracil bicyclonucleoside monomers, oligomerisation, and unprecedented nucleic acid recognition. *Tetrahedron Lett* 54, 3607–3630.
10. Obika, S., Nanbu, D., Hari, Y., Morio, K., In, Y., Ishii, J. K., Imanishi, T. (1997) Synthesis of 2′-*O*, 4′-*C* methyleneuridine and cytidine. Novel bicyclic nucleosides having a fixed C_3-*endo* sugar puckering. *Tetrahedron Lett* 38, 8735–8738.
11. Jacobsen, N., Nielsen, P. S., Jeffares, D. C., Eriksen, J., Ohlsson, H., Arctander, P., Kauppinen, S. (2004) Direct isolation of poly(A)+ RNA from 4 M guanidine thiocyanate-lysed cell extracts using locked nucleic acid-oligo(T) capture. *Nucleic Acids Res* 32, e64.
12. Morrissey, D. V., Lombardo, M., Eldredge, J. K., Kearney, K. R., Groody, E. P., Collins, M. L. (1989) Nucleic acid hybridization assays employing dA-tailed capture probes. I. Multiple capture methods. *Anal Biochem* 181, 345–359.
13. Green, N. M., Toms, E. J. (1972) The dissociation of avidin-biotin complexes by guanidinium chloride. *Biochem J* 130, 707–711.
14. Green, N. M. (1975) Avidin. *Adv Protein Chem* 29, 85–133.
15. Green, N. M. (1990) Avidin and streptavidin. *Methods Enzymol* 184, 51–67.

Chapter 5

Genome Browsers

Elfar Torarinsson

Abstract

Genome browsers are important tools for studying genomes given the vast amounts of data available. This chapter focuses on providing the reader with the skills necessary to perform relatively simple, yet powerful, analysis relating to the structure of the transcription unit. Studying available data should be one of the very first steps taken in designing experiments. This can save considerable time in your research or as expressed by Alan Bleasby "Two months in the lab can easily save an afternoon on the computer."

Key words: Genome browser, UCSC, Ensembl, track, view, comparative genomics, data, bioinformatics, expression, regulation.

1. Introduction

Whole genome data are now available from a number of closely and distantly related vertebrates. Many projects can greatly benefit from relatively simple sequence comparisons of genomic data. Here I describe some of the analysis relating to the structure of the transcription unit that would help in design of experiments.

Genome browsers are great tools to access the vast amount of genomic data available. There are three major genome browsers available, UCSC, Ensembl, and NCBI. Each of these browsers provides their own annotation of the common assembled sequence. Focus will be on the genome browsers at UCSC and Ensembl. I will describe how you easily can start using the browsers (*see* **Note 1**). Once you are comfortable navigating in the browsers it is quite simple to continue on your own to learn more and exploit the power of the browsers. Later we will deal

H. Nielsen (ed.), *RNA*, Methods in Molecular Biology 703,
DOI 10.1007/978-1-59745-248-9_5, © Springer Science+Business Media, LLC 2011

with some more concrete examples to help us find information relevant to the structure of the transcription unit. I recommend that you read this chapter while using a computer to follow the instructions. In selected sections I have added questions (in the Notes section) to make this chapter more interactive and to help you understand the potential of genome browsers.

2. Getting Started

The three major genome browsers at UCSC, Ensembl and NCBI all provide two main entry points to the browsers. These are with a known sequence or querying for known coordinates or some search term. In this chapter we will focus on the case when we know which gene we are interested in. If you want to enter the browsers with an unknown sequence, instead of a known gene, do not worry, the browser navigation described below is exactly the same, the only difference is that you use BLAT (UCSC) or BLAST (Ensembl and NCBI) to compare your sequence to their database and enter browser via the results, and not by accessing it directly with a known gene. Beware, although major updates on the browsers' appearances are rare, some things might have changed since this was written.

2.1. UCSC

1. Point your browser to **http://genome.ucsc.edu**.
2. Choose either "Genomes" or "Genome Browser" in the top left corner.
3. Here you can choose the clade, the genome, and which assembly. In the fields named "position or search term" you can enter different kinds of information. These include the following:
   - Gene names → BRCA1
   - Specific region → chr7:1–10,000 or simply chr7
   - Keywords → kinase, receptor, specific disease
   - IDs → NP, NM, OMIM and more
4. To demonstrate we will use the Human assembly from March 2006 and search for DTNBP1 (*see* **Fig. 5.1**). When you search for this you get a list of data matching your search term. In our case I chose the third listing under UCSC genes

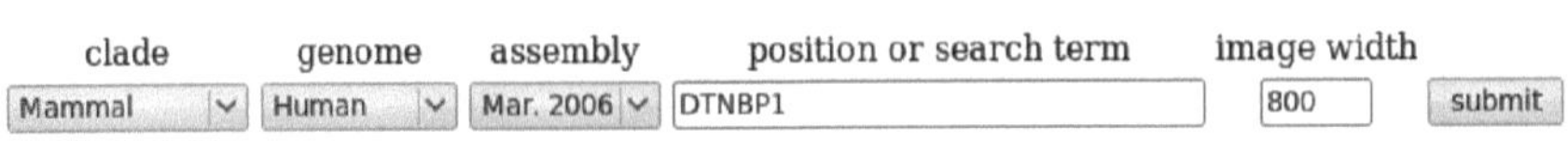

Fig. 5.1. How you search for DTNBP1 in the UCSC genome browser.

"dystrobrevin binding protein 1 isoform a." By following this link you will reach the heart of the browser.

2.1.1. Genome Browser

To better understand how this browser works it is good to know the basic organization of the underlying data. Everything in the browser is organized along the genomic sequence backbone. The data exist in so-called "tracks," which are kept in MySQL databases. For example, there is a track called "UCSC Genes" and the corresponding database for this track holds information like the name and the ID of this gene, which chromosome it belongs to and start and stop positions (also start positions for the UTR and exons, etc.). The positions are all relative to the genomic backbone, so if we are studying a region on chromosome 7 between position 10 and 10,000, the genome browser will check the "UCSC Genes" track and plot a gene on the image if it finds an overlap. The tracks are grouped together so that each group contains similar type of information, i.e., the "Genes and Gene Prediction Tracks." Continuing with our DTNBP1 example we are now at the heart of the genome browser.

5. If you have never used your current computer to access the UCSC genome browser it will display the default tracks; otherwise, if you used it before, it will remember if you removed or added tracks and show these tracks again. If you are interested in the default tracks, simply click the "default tracks" button. To begin with all this information can be overwhelming so we start by removing all the tracks by selecting the "Hide all" button, located just below the image.
6. Now the image is almost empty, only displaying where we are on chromosome 6. Now let us add the "UCSC Genes" track. The "UCSC Genes" track is located under the "Genes and Gene Prediction Tracks." By clicking on the pulldown menu you will usually have five options:
   - "Hide": completely removes a track from your image.
   - "Dense": all items become collapsed into a single line – fuses all the rows of data into one.
   - "Squish": each item is on a separate line, but at 50% of its regular height.
   - "Pack": each item is separate, but efficiently stacked like sardines. However, they are full height, which makes it different from squish.
   - "Full": each item, e.g., gene, is on a separate line.

By selecting the link above the pulldown menus you can read the information about the track, how it was generated, and so on. Sometimes you may further specify how the track should be displayed. In the "UCSC Genes" case you can, for example, change which ID it displays and read that this track is based on Ref-

Seq, UniProt, CCDS, and Comparative Genomics. Let us choose "full" for the "Known Genes" track. Then we update the image by clicking on the "Refresh" button, either just below the image or at the bottom of the page.

7. The image now displays a few versions of the DTNBP1 gene, with different colors (*see* **Note 2**). Our selected isoform is the one with dark blue background in its name (DTNBP1). The full-size boxes indicate exons, the half size-boxes indicate UTRs, and the arrows indicate the direction of the transcription.
8. To get more information about the gene you click on one of the genes (*see* **Note 3**), (*see* **Note Q-1**). This takes you to a new page with many information and links to other databases with further details about this gene. On the top of this page there are links to all the information within this page, and just below there are links to external databases and other resources within UCSC like the "Proteome Browser" and "VisiGene."
9. Finally, it is worth mentioning that when you are on the page with the *browser image*, there are a couple of useful links on the top in the blue horizontal bar. Selecting "DNA" will able you to get the DNA sequence for the region where the browser is located; furthermore, there are several options to manipulate the DNA output, like repeats in lower case and coloring of some features. The "PDF/PS" link gives you a PDF and a PS-formatted file of the image, which is useful for publications or presentations.

Although we only used two tracks and one genome in this little example, the beauty of the genome browser is that the procedure is exactly the same for all the tracks and genomes. The procedure is always the same but the information available varies between tracks and genomes. So if you can follow this small example, you should be able to study every track and genome in the browser. The best way to learn to navigate the browser is by experimenting on your own.

2.2. Ensembl

1. Point your browser to **http://www.ensembl.org/index.html**.
2. We stay true to our species and select Homo sapiens, assembly GRCh37 (the link to the right next to the "Michelangelo" icon).
3. Here we can search with some search term similar to UCSC. Here we search for DTNBP1 again. Below "By Species" click on "Homo sapiens" and then "Gene" to go to the search results (you can also enter "By Feature type" here, it does not matter).

4. This gives us two matches, either the Havana or the Ensembl protein coding gene. We choose Ensembl since it contains more information. There are two different links, a long one with the name and a shorter one named "Region in detail." The long link will take you directly to gene report for this gene, whereas the "Region in detail" link will take you to the Ensemble equivalent of the UCSC genome browser. Let us start by selecting the "Region in detail." Like UCSC everything in the viewer is organized along the genomic sequence backbone in different tracks.
5. This View displays three image boxes. The top image titled "Chromosome 6" shows chromosome 6 with a red box surrounding the region where we are. The next image zooms in on chromosome 6, again with a red box surrounding the region where we are. The information in this view includes "Contigs," Ensembl/Havana genes, non-coding RNA (ncRNA) genes, and ncRNA pseudogenes. Here, the red box surrounds our DTNBP1 gene and we can see that there is a gene named JARID2 upstream of our gene. Furthermore, the ncRNA gene U6 is upstream of our gene. Here, we can click on every gene to obtain further information about each gene, but before you do that, study the bottom image.
6. The bottom image is where we can hide and show all the tracks available at Ensembl. To add or remove tracks, follow the "Configure this page" link in the left-side navigation. You select the group of tracks on the left and click on the box in front of the track you are interested to select if and how it should be displayed. Finally, click on the "Save and Close" icon in the top right corner of the popup window, this will update the image (*see* **Note 4**). If you have a look at the "Ensembl/Havana gene" track you see that exons are indicated with filled boxes and UTRs with non-filled boxes.
7. In the bottom view, if you click on one of the transcripts in the "Ensembl/Havana genes" track like the "DTBP1-001" you will see a popup box. In this box you can choose between accessing the gene, transcript, or peptide information page. Choose the gene ("Gene:ENSG00000047579"). This will take you to the gene report for that gene.
8. This page displays the usual information at the top like the ID and a description. Below that, there are some data concerning the transcripts. In the left menu you can find links to features that are often very relevant in understanding the gene structure and potential regulation of this gene (*see* **Note Q-2**). These will be discussed in more details in the next section.

3. Comparative Genomics

UCSC and Ensembl are useful in different ways, when studying the conservation of a given gene in different organisms. UCSC is very good to quickly locate highly conserved regions, for example, high conservation upstream, downstream, or in the UTR of a given gene, indicating a possible regulatory role of that region. With UCSC you can find links, from the gene information page, to a few orthologues and view them separately in the browser. Ensembl on the other hand has many more orthologue predictions, with emphasis on predictions, and it is possible to view them simultaneously. This makes things like comparing exon structure and genomic context much easier with Ensembl. Furthermore, it is easy to retrieve pairwise or multiple alignments. Let us work with an example to illustrate these different strengths.

3.1. UCSC

1. Like in our earlier example, we go to **http://genome.ucsc.edu** and choose "Genome Browser" in the top left corner.
2. In this example we will work with homeobox C8. Select Human, assembly March 2006, and either search for hoxc8 (and choose the first match in "UCSC Genes") or go directly to the location "chr12:52,689,157-52,692,812."
3. To ease the visual inspection of the tracks, start by clicking on "hide all" tracks just below the image. Now select "pack" in the pulldown menus for the "UCSC Genes," "Conservation" and "28-way Most Cons" (*see* **Note 5**) tracks (in the "Comparative Genomics" group). Click on "Refresh" to apply your changes.
4. The image (*see* **Fig. 5.2**) now displays the HOXC8 gene. Below the "UCSC Genes" track you see the "Conservation" track. The histogram indicates the level of conservation and below you can see where the conserved regions lie in the respective organisms. Finally, at the bottom you see the "28-way Most Cons" track. It is often interesting to study the gene and the conservation simultaneously, like for this gene we can see that the 3′UTR is extremely well conserved. It is often good to be aware of simple things like this when studying the transcription and regulation of this gene (*see* **Note Q-3**).

3.2. Ensembl

1. Point your browser to **http://www.ensembl.org/index.html**.
2. Select the human genome and then search for HOXC8. Click on "Homo sapiens" and the "Gene." Go to the gene report page for the Ensembl gene (the Ensembl ID is

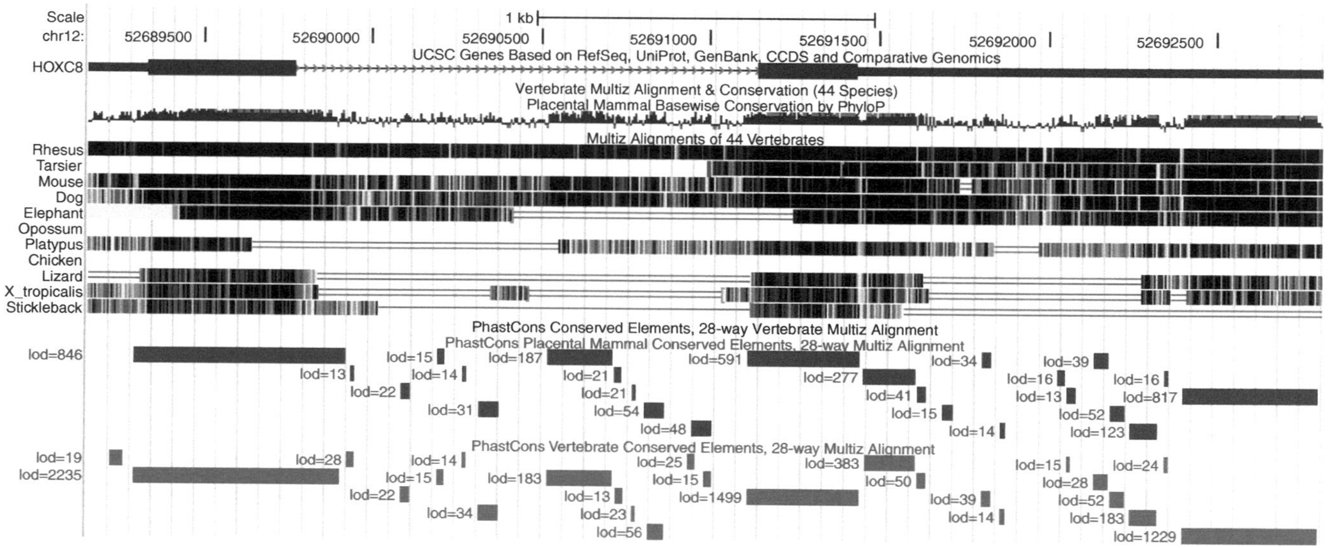

Fig. 5.2. A UCSC genome browser image displaying three tracks, the "UCSC Genes" track, the "Conservation" track, and the "28-way Most Cons" track.

ENSG00000037965). Do NOT click on "Region in detail" click on the long link with the gene name; this will take you directly to the gene report page.

3. The gene report for HOXC8 reveals, amongst other things, that there is only one known transcript, several putative orthologues, and several putative paralogues in human (Orthologues and Paralogues links are in the left-side menu).
4. From the link "Genomic alignments" you can view this gene in genomic alignments to other species. Select the "11 eutherian mammals EPO" from the "Select an alignment" pulldown. Right click on "Go to a graphical view" and open it in a new window/tab. This window makes it quite easy to compare genomic contexts in several species simultaneously. Now we are at zoom level two (the bar on the right surrounded by "+" and "–" icons is at position 2) and only see HOXC8; let us change to zoom level five by clicking on bar number five (corresponds to region 54354719–54454718) (*see* **Fig. 5.3**).
5. Studying the genomic context of a given gene in several organisms can often be very useful. For example, when studying how the gene might be regulated, but also to do things like annotate genes. One could say that it is quite likely that the "Novel RNA genes" ENSBTAG00000029788 and ENSECAG00000026361 in cow and horse, respectively, is the micro RNA miR-196, considering the annotation in the other mammals. So here we have a relatively easy way of using well-annotated organisms, to help annotate other less annotated organisms.
6. Now go back to the gene report for human HOXC8; if you do not have it open, just go back or click on the HOXC8 gene in the human box. In the Orthologues view you can do four things: (i) click on the first link and view the gene report for the orthologue, (ii) click on "Multi-species view" where you can view the orthologue, together with your gene, in a similar way to what we just did in Step 5, (iii) click on "Align" to obtain the alignment between your gene and your orthologue. Via "Configure this page" you can choose between DNA or peptide, several output formats, and species, and (iv) view the gene tree.
7. Still in the Orthologue view, we can click on "View sequence alignments of these homologues." As the name implies, this will show all the pairwise orthologues and paralogues alignments (the same as clicking on "Align" for every orthologue).
8. Finally in the transcript view, in accessed through the "Transcript: HOXC8-201" link at the top (next to "Gene:

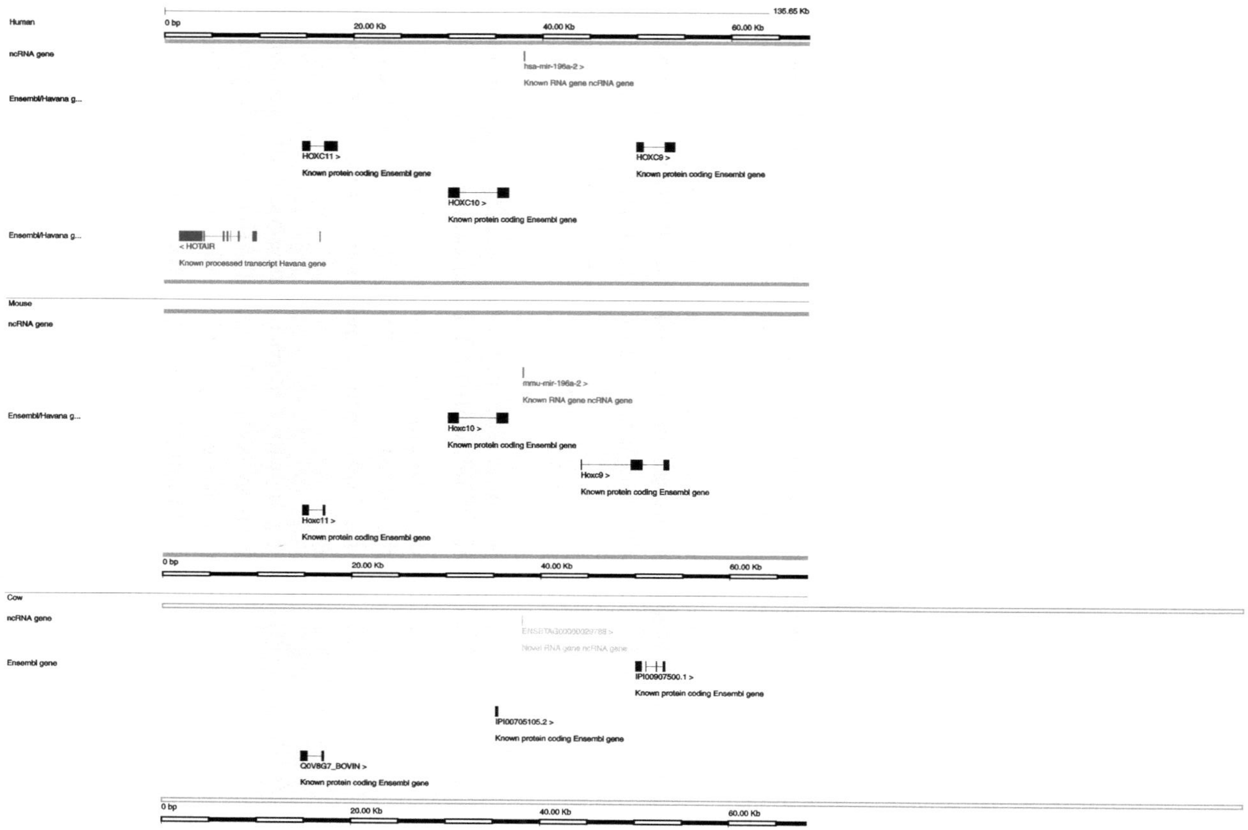

Fig. 5.3. A simplified image of the region surrounding the HOXC8 gene in humans. This image only shows the Ensembl/Havana genes and ncRNAs for human, mouse and cow (i.e., in "Configure this page" I removed some default tracks and species).

HOXC8") there is more interesting data to be obtained. These include the following links:

- "Gene Ontology": where you can see which GO terms (see Note 6) have been mapped to this gene, and by following the links there you can further information concerning the GO term.
- "Domains & features": where you see which domains the gene has and view all the genes with the same domain.
- "Population comparison": where you can see variations in this transcript (i.e., to Watson and Venter).

4. Expression and Regulation

Again, when studying expression and regulation, the strengths of UCSC and Ensembl are different. UCSC, with its simple way of viewing many tracks simultaneously, makes it very easy to compare your gene with various expression and regulation tracks. To some extent this is also possible in Ensembl, though it is more difficult and time consuming. What Ensembl has is a nice view of the regulatory factors from the cisRED database, predicted miRNA target sites from miranda analysis, and regulatory features from the Ensembl Regulatory Build, to mention a few. Again, with custom tracks, this is also possible in UCSC but more difficult. Here is a simple example.

4.1. UCSC

1. We continue studying the HOXC8 gene. If you do not have it open from **Section 3.1**, repeat Steps 1–3.
2. Select "configure" under the browser image. Scroll down to the "Expression" and "Regulation" groups, click on "show all" for both groups and then "submit" at the top of the page.
3. Here you can compare our gene with several expression tracks from GNF, Yale and Affymetrix. Regulatory tracks include, for example, CpG islands, conserved transcription factor binding sites, regulatory elements from the ORegAnno database, and a track displaying ESPERR regulatory potential scores computed from alignments of seven organisms (the darker the color, the higher regulatory potential) (*see* **Note Q-4**).

4.2. Ensembl

1. We continue studying the HOXC8 gene. If you do not have it open from **Section 3.2**, repeat Steps 1 and 2.

2. In the left menu, select "Regulation." This page contains a graphical display and a listing with relevant links of regulatory features from the cisRED database and regulatory features from the Ensembl Regulatory Build, among others (*see* **Note Q-5**).

5. Other

I have just covered a fraction of the functionality at UCSC and Ensembl. There are several other interesting features in both browsers.

5.1. UCSC

Other interesting features at UCSC include the VisiGene browser and the ENCODE tracks. The VisiGene browser is a virtual microscope for viewing in situ images. These images show where a gene is used in an organism, sometimes down to cellular resolution. With VisiGene users can retrieve images that meet specific search criteria, then interactively zoom and scroll across the collection. A link to the VisiGene browser is available from the UCSC browser home page.

If your gene of interest happens to be located in the ENCODE regions, there are many ENCODE specific tracks available. This reveals all the ENCODE tracks, which can be viewed like any other track we have looked at so far. The ENCODE tracks include groups like "Transcription," "Chromatin Immunoprecipitation," "Chromatin Structure," and additional "Comparative Genomics and Variation" tracks.

5.2. Ensembl

No matter if you are looking at a gene report or the "Region in detail" many of the most interesting features at Ensembl are often located in the left menu. For example, when viewing a gene report you can view a multiple alignment of the genomic sequence with several organisms, or get a nice graphical phylogenetic tree (the "Gene Tree (image)" link) or variations (the "Variation table" and "Variation image" links). The "Variation table" site has a listing over the variations, where they are, which alleles are involved and if they are synonymous or non-synonymous when located in a coding region.

6. Notes

1. Much more detailed information and good online tutorials are available for the browsers. OpenHelix (**http://www.openhelix.com/downloads/ucsc/ucsc_home.shtml**)

have developed an online tutorial, slides, and exercises for the UCSC genome browser. UCSC also contains a user guide at **http://genome.ucsc.edu/goldenPath/help/hgTracksHelp.html**. Ensembl also includes extensive documentation. At **http://www.ensembl.org/info/website/help/index.html** you can find animated tutorials, examples, mini-courses, and glossary. I especially recommend the online tutorials for both UCSC and Ensembl as excellent ways to get started.

2. Black: feature has a corresponding entry in the Protein Data Bank (PDB). Dark blue: transcript has been reviewed or validated by either the RefSeq or SwissProt or CCDS staff. Medium blue: other RefSeq transcripts. Light blue: non-RefSeq transcripts.
3. Sometimes, if you click on an annotation track that is actually a compressed track (e.g., "dense"), instead of going to a new web page the track will spread out. You have to click a second time to see the new web page in cases like that.
4. Other options include "Export image," which allows you get the image in various formats if you need that for a publication or a presentation.
5. This track shows predictions of highly conserved elements produced by the phastCons program. PhastCons is part of the PHAST (PHylogenetic Analysis with Space/Time models) package. The predictions are based on a phylogenetic hidden Markov model (phylo-HMM), a type of probabilistic model that describes both the process of DNA substitution at each site in a genome and the way this process changes from one site to the next.
6. GO stands for Gene Ontology and is a project that provides a controlled vocabulary to describe gene and gene product attributes in any organism. The GO project has developed three structured controlled vocabularies (ontologies) that describe gene products in terms of their associated biological processes, cellular components, and molecular functions in a species-independent manner. See **http://www.geneontology.org/** for more details.

Q-1. Select the gene, which name is in a blue box.

(a) How long is the gene (including introns) and how many exons does it contain?

(b) Which disease is caused by defects in this gene?

(c) How long is protein (in amino acids)?

(d) How many orthologous genes does UCSC link to?

Q-2. On the Ensembl gene information page for DTNBP1 can you find out:

(a) What is the genomic location of this gene?

(b) Is there a predicted orthologue in *Xenopus tropicalis* (frog)?

Q-3. Here there are three tracks being displayed.

(a) Studying all the tracks together. Which two regions of the gene are least conserved (5′-UTR, 3′-UTR, intron, exons)?

Q-4. Here, all the expression and regulation tracks for HOXC8 are displayed.

(a) Is there a CpG island overlapping HOXC8?

Q-5. On the "Regulation" site for HOXC8, can you find out:

(a) Are there DNaseI CD4 enriched sites in the intron of HOXC8?

A. Here are the answers to the questions. Beware that although your answers might not be identical they might still be correct since the genome browsers are dynamic with more data being added frequently. The questions are intended to provide you with examples of what kind of information you can find.

A-1. (a) 140233 long and 10 introns, (b) Hermansky-Pudlak syndrome, (c) 351, (d) 3.

A-2. (a) Chromosome 6 at location 15,523,032–15,663,289, (b) Yes.

A-3. (a) The 5′-UTR and the intron (You can see that the conservation scores are lower there. Actually there is a highly conserved region in the intron, but in general it is less conserved).

A-4. (a) Yes.

A-5. (a) Yes (see the gray regulatory feature box in the intron region).

Acknowledgments

The author would like to thank Jan Gorodkin for useful comments.

Chapter 6

Web-Based Tools for Studying RNA Structure and Function

Ajish D. George and Scott A. Tenenbaum

Abstract

Like protein coding sequences, functional motifs in RNA elements are frequently conserved, but this conservation is most often at the structure level rather than sequence based. Proper characterization of these structural RNA motifs is both the key and the limiting step to understanding the nature of RNA–protein interactions. The discovery of elements targeted by RNA-binding proteins and how they function remains one of the most active, yet elusive areas of RNA biology. Only a limited number of these elements have been well characterized with many of the fundamental rules yet to be discovered. Here we present a comprehensive list of web based resources that can be used in the study and identification of RNA-based structural and regulatory motifs and provide a survey of the informatic resources that can have been developed to facilitate this research.

Key words: RNA, RNA-binding Protein (RBP), RNA motifs, RNA binding sites.

1. Introduction

Post-transcriptional regulation of genes and transcripts is an essential aspect of cellular processes, which remains a largely unexplored area of biology. One of the most obvious and central areas of focus is the discovery of functional RNA elements. RNA elements identified thus far include motifs within mRNAs and non-coding RNAs such as pre-miRNAs, snRNAs, snoRNAs, tRNAs, rRNAs, as well as assorted ribozymes. Like protein coding genes, the functional motifs of these RNA elements are highly conserved, but unlike protein coding genes, it is most often structure and not sequence that is conserved. Proper characterization of these structural RNA motifs is essential to understanding the post-transcriptional aspects of the genomic world yet tools to perform this complicated task have only recently begun to be developed.

H. Nielsen (ed.), *RNA*, Methods in Molecular Biology 703,
DOI 10.1007/978-1-59745-248-9_6,

In this chapter we focus on web-based informatics resources and tools that are aimed at discovering structural RNA motifs. First we present existing databases of RNA structures and their known instances (*see* Table **6.1**). These range from databases of directly imaged 3D structures to ones where consensus structures have been compiled either manually from literature or by using a computational approach. They also include databases that catalog the result of genome-wide searches for conserved structures. Complementing these structure databases is a collection of tools for searching out instances of known structures in new sequences (*see* Table **6.2**).

Table 6.1
Databases of RNA structural motifs

1. *Rfam* **http://rfam.janelia.org** 2007	A comprehensive collection of non-coding RNA (ncRNA) families, represented by multiple sequence alignments and profile stochastic context-free grammars (compiled with INFERNAL) that aims to facilitate the identification and classification of new members of known sequence families, and distributes annotation of ncRNAs in over 400 complete genome sequences. Allows querying against motif instances by EMBL ID or by de novo search of up to 2 kb of sequence	Reference (1)
2. *UTRSite* **http://bighost.ba.itb.cnr.it/UTR** 2008	A database of approximately 60 structural and sequential *cis*-regulatory functional motifs in RNA. Patterns are annotated with a modified version of the PatScan pattern definition syntax and are directly parseable only by UTRScan	Reference (2)
3. *UTRdb* **http://www.ba.itb.cnr.it/UTR** 2006	A curated database of 5′- and 3′-untranslated sequences of eukaryotic mRNAs, derived from several sources of primary data that allows selection and extraction of UTR subsets based on their genomic coordinates and/or features of the protein encoded by the relevant mRNA (e.g., GO term, PFAM domain, etc.). Experimentally validated and predicted instances of UTRsite patterns are annotated and cross-linked	Reference (2)
4. *RegRNA* **http://regrna.mbc.nctu.edu.tw** 2006	RegRNA is an integrated web server for identifying the homologs of regulatory RNA motifs and elements against an input mRNA sequence. Either sequence homologs or structural homologs of regulatory RNA motifs can be identified	Reference (3)
5. *TransTerm* **http://uther.otago.ac.nz/Transterm.html** 2007	An interactive database providing access to mRNA sequences and associated regulatory elements. The mRNA sequences are derived from all gene sequence data in Genbank, including complete genomes, divided into putative 5′-UTRs and 3′-UTRs, initiation and termination regions and the full CDS sequences. This data can be searched for defined regulatory elements	Reference (4)

Table 6.1 (Continued)

6. *RibEx* **http://www.ibt.unam.mx/biocomputo/ribex.htm** 2005
A web server capable of searching any sequence for known riboswitches as well as other predicted, but highly conserved, bacterial regulatory elements. It allows the visual inspection of the identified motifs in relation to attenuators and open reading frames (ORFs). Any of the ORF's or regulatory elements' sequence can be obtained with a click and submitted to NCBI's BLAST. Alternatively, the genome context of all other genes regulated by the same element can be explored with our genome context tool
Availability: web service, basic documentation, examples
Reference (5, 6)
7. *EvoFold* **http://www.cbse.ucsc.edu/~jsp/EvoFold** 2006
Phylogenetic stochastic context-free grammars for identifying functional RNAs were used to survey an eight-way genome-wide alignment of the human, chimpanzee, mouse, rat, dog, chicken, zebrafish, and puffer-fish genomes for deeply conserved functional RNAs. The result was a large set of candidate RNA structures including many known functional RNAs such as miRNAs, histone 3′-UTR stem loops, and various types of known genetic recoding elements. The new predictions include heretofore unknown members of known classes such as novel miRNAs and SECIS elements
Reference (7)
8. *CMFinder-ENCODE* **http://genome.ku.dk/resources/cmf_encode** 2009
Used CMfinder, a structure-oriented local alignment tool, to search the ENCODE regions of vertebrate multiple alignments for conserved RNA structures. Will be updated to full genome scan in 2009
Reference (8)
9. *RNAz ncRNAs* **http://www.tbi.univie.ac.at/papers/SUPPLEMENTS/ncRNA** 2006
A comparative screen (using RNAz) of vertebrate genomes for structural non-coding RNAs, which evaluates conserved genomic DNA sequences for signatures of structural conservation of base-pairing patterns and exceptional thermodynamic stability, predicted thousands of highly conserved structured RNA elements. Only a small fraction of these sequences has been described previously but more than 40% of the predicted structured RNAs overlap with experimentally detected sites of transcription
Reference (9)
10. *NDB* **http://ndbserver.rutgers.edu** 2008
The goal of the Nucleic Acid Database Project (NDB) is to assemble and distribute structural information about nucleic acids. The NDB processes data for the crystal structures of nucleic acids. Uses PDB formats
Reference (10)
11. *MiRBase* **http://microrna.sanger.ac.uk** 2008
miRBase Sequences is the primary online repository for miRNA sequence data and annotation. miRBase Targets is a comprehensive new database of predicted miRNA target genes
Reference (11)
12. *RNAJunction* **http://rnajunction.abcc.ncifcrf.gov** 2007
More than 12,000 extracted three-dimensional junction and kissing loop structures as well as detailed annotations for each. If you are interested in RNA as a building block for nano-scale design or if you are analyzing the properties of specific RNA motifs you should find utility in this site. The junctions in this database were extracted using a junction scanning algorithm from a number of structures from the Protein Data Bank
Reference (12)

Table 6.1
(Continued)

13. *SnoRNA-LBME-db* **http://www-snorna.biotoul.fr** 2007
Dedicated database containing human C/D box and H/ACA box small nucleolar RNAs (snoRNAs), and small Cajal body-specific RNAs (scaRNAs) as well as the target sites of their predicted action
Reference (13)
14. *Sno/scaRNAbase* **http://gene.fudan.sh.cn/snoRNAbase.nsf** 2007
Provides an easy-to-use gateway to important sno/scaRNA features such as sequence motifs, possible functions, homologues, secondary structures, genomics organization, sno/scaRNA gene's chromosome location, and more
Reference (14)
15. *SubViral Motifs* **http://subviral.med.uottawa.ca/cgi-bin/motifs.cgi** 2006
Provides secondary structures and sequences of ribozymes and other ncRNAs from viral genomes
Reference (15)
16. *GISSD* **http://www.rna.whu.edu.cn/gissd** 2008
Group I Intron Sequence and Structure Database (GISSD) is a specialized and comprehensive database for group I introns, including sequences, secondary structures, 3D structures, and internal CDSes where available for known and predicted members of the 14 Group I intron subgroups
Reference (16)
17. *CRW* **http://www.rna.ccbb.utexas.edu** 2009
Higher-order structure, and patterns of conservation and variation for organisms that span the phylogenetic tree, has been collected and analyzed for the three ribosomal RNAs (5S, 16S, and 23S rRNA), transfer RNA (tRNA), and two of the catalytic intron RNAs (group I and group II)
Reference (17)

Table 6.2
Search tools for known RNA structural motifs

1. *Infernal* **http://infernal.janelia.org** 2007
Infernal is an implementation of "covariance models" (CMs), which are statistical models of RNA secondary structure and sequence consensus. It is the primary tool used for the Rfam project. Give Infernal a multiple sequence alignment of a conserved structural RNA family, annotated with the consensus secondary structure. The "cmbuild" program builds a statistical profile of your alignment. That CM can be used as a query in a database search to find more homologs of your RNAs (the "cmsearch" program). The latest version also includes the QDB optimization algorithm
Availability: source, extensive documentation, examples
Reference (18)
2. *RSEARCH* **http://selab.janelia.org/software.html#rsearch** 2003
RSEARCH aligns an RNA query to target sequences, using SCFG algorithms to score both secondary structure and primary sequence alignment simultaneously. It's slow, but somewhat more capable of finding significant remote RNA structure homologies than sequence alignment methods like BLAST
Availability: source, extensive documentation
Reference (19)

Table 6.2 (Continued)

3. *RNABOB* **http://selab.janelia.org/software.html#rnabob** 1996

Fast Pattern searching for RNA secondary structures. RNABOB is an implementation of D. Gautheret's RNAMOT, but with a different underlying algorithm using a nondeterministic finite state machine with node rewriting rules. An RNABOB motif is a consensus pattern a la PROSITE patterns, but with base-pairing

Availability: source, limited documentation

References: none

4. *UTRScan* **http://www.pesolelab.it/** 1999

UTRScan is a web service for finding matches to all secondary structure patterns from the UTRSite database in a set of FASTA sequences. It is backed by the PatSearch program which is also provided as a web service for searching custom patterns against Fasta sequences or collected UTR sequences

Availability: web-services, basic documentation

Reference (20)

5. *PatScan* **http://www-unix.mcs.anl.gov/compbio/PatScan** 1997

PatScan is a pattern matcher which searches protein or nucleotide (DNA, RNA, tRNA, etc.) sequence archives for instances of a pattern which you input. Pattern definition rules are provided

Availability: web-service, source, basic documentation

Reference (21)

6. *RegRNA* **http://regrna.mbc.nctu.edu.tw** 2006

RegRNA is an integrated web server for identifying the homologs of Regulatory RNA motifs and elements against an input mRNA sequence. Both sequence homologs and structural homologs of regulatory RNA motifs can be identified

Availability: web service, basic documentation, examples

Reference (3)

7. *TransTerm* **http://uther.otago.ac.nz/Transterm.html** 2007

An interactive database providing access to mRNA sequences and associated regulatory elements. The mRNA sequences are derived from all gene sequence data in Genbank, including complete genomes, divided into putative 5′-UTRs and 3′-UTRs, initiation and termination regions and the full CDS sequences. This data can be searched for defined regulatory elements

Availability: web service, extensive documentation, examples

Reference (4)

8. *RibEx* **http://www.ibt.unam.mx/biocomputo/ribex.htm** 2005

A computational approach that identifies regulatory elements conserved across phylogenetically distant organisms. Intergenic regulatory regions were clustered by orthology of the adjacent genes, and an iterative process was applied to search for significant motifs, enabling new elements of the putative regulon to be added in each cycle. With this approach, we identified highly conserved riboswitches and the Gram-positive T-box. Interestingly, we identified many other regulatory systems that appear to depend on conserved RNA structures

Availability: web service, basic documentation, examples

Reference (5, 6)

Table 6.2
(Continued)

9. *Locomotif* **http://bibiserv.techfak.uni-bielefeld.de/locomotif** 2007

Locomotif: Localization of RNA motifs with generated thermodynamic matchers. Its GUI-based program that allows for the visual design of RNA motifs. The graphical structures are then translated into executable programs to be used for searching a motif in a sequence (plain text or FASTA format)

Availability: JavaWS GUI, extensive documentation

Reference (22)

10. *PHMMTS* **http://phmmts.dna.bio.keio.ac.jp/** 2004

PHMMTS provides a unifying framework and an automata-theoretic model for alignments of trees, structural alignments and pair stochastic context-free grammars. By structural alignment, we mean a pairwise alignment to align "an unfolded RNA sequence" into "an RNA sequence of known secondary structure." PHMMTS takes a folded RNA sequence and searches for a structural alignment to it in an unfolded sequence

Availability: web service, sources, basic documentation

Reference (23)

11. *Stem Kernel* **http://stem-kernel.dna.bio.keio.ac.jp/** 2004

Several computational methods based on stochastic context-free grammars have been A novel kernel function, stem kernel, for the discrimination and detection of functional RNA sequences using support vector machines (SVMs) is proposed. The stem kernel is tailored to measure the similarity of two RNA sequences from the viewpoint of secondary structures. The stem kernels are then applied to discriminate members of an RNA family from nonmembers using SVMs. The study indicates that the discrimination ability of the stem kernel is strong compared with conventional methods

Availability: sources, limited documentation

Reference (24)

12. *PSTAG* **http://pstag.dna.bio.keio.ac.jp/** 2004

This software provides an implementation of the "pair stochastic tree adjoining grammars (PSTAGs)" for modeling "pseudoknot" RNA structures, which is an extension of the "pair hidden Markov models on tree structures (PHMMTSs)." Used to align and predict RNA secondary structures including pseudoknots in unfolded sequences using a folded sequence

Availability: web service, sources, basic documentation

Reference (25)

13. *RNAinverse* **http://rna.tbi.univie.ac.at** 2000

RNAinverse searches for sequences folding into a predefined structure, thereby inverting the folding algorithm. Target structures (in bracket notation) and starting sequences for the search are read alternately from stdin. For each search the best sequence found and its Hamming distance to the start sequence are printed to stdout

Availability: web service, sources, full documentation, examples

Reference (26, 27)

14. *HomoStRscan* **http://protein3d.ncifcrf.gov/shuyun/homostrscan.html** 2004

Homologous Structural RNA Scan: A program for discovering homologous RNAs in complete genomes by taking a single RNA sequence with its secondary structure. It takes account of information of both the primary sequence and the secondary structural constraints of the query RNA in detail including each base-pairs in the duplexes and each nucleotide in the single strand. The homologous RNA structures are strictly inferred from a robust statistical distribution of a quantitative measure, maximal similarity score of RNA structures

Availability: sources, basic documentation, examples

Reference (28)

Table 6.2 (Continued)

15. *RNAMotif* **http://www.scripps.edu/mb/case/casegr-sh-3.5.html** 2008

The rnamotif program searches a database for RNA sequences that match a "motif" describing secondary structure interactions. A match means that the given sequence is capable of adopting the given secondary structure, but is not intended to be predictive. Matches can be ranked by applying scoring rules that may provide finer distinctions than just matching to a profile. It is an extension of RNAMOT and RNABOB

Availability: sources, extensive documentation
Reference (29)

16. *ERPIN* **http://tagc.univ-mrs.fr/erpin** 2005

Easy RNA Profile IdentificationN is an RNA motif search program. Unlike most RNA pattern matching programs, ERPIN does not require users to write complex descriptors before starting a search. Instead ERPIN reads a sequence alignment and secondary structure, and automatically infers a statistical "Secondary Structure Profile" (SSP). An original Dynamic Programming algorithm then matches this SSP onto any target database, finding solutions and their associated scores. Web service allows search with precompiled RNA motifs

Availability: web service, sources, binaries, full documentation, examples
Reference (30, 31)

17. *RSmatch/RADAR* **http://datalab.njit.edu/biodata/rna/RSmatch/server.htm** 2007

RADAR can align structure-annotated RNA sequences so that both sequence and structure information are taken into consideration. This server is capable of performing database search, multiple structure alignment, and pairwise structure comparison. In addition, RADAR provides two salient features (1): constrained alignment of RNA secondary structures, and (2) prediction of the consensus structure for a set of RNA sequences

Availability: web service, binaries, full documentation, examples
Reference (32)

18. *Rscan* **http://bioinfo.au.tsinghua.edu.cn/member/cxue/rscan/RScan.htm** 2007

RScan is designed to quickly find structural similarities for a query sequence with known or predicted secondary structure in a genomic database. The input format of a structured query is the output of Vienna's RNAfold

Availability: sources, basic documentation, examples
Reference (33)

19. *INFO-RNA* **http://www.bioinf.uni-freiburg.de/Software/INFO-RNA/start.html** 2007

We consider the inverse RNA folding problem, which is the design of RNA sequences that fold into a desired structure. Given a set of base pairs, we aim at finding an RNA sequence that is going to adopt these pairs. Additionally, restrictions on the sequence level can be specified by constraints given in IUPAC symbols. The resulting sequences could be used in a BLAST or related sequence search

Availability: web service, full documentation
Reference (34)

Table 6.2
(Continued)

20. *FastR/PFastR* **http://ribozyme.ucsd.edu/fastr** 2007
Given an RNA sequence with a known secondary structure, efficiently compute all structural homologs (computed as a function of sequence and structural similarity) in a genomic database. Structural filters that eliminate a large portion of the database allow us to search a typical bacterial database in minutes on a standard PC with high sensitivity and specificity. The web service allows querying of input sequences against a number of RNA structure profiles
Availability: web service, limited documentation
Reference (35, 36)
21. *MilPat* **http://cat.toulouse.inra.fr/~rnaworld/MilPat/MilPat.pl** 2006
Searching for RNA structure profiles in target sequences according to a constraint network model. Also allows searching for target sequences in interaction with RNA structures
Availability: web service, basic documentation
Reference (37)
22. *STRMS* **http://www.cs.bgu.ac.il/~vaksler/STRMS.htm** 2006
STRMS is an RNA motif search tool. Prefolds the target RNA sequences using an mfold based approach and convert them into structure trees. Then takes a query structure, builds a tree-structure from it and runs tree-alignment of the query tree against the target database
Availability: sources, basic documentation, examples
Reference (38)
23. *RNAMST* **http://bioinfo.csie.ncu.edu.tw/~rnamst** 2006
RNAMST is an efficient and flexible RNA Motif Search Tool for RNA structural homologs. RNAMST web server accepts four different kinds of input formats to facilitate the user to describe a RNA structure easily. Besides, several databases are provided and have been processed by our algorithm. Therefore, the user can easily and quickly search the RNA structural homologs against the huge amount of sequences. In addition, RNAMST is able to search structures with asymmetric mispairs and bulges that makes the search more comprehensive and practical
Availability: web service, full documentation
Reference (39)

Next we provide a list of to tools developed for the discovery of new structural motifs contained in a set of related sequences. These are divided into two main families – ones that rely on pre-aligning the sequences (*see* Table **6.3**) and those that can work with unaligned sequences (*see* Table **6.4**). The first group includes notable covariance model-based approaches as well as several classifier-driven, Bayesian, thermodynamic, and aggregate approaches. The latter table contains many improvements for simultaneous sequence/structure alignment along with novel approaches such as shape-abstraction, suffix-arrays, genetic programming, and formal grammars. To aid in the comparison and benchmarking of these motif prediction algorithms, we have also provided two of the known attempts at compiling standardized

Table 6.3
Programs for aligning sequences of RNA consensus structures

1. *Infernal* **http://infernal.janelia.org** 2007

Infernal is an implementation of "covariance models" (CMs), which are statistical models of RNA secondary structure and sequence consensus. It is the primary tool used for the Rfam project. Give Infernal a multiple sequence alignment of a conserved structural RNA family, annotated with the consensus secondary structure. The "cmbuild" program builds a statistical profile of your alignment. That CM can be used as a query in a database search to find more homologs of your RNAs (the "cmsearch" program). The latest version also includes the QDB optimization algorithm. Pre-aligned input

Availability: source, extensive documentation, examples

Reference (18)

2. *RNAz* **http://www.tbi.univie.ac.at/~wash/RNAz/** 2006

RNAz is a program for predicting structurally conserved and thermodynamically stable RNA secondary structures in multiple sequence alignments. It can be used in genome wide screens to detect functional RNA structures, as found in non-coding RNAs and *cis*-acting regulatory elements of mRNAs. Pre-aligned input

Availability: sources, windows binary, extensive documentation, examples

Reference (9)

3. *EvoFold* **http://www.cbse.ucsc.edu/~jsp/EvoFold/** 2004

EvoFold is a comparative method for identifying functional RNA structures in multiple-sequence alignments. It is based on a probabilistic model-construction called a phylo-SCFG and exploits the characteristic differences of the substitution process in stem-pairing and unpaired regions to make its predictions. Each prediction consists of a specific secondary structure and a folding potential score. Pre-aligned input

Availability: linux binary, basic documentation, static results

Reference (7)

4. *ddbRNA* **http://dibernardo.tigem.it/wiki/index.php/DdbRNA** 2003

An algorithm able to detect conserved secondary structures in both pairwise and multiple DNA sequence alignments with computational time proportional to the square of the sequence length. Pre-aligned input

Availability: cross-platform jar, limited documentation

Reference (40)

5. *RNAalifold (Vienna)* **http://rna.tbi.univie.ac.at/** 2007

RNAalifold reads aligned RNA sequences from stdin or *file.aln* and calculates their minimum free energy (mfe) structure, partition function (pf), and base pairing probability matrix. Currently, the input alignment has to be in CLUSTAL format. It returns the mfe structure in bracket notation, its energy, the free energy of the thermodynamic ensemble, and the frequency of the mfe structure in the ensemble. Pre-aligned input

Availability: web service, sources, full documentation, examples

Reference (41)

6. *SCA* **http://rna.tbi.univie.ac.at/cgi-bin/SCA.cgi** 2008

Allows the use of a variety of methods including distance between individually folded structures, global minimum free energies, and folding space searches. Wraps RNAalifold, RNAdistance, RNApdist and other algorithms using cost, distance, and clustering methods to approximate a conserved structure. Pre-aligned input

Availability: web service, limited documentation, examples

References: none

Table 6.3
(Continued)

7. *QRNA* **http://selab.janelia.org/software.html** 2003

A prototype non-coding RNA gene finder, based on comparative genome sequence analysis. QRNA uses comparative genome sequence analysis to detect conserved RNA secondary structures, including both ncRNA genes and *cis*-regulatory RNA structures. Pre-aligned input

Availability: sources, extensive documentation, examples

Reference (42)

8. *McCaskill-MEA* **http://www.ncrna.org/papers/McCaskillMEA** 2005

The McCaskill-MEA method first computes the base-pairing probability matrices for all the sequences in the alignment and then obtains the base-pairing probability matrix of the alignment by averaging over these matrices. The consensus secondary structure is predicted from this matrix such that the expected accuracy of the prediction is maximized. We show that the McCaskill-MEA method performs better than other methods, particularly when the alignment quality is low and when the alignment consists of many sequences. Pre-aligned input

Availability: sources, limited documentation, examples, benchmarks

Reference (43)

9. *ERPIN* **http://tagc.univ-mrs.fr/erpin/** 2006

Unlike most RNA pattern matching programs, ERPIN does not require users to write complex descriptors before starting a search. Instead ERPIN reads a sequence alignment and secondary structure, and automatically infers a statistical "secondary structure profile" (SSP). An original Dynamic Programming algorithm then matches this SSP onto any target database, finding solutions, and their associated scores. In the latest version (unpublished) Erpin computes E-values for matches. Pre-aligned input

Availability: web service, sources, full documentation, examples

Reference (30, 31)

10. *MSARI* **http://groups.csail.mit.edu/cb/MSARi/** 2004

A highly accurate method for identifying genes with conserved RNA secondary structure by searching multiple sequence alignments of a large set of candidate orthologs for correlated arrangements of reverse-complementary regions. This approach is growing increasingly feasible as the genomes of ever more organisms are sequenced. A program called msari implements this method and is significantly more accurate than existing methods in the context of automatically generated alignments, making it particularly applicable to high-throughput scans. Subsequently lists RNAz and Pfold as more accurate. Pre-aligned input

Availability: sources, limited documentation, examples

Reference (44)

11. *PFold* **http://www.daimi.au.dk/~compbio/rnafold** 2003

A practical way of predicting RNA secondary structure that is especially useful when related sequences can be obtained. The method improves a previous algorithm based on an explicit evolutionary model and a probabilistic model of structures. Pre-aligned input

Availability: web service, limited documentation, examples

Reference (45)

Table 6.3 (Continued)

12. *ILM* **http://www.cse.wustl.edu/~zhang/projects/rna/ilm/** 2003
Iterative loop matching (ILM) is an extended dynamic programming algorithm that is able to predict RNA secondary structures including pseudoknots. ILM can not only predict consensus structures for aligned homologous sequences, using combined thermodynamic and covariance scores, but can also be applied to individual sequences, using thermodynamic information alone. Pre-aligned input
Availability: sources, limited documentation, examples
Reference (46)
13. *BayesFold* **http://bayes.colorado.edu/Bayes/** 2003
BayesFold is a web application that finds, ranks, and draws the likeliest structures for a sequence alignment. Foldings are based on the predictions of the Bayesian statistical method. BayesFold provides convenient structure comparison and formatting functionality, and produces publication-quality graphics. Pre-aligned input
Availability: web service (MSIE only), extensive documentation
Reference (47)
14. *KnetFold* **http://knetfold.abcc.ncifcrf.gov** 2006
KNetFold is a new software for predicting the consensus RNA secondary structure for a given alignment of nucleotide sequences. It uses an innovative classifier system (a hierarchical network of *K*-nearest neighbor classifiers) to compute for each pair of alignment positions a "base-pair" or "no base-pair" prediction. We evaluated the accuracy of the KNetFold algorithm with a set of 49 RNA sequence alignments obtained from the RFAM database. In our recent publication, we show that for this test set, the performance of the method is higher compared to the programs PFOLD and RNAalifold. Pre-aligned input
Availability: web service, sources, basic documentation, examples
Reference (48)
15. *ConStruct* **http://www.biophys.uni-duesseldorf.de/construct3/** 2008
ConStruct is an RNA alignment editor and consensus structure prediction tool. It combines multiple sequence alignment, thermodynamic structure prediction, and statistics in a semiautomatical fashion. Its sophisticated GUI guides the user through correcting an initial sequence alignment with respect to a consensus structure. Its built-in structure prediction routines allow for optimal secondary structures, suboptimal secondary structures and also tertiary interactions, e.g., pseudoknots. Pre-aligned input
Availability: sources, debian package, extensive documentation, examples
Reference (49)
16. *SimulFold* **http://www.cs.ubc.ca/~irmtraud/simulfold/** 2007
SimulFold 1.0 is a computer program for co-estimating an RNA structure including pseudoknots, a multiple-sequence alignment and an evolutionary tree, given a set of evolutionarily related RNA sequences as input. In other words, you give SimulFold an initial alignment of RNA sequences as input and it will predict a consensus RNA structure which may include pseudoknots while simultaneously estimating the sequence alignment and the evolutionary tree relating the RNA sequences. Pre-aligned input
Availability: sources, limited documentation, examples
Reference (50)

Table 6.3
(Continued)

17. *STRAL* **http://www.biophys.uni-duesseldorf.de/stral/** 2006
StrAl is an alignment tool designed to provide multiple alignments of non-coding RNAs following a fast progressive strategy. It combines the thermodynamic base-pairing information derived from RNAfold calculations in the form of base-pairing probability vectors with the information of the primary sequence. Thus the scoring system is composed of two major parts evaluating the given structural and the sequence information, respectively. Pre-aligned input
Availability: web service, sources, benchmarks
Reference (51)
18. *R-Coffee/RM-Coffee* **http://www.tcoffee.org/** 2000
R-Coffee is a multiple RNA alignment package, derived from T-Coffee, designed to align RNA sequences while exploiting secondary structure information. R-Coffee uses an alignment-scoring scheme that incorporates secondary structure information within the alignment. It works particularly well as an alignment improver and can be combined with any existing sequence alignment method. Uses any of a number of sequence aligners alongside pairwise structure aligners incorporating a novel score-improving scheme. Alignment step first
Availability: web service, limited documentation
Reference (52, 53)

Table 6.4
Tools for identifying consensus structures in unaligned RNA sequences

1. *RNAShapes* **http://bibiserv.techfak.uni-bielefeld.de/rnashapes/** 2008
RNA shape abstraction maps structures to a tree-like domain of shapes, retaining adjacency and nesting of structural features, but disregarding helix lengths. Shape abstraction integrates well with dynamic programming algorithms, and hence it can be applied during structure prediction rather than afterwards. This avoids exponential explosion and can still give us a non-heuristic and complete account of properties of the molecule's folding space. RNAshapes offers three powerful RNA analysis tools in one single software package: Computation of a small set of representative structures of different shapes, complete in a well-defined sense, computation of accumulated shape probabilities, comparative prediction of consensus structures, as an alternative to the over-expensive Sankoff Algorithm
Availability: SOAP service, sources, binaries, full documentation, examples
Reference (54)
2. *RNAmine* **http://rnamine.ncrna.org/** 2006
Frequent stem pattern miner from unaligned RNA sequences (RNAmine) is a software tool to extract the structural motifs from a set of RNA sequences. The potential secondary structures of the RNA sequences are represented by directed labeled graphs with label taxonomy, and the common secondary structures are extracted by using graph mining technique. RNAmine is used for motif finding, cluster detection and common secondary structure prediction from a set of RNA sequences
Availability: web service, limited documentation, benchmarks
Reference (55)

Table 6.4 (Continued)

3. *MX-SCARNA* **http://mxscarna.ncrna.org/** 2008

MXSCARNA (Multiplex Stem Candidate Aligner for RNAs) is a multiple alignment tool for RNA sequences using progressive alignment based on pairwise structural alignment algorithm of SCARNA. This software is fast enough for large scale analyses, while the accuracies of the alignments are better than or comparable with the existing algorithms which are computationally much more expensive in time and memory

Availability: web service, sources, full documentation, benchmarks

Reference (56)

4. *SOCOS/CAN* **http://www.cbrc.jp/sokos/** 2007

An experimental implementation of stochastic or probabilistic context-free grammar (SCFG) for RNA sequence analysis with capability of computing the marginalized count kernel which is a metric similarity between two RNA sequences. The similarity takes into account of potential RNA secondary structures of the RNA's. SOKOS/CAN can be used for generic RNA sequence analysis including secondary structure prediction and homology search

Availability: sources, basic documentation, examples, benchmarks

Reference (57)

5. *RNAGA* **http://protein3d.ncifcrf.gov/shuyun/rnaga.html** 2003

A program for predicting a secondary structure common to a number of phylogenetically related sequences without the need for pre-aligned RNA sequences. One of the remarkable features of RNAGA is that RNA secondary structures are automatically optimized by not only the free energy of the formation of the structure but also the structural similarity among homologous sequences

Availability: web service, sources, full documentation, examples

Reference (58)

6. *Carnac* **http://bioinfo.lifl.fr/carnac** 2004

Predicts if sequences share a common secondary structure. When this structure exists, Carnac is then able to correctly recover a large amount of the folded stems. The input is a set of single-stranded RNA sequences that need not to be aligned. The folding strategy relies on a thermodynamic model with energy minimization. It combines information coming from locally conserved elements of the primary structure and mutual information between sequences with covariations too

Availability: web service, sources, basic documentation, examples

Reference (59)

7. *comRNA* **http://ural.wustl.edu/~yji/comRNA** 2004

The algorithm applies graph-theoretical approaches to automatically detect common RNA secondary structure motifs in a group of functionally or evolutionarily related RNA sequences. The advantages of this method are that it: does not require the presence of global sequence similarities (but can take advantage of it) does not require prior structural alignment and is able to detect pseudoknot structures. It finds sets of stable stems conserved across multiple sequences, and assembles compatible conserved stems to form consensus secondary structure motifs

Availability: sources, basic documentation, examples, benchmarks

Reference (60)

Table 6.4 (Continued)

8. *GPRM* **http://bioinfo.life.nctu.edu.tw/tools.php** 2003
GPRM is aimed at finding common secondary structure elements, not a global alignment, in a sufficiently large family (e.g., more than 15 members) of unaligned RNA sequences. It is not applicable to finding the possible folding of a single sequence. Besides, owing to the hardware limitation of our current PC server, GPRM is currently limited to finding structure elements with no more than five stems
Availability: web service (server unreachable)
Reference (61)
9. *CMFinder* **http://bio.cs.washington.edu/yzizhen/CMfinder/** 2005
CMfinder is a RNA motif prediction tool. It is an expectation maximization algorithm using covariance models for motif description, carefully crafted heuristics for effective motif search, and a novel Bayesian framework for structure prediction combining folding energy and sequence covariation. This tool performs well on unaligned sequences with long extraneous flanking regions, and in cases when the motif is only present in a subset of sequences. CMfinder also integrates directly with genome-scale homology search and can be used for automatic refinement and expansion of RNA families
Availability: web service, sources, basic documentation, examples, benchmarks
Reference (62)
10. *RNAProfile* **http://www.pesolelab.it/** 2005
We present an algorithm that takes as input a set of unaligned RNA sequences expected to share a common motif, and outputs the regions that are most conserved throughout the sequences, according to a similarity measure that takes into account both the sequence of the regions and the secondary structure they can form according to base-pairing and thermodynamic rules. Only a single parameter is needed as input, which denotes the number of distinct hairpins the motif has to contain. No further constraints on the size, number, and position of the single elements comprising the motif are required
Availability: sources, basic documentation
Reference (63)
11. *RNA Sampler* **http://ural.wustl.edu/~xingxu/RNASampler** 2008
The algorithm applies a probabilistic sampling approach and combines intrasequence base-pairing probabilities and intersequence base alignment probabilities to prediction consensus structure on two sequences. It is extended by using a consistency-based method to incorporates pairwise structural information to predict the common structure conserved among multiple sequences
Availability: sources, basic documentation, benchmarks
Reference (64)
12. *RNAspa* **http://faculty.biu.ac.il/~unger/RNAspa/** 2007
We developed the RNAspa program, which comparatively predicts the secondary structure for a set of ncRNA molecules in linear time in the number of molecules. We observed that in a list of several hundred suboptimal minimal free energy (MFE) predictions, as provided by the RNAsubopt program of the Vienna package, it is likely that at least one suggested structure would be similar to the true, correct one. The suboptimal solutions of each molecule are represented as a layer of vertices in a graph. The shortest path in this graph is the basis for structural predictions for the molecule
Availability: sources, limited documentation, benchmarks
Reference (65)

Table 6.4 (Continued)

13. *X-INS-I/Q-INS-I* **http://align.bmr.kyushu-u.ac.jp/mafft/software/source65.html** 2008
Part of MAFFT. Methods are suitable for a global alignment of highly diverged ncRNA sequences. Q-INS-i: Applicable to up to <200 sequences, <1,000 nt; Uses the Four-way Consistency objective function for incorporating structural information. X-INS-i: Applicable to up to <50 sequences, <1,000 nt; a framework based on the Four-way Consistency objective function to build a multiple structural alignment by combining pairwise structural alignments given by an external program. At present, the external program can be selected from MXSCARNA, LaRA and FOLDALIGN (the local and global options)
Availability: sources, limited documentation, benchmarks
Reference (66)
14. *RAF* **http://contra.stanford.edu/contrafold/** 2008
An efficient algorithm for simultaneous alignment and consensus folding of unaligned RNA sequences. Algorithmically, RAF exploits sparsity in the set of likely pairing and alignment candidates for each nucleotide (as identified by the CONTRAfold or CONTRAlign programs) to achieve an effectively quadratic running time for simultaneous pairwise alignment and folding. RAF's fast sparse dynamic programming, in turn, serves as the inference engine within a discriminative machine learning algorithm for parameter estimation
Availability: yet-to-be-published
Reference (67)
15. *StemLoc* **http://biowiki.org/StemLoc** 2005
An SCFG based algorithm using "alignment envelopes" and "fold envelopes" to simultaneously constrain both the alignment and the secondary structures of the sequences being compared and make the Sankoff algorithm tractable. Stemloc moves from all versus all pairwise alignments into global or local focused searches for alignments/structures
Availability: sources, full documentation, examples
Reference (68)
16. *MASTR* **http://servers.binf.ku.dk/mastr/** 2008
Using Markov chain Monte Carlo in a simulated annealing framework, the algorithm MASTR (Multiple Alignment of STructural RNAs) iteratively improves both sequence alignment and structure prediction for a set of RNA sequences. This is done by minimizing a combined cost function that considers sequence conservation, covariation and base-pairing probabilities. The results show that the method is very competitive to similar programs available today, both in terms of accuracy and computational efficiency. Unaligned input
Availability: web service, sources, limited documentation, examples
Reference (69)
17. *Murlet* **http://software.ncrna.org/** 2005
A variant of the Sankoff algorithm that uses an efficient scoring system to reduce the time and space requirements considerably without compromising on the alignment quality. First, our algorithm computes the match probability matrix that measures the alignability of each position pair between sequences as well as the base-pairing probability matrix for each sequence. These probabilities are then combined to score the alignment using the Sankoff algorithm. By itself, our algorithm does not predict the consensus secondary structure of the alignment but uses external programs for the prediction. Unaligned input
Availability: web service, sources, basic documentation, benchmarks
Reference (70)

Table 6.4 (Continued)

18. *Pmcomp/pmmulti* **http://www.tbi.univie.ac.at/RNA/PMcomp/** 2004
pmcomp and pmmulti are two programs to perform pairwise and progressive multiple alignments of RNA sequences. pmcomp is a variant of Sankoff's algorithm for simultaneous folding and alignment, which takes as input pre-computed base-pair probability matrices from McCaskill's algorithm as produced by RNAfold-p. Thus the method can also be viewed as way to compare base-pair probability matrices. pmmulti is a simple wrapper program that does progressive multiple alignments by repeatedly calling pmcomp
Availability: sources, limited documentation, examples
Reference (71)
19. *LocARNA* **http://www.bioinf.uni-freiburg.de/Software/LocARNA/** 2007
A structure-based clustering approach that is capable of extracting putative RNA classes from genome-wide surveys for structured RNAs. The LocARNA (local alignment of RNA) tool implements a novel variant of the Sankoff algorithm that is sufficiently fast to deal with several thousand candidate sequences. The method is also robust against false positive predictions, i.e., a contamination of the input data with unstructured or non-conserved sequences
Availability: web service, sources, examples, benchmarks
Reference (72)
20. *MARNA* **http://biwww2.informatik.uni-freiburg.de/Software/MARNA/** 2005
MARNA is a multiple alignment of RNAs taking into consideration both the primary sequence and the secondary structure. It is based on pairwise comparisons using costs of edit operations. The edit operations can be divided into edit operations on arcs and edit operations on bases. Additionally, MARNA predicts a consensus sequence as well as a consensus structure
Availability: web service, sources, limited documentation, examples
Reference (73)
21. *Seed* **http://bio.site.uottawa.ca/software/seed/** 2006
Program takes as input a set of unaligned RNA sequences and produces a set of secondary structure motifs. Suffix arrays are used enumerate complementary regions, possibly containing interior loops, as well for matching RNA secondary structure expressions
Availability: sources, limited documentation
Reference (74)
22. *lara* https:/**/www.mi.fu-berlin.de/w/LiSA/Lara** 2008
lara ("lagrangian relaxed structural alignment") is a tool for the sequence-structure alignment of RNA sequences. It employs methods from combinatorial optimization to compute feasible solutions for an integer linear program. lara computes all pairwise sequence-structure alignments of the input sequences and passes this information on to T-Coffee, which computes a multiple sequence-structure alignment given the pairwise alignments. This is in contrast to plara where we compute a multiple sequence-structure alignment in a progressive fashion
Availability: web service (at T-Coffee), sources, full documentation
Reference (75)

datasets of motif-containing sequences (*see* Table **6.5**). The newer of these, the UAlbany-TUTR collection, also contains matched control sets to help properly estimate sensitivity and specificity parameters for each algorithm.

Table 6.5
Benchmark datasets for consensus RNA structure prediction

1. *TUTR* **http://ribonomics.albany.edu/** 2008

To facilitate the development, evaluation, and training of new software programs that identify RNA motifs, we created the UAlbany training UTR (TUTR) database, which is a collection of validated sets of sequences containing experimentally defined regulatory motifs. Presently, eleven training sets have been generated with associated indexes and "answer sets" provided that identify where the previously characterized RNA motif [the iron responsive element (IRE), AU-rich class-2 element (ARE), selenocysteine insertion sequence (SECIS), etc.] resides in each sequence. For each training set, control sets of corresponding size are also provided that are composed not of random sequences but rather sequences from the concatenated space of genomic UTRs

Reference (76)

2. *BRAliBase* **http://people.binf.ku.dk/pgardner/bralibase/** 2004

A dataset for benchmarking secondary structure prediction algorithms. We have compiled RNA sequence alignments consisting of up to 11 sequences derived from reliable sources. These have been used to test several RNA analysis packages. Each alignment contains at least one reference sequence with (preferably) an experimentally verified secondary structure. Experimental verification of a structure may be from a variety of sources: X-ray crystallography, NMR, enzymatic structure probing, or phylogenetic inference. Most new algorithms have already been benchmarked using this set

Reference (77)

We have chosen not to include the many tools for predicting structures in individual sequences, for predicting interactions between RNA structures, or those for folding sequences in a specific association context or with specific thermodynamic constraints. We feel that these tools go beyond the task of motif prediction as it relates to *families* of functionally related mRNAs and ncRNAs and are, therefore, outside the scope of this article.

The information provided in the following tables is intended to provide an ideal starting point for studies of post-transcriptional regulatory elements in mRNA and non-coding RNA. Using the listed tools, it should be possible to survey the known space of functional RNA motifs, to search for known motifs in identified sequences of interest, and to discover new structure families in related sets of aligned and unaligned sequences.

Acknowledgments

We wish to thank the members of the Tenenbaum Lab for helpful suggestions and discussion, especially Chris Zaleski and Frank Doyle. This work was supported in part by NIH grant U01HG004571 to SAT from the NHGRI.

References

1. Griffiths-Jones, S., et al. (2005) Rfam: annotating non-coding RNAs in complete genomes. *Nucleic Acids Res* 33, suppl_1, D121–D124.
2. Mignone, F., et al. (2005) UTRdb and UTR-site: a collection of sequences and regulatory motifs of the untranslated regions of eukaryotic mRNAs. *Nucleic Acids Res* 33 Suppl 1, D141–D146.
3. Huang, H., et al. (2006) RegRNA: an integrated web server for identifying regulatory RNA motifs and elements. *Nucleic Acids Res* 34, Web Server issue, W429–W434.
4. Jacobs, G. H., et al. (2006) Transterm–extended search facilities and improved integration with other databases. *Nucleic Acids Res* 34, Database issue, D37–D40.
5. Abreu-Goodger, C., et al. (2004) Conserved regulatory motifs in bacteria: riboswitches and beyond. *TIGS* 20, 475–479.
6. Abreu-Goodger, C., Merino, E. (2005) RibEx: a web server for locating riboswitches and other conserved bacterial regulatory elements. *Nucleic Acids Res* 33, Web Server issue, W690–W692.
7. Pedersen, J. S., et al. (2006) Identification and classification of conserved RNA secondary structures in the human genome. *PLoS Comput Biol* 2, e33.
8. Torarinsson, E., et al. (2008) Comparative genomics beyond sequence-based alignments: RNA structures in the ENCODE regions. *Genome Res* 18, 242–251.
9. Washietl, S., Hofacker, I. L., Stadler, P. F. (2005) Fast and reliable prediction of noncoding RNAs. *Proc Natl Acad Sci* 102, 2454–2459.
10. Berman, H. M., et al. (1992) The nucleic acid database. A comprehensive relational database of three-dimensional structures of nucleic acids. *Biophys J* 63, 751–759.
11. Griffiths-Jones, S., et al. (2006) miRBase: microRNA sequences, targets and gene nomenclature. *Nucleic Acids Res* 34, suppl_1, D140–D144.
12. Bindewald, E., et al. (2008) RNAJunction: a database of RNA junctions and kissing loops for three-dimensional structural analysis and nanodesign. *Nucleic Acids Res* 36, Database issue, D392–D397.
13. Lestrade, L., Weber, M. J. (2006) snoRNA-LBME-db, a comprehensive database of human H/ACA and C/D box snoRNAs. *Nucleic Acids Res* 34, Database issue, D158–D162.
14. Xie, J., et al. (2007) Sno/scaRNAbase: a curated database for small nucleolar RNAs and cajal body-specific RNAs. *Nucleic Acids Res* 35, Database issue, D183–D187.
15. Rocheleau, L., Pelchat, M. (2006) The Subviral RNA Database: a toolbox for viroids, the hepatitis delta virus and satellite RNAs research. *BMC Microbiol* 6, 24.
16. Zhou, Y., et al. (2008) GISSD: group I intron sequence and structure database. *Nucleic Acids Res* 36, Database issue, D31–D37.
17. Cannone, J. J., et al. (2002) The comparative RNA web (CRW) site: an online database of comparative sequence and structure information for ribosomal, intron, and other RNAs. *BMC Bioinformatics* 3, 2.
18. Eddy, S. R. (2006) Computational analysis of RNAs. *Cold Spring Harb Symp Quant Biol* 71, 117–128.
19. Klein, R. J., Eddy, S. R. (2003) RSEARCH: finding homologs of single structured RNA sequences. *BMC Bioinformatics* 4, 44.
20. Pesole, G., Liuni, S. (1999) Internet resources for the functional analysis of 5′ and 3′ untranslated regions of eukaryotic mRNAs. *TIGS* 15, 378.
21. Dsouza, M., Larsen, N., Overbeek, R. (1997) Searching for patterns in genomic data. *Trends Genet* 13, 597–498.
22. Reeder, J., Reeder, J., Giegerich, R. (2007) Locomotif: from graphical motif description to RNA motif search. *Bioinformatics* 23, i392–i400. data. *TIGS*, 13, 497–498.
23. Sakakibara, Y. (2003) Pair hidden Markov models on tree structures. *Bioinformatics* 19 Suppl 1, i232–i240.
24. Sakakibara, Y., et al. (2007) Stem kernels for RNA sequence analyses. *J Bioinformatics Comput Biol* 5, 1103–1122.
25. Matsui, H., Sato, K., Sakakibara, Y. (2004) Pair stochastic tree adjoining grammars for aligning and predicting pseudoknot RNA structures, in *Proceedings/IEEE Computational Systems Bioinformatics Conference, CSB. IEEE Computational Systems Bioinformatics Conference*, pp. 290–299.
26. Hofacker, I. L. (2003) Vienna RNA secondary structure server. *Nucleic Acids Res* 31, 3429–3431.
27. Hofacker, I. L. (2004) RNA secondary structure analysis using the Vienna RNA package. *Curr Protoc Bioinformatics* (Editoral Board, Andreas D. Baxevanis et al., **Chapter 12**, Unit 12.2).
28. Le, S. Y., Zhang, K., Maizel, J. V. (1995) A method for predicting common structures of

homologous RNAs. *Comput Biomed Res Int J* 28, 53–66.
29. Macke, T. J., et al. (2001) RNAMotif, an RNA secondary structure definition and search algorithm. *Nucleic Acids Res* 29, 4724–4735.
30. Gautheret, D., Lambert, A. (2001) Direct RNA motif definition and identification from multiple sequence alignments using secondary structure profiles. *J Mol Biol* 313, 1003–1011.
31. Lambert, A., et al. (2005) Computing expectation values for RNA motifs using discrete convolutions. *BMC Bioinformatics* 6, 118.
32. Liu, J., et al. (2005) A method for aligning RNA secondary structures and its application to RNA motif detection. *BMC Bioinformatics* 6, 89.
33. Xue, C., Liu, G. (2007) RScan: fast searching structural similarities for structured RNAs in large databases. *BMC Genomics* 8, 257.
34. Busch, A., Backofen, R. (2006) INFO-RNA – a fast approach to inverse RNA folding. *Bioinformatic* 22, 1823–1831.
35. Bafna, V., Zhang, S. (2004) FastR: fast database search tool for non-coding RNA, in *Proceedings/IEEE Computational Systems Bioinformatics Conference, CSB. IEEE Computational Systems Bioinformatics Conference*, pp. 52–61.
36. Zhang, S., et al. (2005) Searching genomes for noncoding RNA using FastR. *IEEE/ACM Trans Comput Biol Bioinformatics* 2, 366–379.
37. Thébault, P., et al. (2006) Searching RNA motifs and their intermolecular contacts with constraint networks. *Bioinformatics* 22, 2074–2080.
38. Veksler-Lublinsky, I., et al. (2007) A structure-based flexible search method for motifs in RNA. *J Comput Biol J Comput Mol Cell Biol* 14, 908–926.
39. Chang, T., et al. (2006) RNAMST: efficient and flexible approach for identifying RNA structural homologs. *Nucleic Acids Res* 34, Web Server issue, W423–W428.
40. diBernardo, D, Down, T. and Hubbard, T (2003) ddbRNA: detection of conserved secondary structures in multiple alignments. *Bioinformatics* 19, 1606–1611.
41. Hofacker, I. L. (2007) RNA consensus structure prediction with RNAalifold. *Methods Mol Biol* 395, 527–544.
42. Rivas, E., Eddy, S. R. (2001) Noncoding RNA gene detection using comparative sequence analysis. *BMC Bioinformatics* 2, 8.
43. Kiryu, H., Kin, T., Asai, K. (2007) Robust prediction of consensus secondary structures using averaged base pairing probability matrices. *Bioinformatics* 23, 434–441.
44. Coventry, A., Kleitman, D. J., Berger, B. (2004) MSARI: multiple sequence alignments for statistical detection of RNA secondary structure. *Proc Natl Acad Sci* 101, 12102–12107.
45. Knudsen, B., Hein, J. (2003) Pfold: RNA secondary structure prediction using stochastic context-free grammars. *Nucleic Acids Res* 31, 3423–3428.
46. Ruan, J., Stormo, G. D., Zhang, W. (2004) An iterated loop matching approach to the prediction of RNA secondary structures with pseudoknots. *Bioinformatics* 20, 58–66.
47. Knight, R., Birmingham, A., Yarus, M. (2004) BayesFold: rational 2 degrees folds that combine thermodynamic, covariation, and chemical data for aligned RNA sequences. *RNA* 10, 1323–1336.
48. Bindewald, E., Shapiro, B. A. (2006) RNA secondary structure prediction from sequence alignments using a network of k-nearest neighbor classifiers. *RNA* 12, 342–352.
49. Wilm, A., Linnenbrink, K., Steger, G. (2008) ConStruct: Improved construction of RNA consensus structures. *BMC Bioinformatics* 9, 219.
50. Meyer, I. M., Miklós, I. (2007) SimulFold: simultaneously inferring RNA structures including pseudoknots, alignments, and trees using a Bayesian MCMC framework. *PLoS Comput Biol* 3, e149.
51. Dalli, D., et al. (2006) STRAL: progressive alignment of non-coding RNA using base pairing probability vectors in quadratic time. *Bioinformatics* 22, 1593–1599.
52. Wilm, A., Higgins, D. G., Notredame, C. (2008) R-Coffee: a method for multiple alignment of non-coding RNA. *Nucleic Acids Res* 36, e52.
53. Moretti, S., et al. (2008) R-Coffee: a web server for accurately aligning noncoding RNA sequences. *Nucleic Acids Res* 36, Web Server issue, W10–W13.
54. Steffen, P., et al. (2006) RNAshapes: an integrated RNA analysis package based on abstract shapes. *Bioinformatics* 22, 500–503.
55. Hamada, M., et al. (2006) Mining frequent stem patterns from unaligned RNA sequences. *Bioinformatics* 22, 2480–2487.
56. Tabei, Y., et al. (2008) A fast structural multiple alignment method for long RNA sequences. *BMC Bioinformatics* 9, 33.
57. Kin, T., Tsuda, K., Asai, K. (2002) Marginalized kernels for RNA sequence data analysis.

Genome Inform Int Conf Genome Inform 13, 112–122.
58. Le, S., Maizel, J. V., Zhang, K. (2004) An algorithm for detecting homologues of known structured RNAs in genomes, in *Proceedings/IEEE Computational Systems Bioinformatics Conference, CSB. IEEE Computational Systems Bioinformatics Conference*, pp. 300–310.
59. Touzet, H. (2007) Comparative analysis of RNA genes: the caRNAc software. *Methods Mol Biol* 395, 465–474.
60. Ji, Y., Xu, X., Stormo, G. D. (2004) A graph theoretical approach for predicting common RNA secondary structure motifs including pseudoknots in unaligned sequences. *Bioinformatics* 20, 1591–1602.
61. Hu, Y. (2003) GPRM: a genetic programming approach to finding common RNA secondary structure elements. *Nucleic Acids Res* 31, 3446–3449.
62. Yao, Z., Weinberg, Z., Ruzzo, W. L. (2006) CMfinder–a covariance model based RNA motif finding algorithm. *Bioinformatics* 22, 445–452.
63. Pavesi, G., et al. (2004) RNAProfile: an algorithm for finding conserved secondary structure motifs in unaligned RNA sequences. *Nucleic Acids Res* 32, 3258–3269.
64. Xu, X., Ji, Y., Stormo, G. D. (2007) RNA Sampler: a new sampling based algorithm for common RNA secondary structure prediction and structural alignment. *Bioinformatics* 23, 1883–1891.
65. Horesh, Y., et al. (2007) RNAspa: a shortest path approach for comparative prediction of the secondary structure of ncRNA molecules. *BMC Bioinformatics* 8, 366.
66. Katoh, K., Toh, H. (2008) Improved accuracy of multiple ncRNA alignment by incorporating structural information into a MAFFT-based framework. *BMC Bioinformatics* 9, 212.
67. Do, C. B., Foo, C., Batzoglou, S. (2008) A max-margin model for efficient simultaneous alignment and folding of RNA sequences. *Bioinformatics* 24, i68–i76.
68. Holmes, I. (2005) Accelerated probabilistic inference of RNA structure evolution. *BMC Bioinformatics* 6, 73.
69. Lindgreen, S., Gardner, P. P., Krogh, A. (2007) MASTR: multiple alignment and structure prediction of non-coding RNAs using simulated annealing. *Bioinformatics* 23, 3304–3311.
70. Kiryu, H., Tabei, Y., et al. (2007) Murlet: a practical multiple alignment tool for structural RNA sequences. *Bioinformatics* 23, 1588–1598.
71. Hofacker, I. L., Bernhart, S. H. F., Stadler, P. F. (2004) Alignment of RNA base pairing probability matrices. *Bioinformatics* 20, 2222–2227.
72. Will, S., et al. (2007) Inferring noncoding RNA families and classes by means of genome-scale structure-based clustering. *PLoS Comput Biol* 3, e65.
73. Siebert, S., Backofen, R. (2005) MARNA: multiple alignment and consensus structure prediction of RNAs based on sequence structure comparisons. *Bioinformatics* 21, 3352–3359.
74. Anwar, M., Nguyen, T., Turcotte, M. (2006) Identification of consensus RNA secondary structures using suffix arrays. *BMC Bioinformatics* 7, 244.
75. Bauer, M., Klau, G. W., Reinert, K. (2007) Accurate multiple sequence-structure alignment of RNA sequences using combinatorial optimization. *BMC Bioinformatics* 8, 271.
76. Doyle, F., et al. (2008) Bioinformatic tools for studying post-transcriptional gene regulation: the UAlbany TUTR collection and other informatic resources. *Methods Mol Biol* 419, 39–52.
77. Gardner, P., Giegerich, R. (2004) A comprehensive comparison of comparative RNA structure prediction approaches. *BMC Bioinformatics* 5, 140.

Chapter 7

Northern Blotting Analysis

Knud Josefsen and Henrik Nielsen

Abstract

Northern blotting analysis is a classical method for analysis of the size and steady-state level of a specific RNA in a complex sample. In short, the RNA is size-fractionated by gel electrophoresis and transferred by blotting onto a membrane to which the RNA is covalently bound. Then, the membrane is analysed by hybridization to one or more specific probes that are labelled for subsequent detection. Northern blotting is relatively simple to perform, inexpensive, and not plagued by artefacts. Recent developments of hybridization membranes and buffers have resulted in increased sensitivity closing the gap to the more laborious nuclease protection experiments.

Key words: Gel electrophoresis, northern blotting, probe, hybridization analysis, mRNA.

1. Introduction

Northern blotting analysis is a method for obtaining information on the size and abundance of a specific RNA in a complex mixture. In short, the RNA sample (e.g. whole cell RNA or a fraction hereof) is size-fractionated by gel electrophoresis. Then, the RNA is transferred onto a membrane ("blotted") and analysed by binding ("hybridization") of one or more labelled probes specific for the RNA in question. Northern blotting (1, 2) is a further development of a similar technique for DNA analysis, Southern blotting, developed by Ed Southern (3). Thus, its name, like western blotting, was introduced as a joke. Because northern blotting, unlike Southern blotting, does not refer to a person's name, it is spelled in lower case.

Northern blotting analysis gives information on the length of the RNA molecule and the possible existence of length variants

H. Nielsen (ed.), *RNA*, Methods in Molecular Biology 703,
DOI 10.1007/978-1-59745-248-9_7, © Springer Science+Business Media, LLC 2011

because the RNA is electrophoresed under denaturing conditions in parallel with a molecular weight marker (an RNA ladder). RNA molecules that are branched or circular have aberrant mobility's and their analysis requires specialized procedures (4). Northern blotting is frequently used simply to demonstrate the presence of a specific RNA in a sample, but the method also allows for quantitative measurements. It is the steady-state level of the RNA, that is the sum of its production and removal, that is being measured. If the transcriptional activity is of interest, this is measured by e.g. nuclear run-on experiments. Decay of the RNA is measured independently by measurement of the transcript level at various timepoints following addition of a transcription initiation inhibitor. When used as a quantitative method, northern blotting is usually used to compare the RNA level in different experimental situations rather than for determination of absolute amounts. This is because hybridization to filter-bound RNA is suboptimal and strongly dependent on experimental parameters, such as degree of covalent attachment to the filter. Addition of known amounts of an internal standard of different length but with similar sequence made by in vitro transcription could be used to estimate absolute transcript levels, but this is rarely seen.

An outline of a northern blotting analysis is shown in Fig. **7.1**. Only the most common variations of the technique are covered by this chapter. For a more comprehensive description, *see* (5–7). The input RNA can be of any kind, but mostly it is whole cell RNA ("total RNA") prepared by the acid phenol/guanidinium thiocyanate extraction method (8) or by purification on an affinity matrix as in many commercial kits for preparation of RNA. If an mRNA is being analysed and the sensitivity of the analysis is an issue, polyA+ RNA (mostly mRNA) is isolated by oligo(dT) chromatography (9) prior to gel electrophoretic fractionation. Cytoplasmic, polyadenylated RNA constitute approximately 3% of whole cell RNA in human cells. By isolating this fraction, more of the relevant RNA can be loaded on the gel and consequently less abundant RNAs can be detected. Another application that calls for fractionation is analysis of the small RNAs. Simple fractionation protocols based on RNA binding columns can be applied to size-fractionate whole cell RNA and thus used to increase the capacity of gels, e.g. for analysis of cellular microRNAs.

Large RNAs (mRNA size) are size-fractionated on agarose gels. The useful range of agarose-concentration is 0.8% (for very large RNAs) to 1.4% (w/v). Outside this range, the gels either become difficult to handle or less efficient in blotting. Small RNAs are size-fractionated on polyacrylamide gels in the range 4–20% acrylamide. The size-fractionation is performed at denaturing conditions. RNA molecules fold into complex structures that involve the formation of double-stranded segments by base pairing. Thus, size-fractionation at native conditions will depend

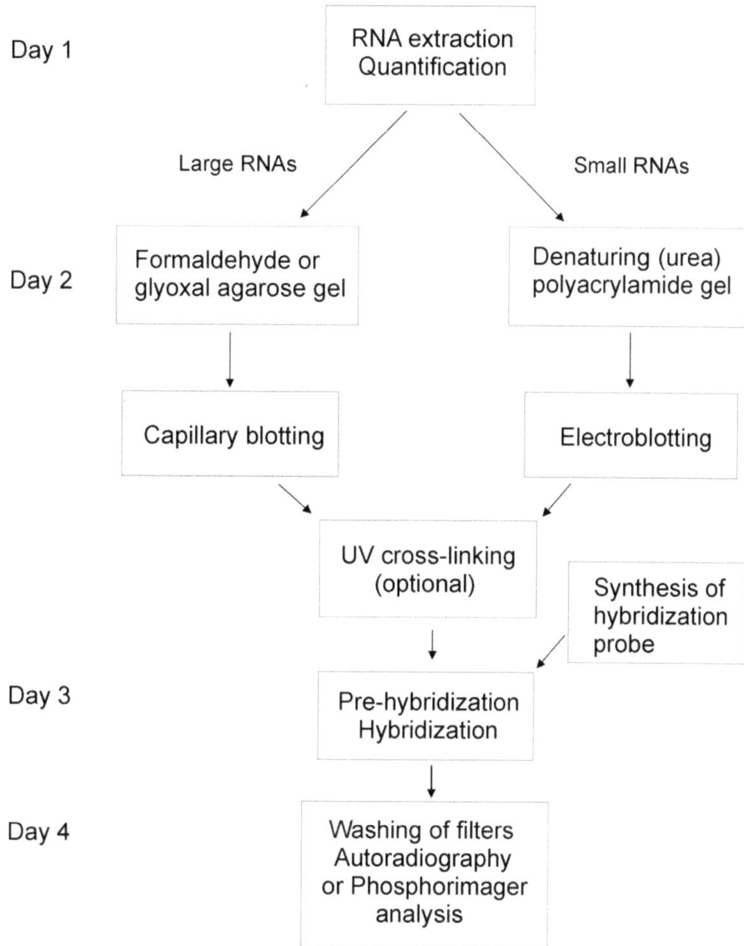

Fig. 7.1. Outline of a northern blotting analysis. The individual steps, except the extraction of RNA, are discussed in the text. A standard experiment takes 3–4 days. The overnight blotting step by upward capillary transfer can be shortened to few hours by application of alternatives discussed in the main text. The overnight hybridization step can similarly be performed in hours using "fast" hybridization solution alternatives.

not only on the length of the RNA chain but also on the folding. The two main methods for denaturing agarose gels are based on formamide/formaldehyde (10) and glyoxal (11), respectively. Formaldehyde is used both in the loading buffer (together with formamide) and in the gel. Formamide denatures the RNA and formaldehyde maintains the denatured state by reacting covalently with the amine groups of adenine, guanine, and cytosine bases. Since these bases are directly involved in formation of the A-U, G-U, and G-C base pairs in RNA, formation of secondary structure during the gel run is prevented. The reaction is reversed by treatment of the filter after blotting to allow the RNA to base pair with the probe. Glyoxal is not used in the gel but only in pre-treatment of the RNA sample. Glyoxal denatures RNA by reacting covalently with guanine bases with the effect of inhibiting base pair formation similarly to formaldehyde. The glyoxal reaction is also reversed by treatment of the filter prior to hybridization analysis. Glyoxal gels tend to produce sharper bands than formaldehyde gels but care must be taken to prevent pH-dependent reversal of the glyoxalation during the gel run. In northern blotting analysis of small RNAs, the sample is denatured by heating in a loading buffer containing formamide or urea and the gels are made with 50% (approximately 8 M) urea (12).

One of the most common uses of northern blotting analysis is comparison of RNA levels in different experimental situations represented by different lanes on the gel. Thus, equal loading of RNA samples is critical. When whole cell RNA is used, equal amounts (up to 20 μg) based on spectrophotometry is applied to the gel. This is robust because cytoplasmic ribosomal RNA make up 70–80% of total RNA and together with other stable RNA are unlikely to be affected by the experimental variable. When enrichment by isolation of polyA+ RNA or by other means is used, the RNA of interest can be enriched to different degrees in the samples that are compared. In this case, the RNA levels are normally expressed in relation to an RNA that is detected in parallel by a different probe and known to be unaffected by the experimental variable. Popular examples of standards for comparing mRNA levels are cytoskeletal mRNAs (actin, tubulin), cyclophilin, and the mRNA coding for the glycolytic enzyme glucose-6-phosphate dehydrogenase (GAPDH). However, there are numerous examples of variations in the expression levels of these and the choice of standard should be considered carefully for each type of experiment. It has been suggested that ribosomal RNA would be a convenient general standard based on the arguments presented above. This runs into technical problems in northern blotting analysis because the ribosomal RNAs are so prominently present that they saturate the filter at the position of the ribosomal RNA bands making quantitative hybridization impossible.

Northern blotting analysis of mRNA is very sensitive to even slight degradation of the sample. For this reason, it is important to visually inspect the RNA after gel electrophoresis. RNA is conveniently stained by ethidium bromide similarly to DNA albeit less efficiently because binding is by ionic interaction rather than intercalation. One approach is to include ethidium bromide in the loading buffer for formaldehyde gels. This allows inspection during the gel run. This is not an option with glyoxal gels because ethidium bromide reacts with glyoxal and interferes with its ability to denature the RNA. Alternatively, the gel is stained after the run. Inclusion of ethidium bromide in the gel is not recommended because it migrates towards the cathode creating a front of stain in the gel. An important point in staining of whole cell RNA is that the quality of the RNA can be assessed from inspection of the ribosomal RNA bands. The large (LSU) and small (SSU) subunit ribosomal RNAs are found in 1:1 ratio in cells. Thus, the differences in size (5,035 nt for LSU (28S) and 1,871 nt for SSU (18S) in humans) which is 2.7:1 should be reflected in the intensities of the two bands in ethidium bromide staining. If the ratio is less, this indicates that the RNA is partially degraded. Binding of ethidium to the RNA may reduce transfer and hybridization efficiency. If this is a problem, a test lane of the RNA should be run in parallel and stained after the gel run.

There are essentially two approaches to the blotting step of northern blotting analysis, capillary blotting and electroblotting. The traditional passive capillary transfer method (similar to the classical Southern blotting method (3)) is beautiful in its simplicity and can be performed with materials that are generally at hand in the laboratory. In essence, a stack of paper towels or similar absorbent material is used to draw a high salt transfer buffer from a reservoir, through the gel and an RNA binding membrane placed on top of the gel. The flow of buffer carries the RNA out of the gel so that it is trapped on the membrane. The setup for such an upward capillary transfer is shown in Fig. **7.2a**. Transfer takes several hours and is usually performed over night. A faster and equally simple setup is the downward capillary transfer depicted

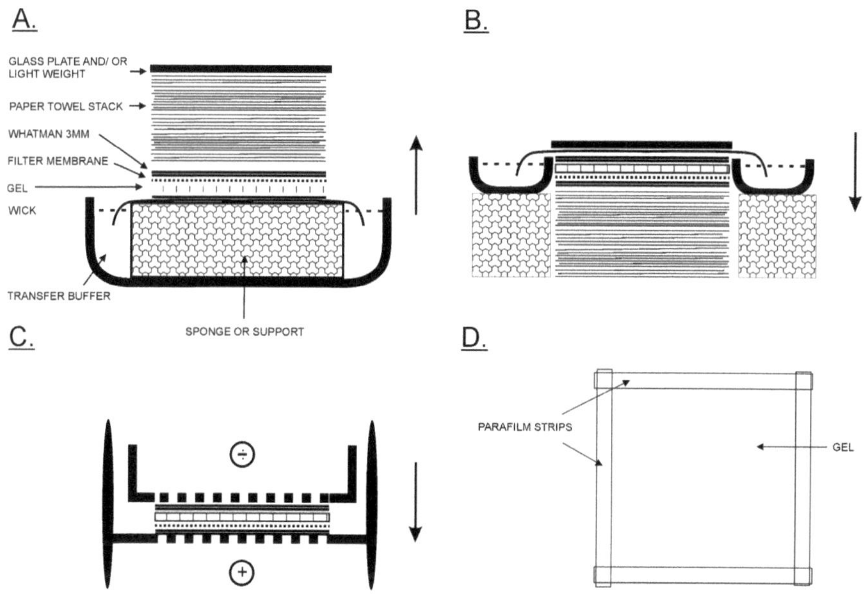

Fig. 7.2. Three different configurations for the gel transfer (blotting) step in northern blotting analysis. **a** Standard upward capillary transfer. This is used with agarose gels and takes several hours (3–6) to overnight to go to completion. **b** Downward capillary transfer. This is used with agarose gels and is completed in 1 h but can be left for longer times. **c** Electroblotting. This is used with polyacrylamide gels and (rarely) with agarose gels. The time depends on the type of gel but is usually complete in 1 h. **d** In all three transfer configurations is the gel lined with overlapping strips of parafilm to avoid short circuiting.

in Fig. **7.2b** (13). Transfer in this setup takes around 1 h but can be left for longer periods of time without problems. Other ways to speed up the transfer is to use a vacuum below the gel or to apply a pressure above the gel. Specialized equipments for these configurations are commercially available. Capillary transfer is the method of choice for agarose gels, whereas electroblotting (14) is used for polyacrylamide gels. In this method, the gel and the membrane are placed in a special cassette placed in an electrophoresis cell (*see* Fig. **7.2c**). An electrical field is applied perpendicular to the cassette and the RNA is transferred onto the membrane. This approach is also used for western blotting of proteins and a variety of electroblotting apparatus are available. One particular useful variety is the semi-dry type in which gel and membrane are sandwiched between Whatman 3MM paper wetted with transfer buffer. Electroblotting can be performed in less than an hour in some cases.

The choice of filter membrane is critical in northern blotting analysis. There are several different options, but for most purposes, positively charged nylon membranes are optimal. The positive charge is due to amino groups on the membrane surface and has the dual effect of increasing the affinity for nucleic acid and facilitating the immobilization of the RNA on the membrane. For the very same reasons, nylon membranes tend to produce more background due to unspecific binding of the probe unless care is taken to avoid this. Nylon membranes are characterized by high tensile strength and high binding capacity (4–500 $\mu g/cm^2$). They are produced with different pore sizes (0.22 and 0.45 μm) and some are optimized, e.g. for repeated use or for binding of small RNAs. It is important to notice the manufacturer's recommendations on handling of the membrane. For convenience, it is advisable to use membranes that are charged on both sides.

If the gel was stained with ethidium bromide, successful transfer can be confirmed by inspection of the filter in UV light (or even in daylight). It is also possible to inspect the gel to check that transfer was complete and not biased towards transfer of small RNAs. Alternatively, filter-bound RNA can be stained at later stages by methylene blue (15). RNAs transferred to positively charged nylon membranes at alkaline conditions become covalently bound to the filter. In all other situations, it is important to ensure that the sample RNA is immobilized on the filter in order to prevent loss during subsequent hybridization steps. It is advisable to follow the recommendations of the manufacturer of the membrane, but some general rules apply. The most common method is to use a UV light source to immobilize the RNA (16). UV light activates pyrimidine bases (thymine and uracil) to form covalent bonds with the surface amines of positively charged membranes (17). Wet nylon membranes generally require a dose of 1.6 kJ/m^2, whereas air-dried membranes require only 0.16 kJ/m^2. This is most conveniently delivered by specialized equipment with a calibrated UV light source (e.g. Stratalinker from Stratagene). These instruments generally have an "auto-crosslink" setting at a dose optimized for damp membranes. Other UV sources can be used, but this may take considerable optimization and care should be taken to avoid extensive radiation at short (254 nm) wavelengths that will damage the RNA. An alternative way to immobilize RNA on nylon membranes is to incubate the filter at 65°C for 1–2 h. The exact principle behind fixation is not known in this case. Some manufacturers claim that the ionic binding to the membrane is sufficient and that there is no need for further steps.

Membranes with immobilized RNA can be stored for extended periods of time and can be used as a record of RNA experiments for later analysis. The next step in the procedure is hybridization analysis. This involves application of a probe to the membrane bound material at conditions that favours specific binding to complementary target molecules on the membrane. The probe is labelled to allow subsequent detection of its complex with the target. The most common approach is radioactive labelling (preferably ^{32}P), but non-isotopic alternatives exist. These probes are generally much less sensitive than ^{32}P-labelled probes. The probes can be DNA or RNA that can be labelled in various ways. The most popular type of probes has been double-stranded plasmids or PCR-products labelled with the random-priming labelling method (18), but asymmetric (single-stranded) PCR-products and RNA probes made by in vitro transcription are more sensitive due to the absence of a competing strand in the probe. Oligonucleotide probes are much less sensitive and are used when discrimination between target molecules of closely related sequences is the issue.

A large variety of hybridization buffers or systems are available and can be used with success for membrane hybridizations. Hybridization theory does not apply to nucleic acids immobilized on a membrane and the success is mainly empirically based. However, some rules of thumb exist. The critical value is the melting temperature (T_m), defined as the temperature at which 50% of all hybrids are dissociated into single strands. The maximum rate of hybridization is typically found around 20°C below the melting temperature. For long probes, the T_m for DNA:RNA hybrids is $T_m = 79.8°C + 18.5 \log[Na^+] + 58.4 (\%G+C) + 11.8 (\%G+C)^2 - 0.5 (\% \text{ formamide}) - (820/L)$ (19). (%G+C) is the percentage of guanine+cytosine content expressed as a mole fraction, [Na^+] is the molar sodium concentration, and L is the effective length of the probe taking part in base pairing with the target. The constants are slightly different for DNA:DNA and RNA:RNA hybrids. Formamide is used to reduce the hybridization temperature, which is particularly important when using RNA probes and when filters are to be used for multiple hybridizations (20). For most purposes, a practical approach using "standard" conditions is taken. The temperature of hybridization and the stringency of washing are then varied according to the requirements of the experiment on an empirical basis. The composition of the hybridization buffer is important. Single-stranded DNA of an origin that is unrelated to the experimental samples (mostly salmon sperm DNA) is used to block unspecific nucleic acid binding to the membrane. An additive known as Denhardts solution (Ficoll, polyvinylpyrrolidone, and bovine serum albumin) is frequently used to block the membrane and to increase the probe concentration by reducing the active volume. An important development is hybridization buffers with a high concentration of SDS. Unfortunately, many high-quality hybridization solutions are sold without a declaration of their content and based on the convincing argument that they are optimized for use with the hybridization membranes sold by the same company.

Northern blotting analysis is not the only technique to demonstrate the presence and quantitate the amount of a specific RNA. In nuclease protection experiments, a labelled probe strand is hybridized in solution to the sample RNA. Then, unhybridized single strands are removed by nuclease digestion and the hybrids are analysed on gels. This technique is traditionally considered an order of magnitude more sensitive (detection limit 0.1–1 pg) than classical northern blotting analysis (detection limit 1–10 pg) and more reliable because the hybridization step is in solution. More recent improvements of northern blotting analysis have closed the gap and it is claimed that in some systems as little as 10,000 molecules on a membrane can be detected. This corresponds to a medium abundant mRNA in less than 100 ng of whole cell RNA. Nuclease protection experiments are relatively laborious and require some optimization. RT-PCR based methods are an additional order of magnitude more sensitive than northern blot analysis. The main problem is that they require stringent controls to avoid artefacts and that the quantitative versions of the technique require expensive instrumentation. Thus, the northern blotting analysis remains an important technique that is simple to perform, low-cost, and provide qualitative and quantitative information in the same experiment. In many cases, a northern blot is still required to convince the referees of the validity of an observation.

One final note of caution concerns the interpretation of the result of the experiment. It is becoming increasingly clear that the level of mRNA is correlated with the level of protein for less than 20% of the genes in humans. This observation obviously emphasizes the importance of post-transcriptional regulation. Thus, the quantitation of an mRNA in most cases provides little information on gene expression in protein sense. On the other hand, the increased interest in RNA makes the observation of the transcript level important in its own right!

2. Materials

Northern blotting analysis is very sensitive to even slight degradation of RNA. For this reason, it is important to wear gloves during all manipulations. Equipment should be free of RNases. One simple precaution is to reserve equipment for RNA work. Alternatively, equipment should be thoroughly cleaned prior to use, especially if it has been in contact with RNases, e.g. during preparation of plasmid DNA. Electrophoresis tanks should be cleaned with detergent, rinsed in water, and dried with ethanol. They are then treated with 3% H_2O_2 for 10 min at RT, followed by rinsing with DEPC-water.

2.1. Formaldehyde Agarose Gel

1. Agarose (Seakem GTG; FMC BioProducts, or similar).
2. 20× MOPS: 0.4 M MOPS ((3-(*N*-morpholino) propanesulfonic acid), 0.1 M sodium acetate, 20 mM EDTA. For 1 L, dissolve 83.6 g of MOPS and 8.2 g of sodium acetate in 800 mL of DEPC-treated water. Adjust the pH to 7.0 with 2 N NaOH. Add 40 mL of 0.5 M EDTA, pH 8.0, and adjust the final volume to 1 L. Autoclave or sterilize by filtration through a 0.2-μm Millipore filter (*see* **Note 1**).
3. 37% formaldehyde (standard 12.3 M solution known as "formalin") (*see* **Note 2**).
4. Loading buffer: 250 μL deionized formamide (*see* **Note 3**), 88 μL 37% formaldehyde, 25 μL MOPS, 2 μL EtBr (3 mg/mL).
5. RNA ladder (commercially available from several companies).
6. Positively charged nylon membrane (Hybond N+ (GE Healthcare Life Sciences), BrightStar Plus (Ambion), Gene Screen Plus (PerkinElmer), or similar).
7. 20× SSC: 3 M NaCl, 0.3 M sodium citrate, pH 7.0. For 1 L dissolve 175.3 g of NaCl and 88.2 g of sodium citrate in 800 mL of DEPC-treated water. Adjust the pH to 7.0 with a few drops of 10 N NaOH. Adjust the volume to 1 L with DEPC-treated water and autoclave.
8. 2× SSC made from 20× SSC by dilution with DEPC-treated water.
9. Parafilm.

10. Whatman 3MM paper.
11. Flat paper towels.
12. Plastic wrap.

2.2. Glyoxal Agarose Gel

1. Agarose (Seakem GTG; FMC BioProducts).
2. 20× MOPS: 0. 4 M MOPS ((3-(*N*-morpholino) propane-sulfonic acid), 0.1 M sodium acetate, 20 mM EDTA. For 1 L, dissolve 83.6 g of MOPS and 8.2 g of sodium acetate in 800 mL of DEPC-treated water. Adjust the pH to 7.0 with 2 N NaOH. Add 40 mL of 0.5 M EDTA, pH 8.0 and adjust the final volume to 1 L. Sterilize by filtration through a 0.2-μm Millipore filter (*see* **Note 1**)
3. 6 M glyoxal (40%; freshly deionized) (*see* **Note 4**).
4. DMSO.
5. Loading buffer: 50% glycerol, 1× MOPS, 0.25% bromophenol blue.
6. RNA ladder (commercially available from several companies).
7. Positively charged nylon membrane (Hybond N+ (GE Healthcare Life Sciences), BrightStar Plus (Ambion), Gene Screen Plus (PerkinElmer), or similar).
8. Parafilm.
9. Whatman 3MM paper.
10. Flat paper towels.
11. Plastic wrap.

2.3. Denaturing Polyacrylamide Gel

1. 40% acrylamide (40:1.3): 400 g of acrylamide, 13 g of bisacrylamide. Dissolve in water; adjust to 1 L. Filter through a 3MM filter and store in a dark bottle in the cold (*see* **Note 5**).
2. 10× TBE electrophoresis buffer: 0.9 M Tris, 0.9 M boric acid, 0.02 M EDTA. For 5 L dissolve 544 g of Tris, 278 g of boric acid, and 37.2 g of EDTA. Dissolve in water and adjust to 5 L. Autoclave for long-term storage.
3. 5% UPAG-mix/1× TBE: 125 mL 40% acrylamide-stock (40:1.3), 500 g urea, 100 mL 10× TBE. Add water to dissolve urea; adjust to 1 L. Filtrate through a 3MM filter and store in a dark bottle in the cold.
4. 10% (w/v) ammonium persulfate. Store as frozen aliquots. Once thawed, the solution is stored at 4°C and can be used for a few weeks.
5. TEMED (N, N, N', N'-tetramethylethylenediamine)

6. Loading buffer: 1× TBE, 50% urea, 1 mg/mL bromophenol blue, 1 mg/mL xylene cyanol FF.
7. RNA ladder (commercially available from several companies).
8. Positively charged nylon membrane (Hybond N+ (GE Healthcare Life Sciences), BrightStar Plus (Ambion), Gene Screen Plus (PerkinElmer), or similar).
9. Parafilm.
10. Whatman 3MM paper.

2.4. Hybridization Analysis

1. 20× SSPE: 3 M NaCl, 200 mM NaH_2PO_4, 20 mM EDTA, pH 7.4. For 1 L dissolve 175.3 g of NaCl and 27.6 g of NaH_2PO_4 and 7.4 g of EDTA in 800 mL of DEPC-treated water. Adjust the pH to 7.4 with NaOH approximately 6.5 mL of a 10 N solution). Adjust the volume to 1 L with DEPC-treated water and autoclave.
2. 20× SSC: 3 M NaCl, 0.3 M sodium citrate, pH 7.0. For 1 L dissolve 175.3 g of NaCl and 88.2 g of sodium citrate in 800 mL of DEPC-treated water. Adjust the pH to 7.0 with a few drops of 10 N NaOH. Adjust the volume to 1 L with DEPC-treated water and autoclave.
3. 50× Denhardt's solution: 0.05% (w/v) BSA, 0.05% (w/v) polyvinyl pyrolidone, and 0.05% (w/v) Ficoll 400. For 50 mL dissolve 0.5 g of Ficoll (type 400 Pharmacia), 0.5 g of polyvinylpyrrolidone, and 0.5 g of bovine serum albumine (Fraction V; Sigma) in DEPC-treated water. Filter and store in small aliquots at –20°C.
4. Hybridization buffer: 50% formamide, 5× SSPE, 0.5% SDS, 2× Denhardts solution, 100 μg/mL of denatured carrier DNA. For 10 mL of hybridization buffer, combine 5 mL of deionized formamide (*see* **Note 3**), 4 mL of 20 × SSPE, 0.25 mL of a 20% solution of SDS, 0.2 mL of 50× Denhardts solution, 50 μL of a 20 mg/mL solution of denatured carrier DNA (e.g. salmon sperm DNA) (*see* **Note 6**).
5. Radioactively labelled probe (*see* **Note 7**).
6. Low stringency washing buffer: 2× SSC, 1% sodium pyrophosphate, 0.1% SDS.
7. Medium stringency washing buffer: 0.2× SSC, 1% sodium pyrophosphate, 0.5% SDS.
8. High stringency washing buffer: 0.1× SSC, 1% sodium pyrophosphate, 0.5% SDS.
9. Probe removal buffer: 50% formamide, 2× SSPE.

3. Methods

3.1. Northern Blotting of a Formaldehyde Agarose Gel

The example is a 1.2% agarose gel blotted by traditional upward, passive capillary blotting to a positively charged nylon membrane. This type of gel is used for analysis of small to medium sized mRNA from mammalian cells.

1. Pour 100 mL of dH_2O into a 250-mL Ehrlenmeyer flask. Use a marker pen to indicate the water level on the flask. Remove approximately 10 mL of the water and add 1.2 g of agarose to the flask. Dissolve the agarose by boiling, e.g. in a microwave oven. When the agarose is completely dissolved, cool it to 60°C under running tap water. Add 5 mL of 20× MOPS and 5.36 mL of 37% formaldehyde in a fume hood (*see* **Note 2**). Adjust the volume to the original 100 mL with dH_2O using the mark on the flask. Mix and cast the gel in the tray. Insert the slot former and allow the gel to solidify for at least half an hour.
2. Insert the gel tray into the gel apparatus. Fill up with electrophoresis buffer (1× MOPS) and carefully remove the slot former.
3. Take an aliquot of approximately 10 μg of RNA in 6 μL or less. Add 2.7 vols of loading buffer per μL of RNA to the sample and heat it for 10 min at 70°C. Flush the sample wells with electrophoresis buffer and load the samples. Load a molecular weight marker next to the samples.
4. Run the gel at 2 V/cm for approximately 3 h. The loading buffer contains ethidium bromide allowing the electrophoresis to be followed by inspection of the gel under UV light. Note that not all gel electrophoresis trays are UV transparent.
5. Photograph the gel in the UV transilluminator. Notice the intensity of the two major RNA bands. The upper one (LSU rRNA) should be approximately twice the intensity of the lower (SSU rRNA).
6. Prepare the gel for northern blotting by cutting away those parts of the gel that are not to be transferred (below 5S rRNA, above the wells, and the sides of the gel). Cut off lower left corner of the gel for orientation.
7. Wash out excess formaldehyde and ethidium bromide by soaking the gel in 2× SSC three times 5 min.
8. Optional. Irradiate the gel on a UV transilluminator to introduce random breaks in the RNA backbone. This is used to improve transfer efficiency of large RNA molecules and may improve hybridization efficiency (*see* **Note 8**).

9. Cut a piece of membrane to the size of the gel while leaving the membrane between the two sheets of protective paper. Do not handle the membrane without gloves (*see* **Note 9**).
10. Pre-wet the membrane by floating on the surface of RNase-free water for a few minutes. Once the membrane is wet (usually within less than a minute), inspect the membrane for even wetting (*see* **Note 10**). Submerge the membrane and transfer it to the 10× SSC transfer buffer and equilibrate for 1–2 min.
11. Cut two pieces of Whatman 3MM filter paper to the size of the gel.
12. Make sure that the sponge and the Whatman 3MM filter in the transfer unit is saturated with buffer and that the buffer level in the unit is at least 1 cm from the bottom.
13. Assemble the transfer sandwich: Without trapping air bubbles, place the gel in the transfer unit. Place strips of parafilm along the edges of the gel in order to avoid "short-cutting" of the buffer-flow. Carefully position the filter membrane on top of the gel. Then layer the two pieces of Whatman 3MM filter paper on top of the sandwich, one at a time. Cut a stack of paper towels (5 cm when compressed) and place on top of the sheets of Whatman paper. Finally, put a light weight (e.g. a glass plate) on top of the paper towels to compress the stack. Allow the transfer to proceed for at least 6 h, preferably over night.
14. Disassemble the transfer sandwich. Blot excess liquid using kitchen roll or paper towels and place the filter on top of a sheet of 3MM paper. Do not allow the filter to dry out completely. If the gel was stained with ethidium bromide, successful transfer can be confirmed by inspection of the filter in UV light (or even in daylight). Filter-bound RNA can also be stained at later stages by methylene blue (*see* **Note 11**).
15. Optional. Place the filter in the UV cross-linker with the side of the membrane that was in contact with the gel facing up. Use the "auto-crosslink" setting as a starting point (*see* **Note 12**).

3.2. Northern Blotting of Glyoxalated RNA Separated on an Agarose Gel

The example is a 1% agarose gel blotted by traditional upward, passive capillary blotting to a positively charged nylon membrane. This type of experiment is used for analysis of most mRNA from mammalian cells.

1. Pour 100 mL of dH_2O into a 250-mL Ehrlenmeyer flask. Use a marker pen to indicate the water level on the flask. Remove approximately 10 mL of the water and add 1.0 g

of agarose to the flask. Dissolve the agarose by boiling, e.g. in a microwave oven. When the agarose is completely dissolved, cool it to 60°C under running tap water. Add 5 mL of 20× MOPS. Adjust the volume to the original 100 mL with dH_2O using the mark on the flask. Mix and cast the gel in the tray. Insert the slot former and allow the gel to solidify for at least half an hour (*see* **Note 13**).

2. Insert the gel tray into the gel apparatus. Fill up with electrophoresis buffer (1× MOPS) and carefully remove the slot former.
3. Mix in a RNA quality microfuge tube 5.4 μL of 6 M glyoxal, 16.0 μL of DMSO, 1.5 μL of 20× MOPS, and 7.1 μL of RNA (up to 10 μg). Place the sample at 50°C for 1 h to denature. Remember to treat the RNA ladder in parallel.
4. Cool the RNA sample on ice, add 4 μL of loading buffer, and immediately load the samples into the wells of the gel.
5. Run the gel while ensuring that no pH-gradient is formed during the run. Glyoxal dissociates from RNA at pH>8.0.
6. At the end of the run (typically when the bromophenol blue has migrated 8 cm), the lane(s) containing the RNA ladder can by cut out and stained. Alternatively, proceed to the blotting steps.
7. Transfer the RNA to a positively charged nylon membrane as described in **Section 3.1, Steps 6, and 8–15.**
8. Optional. After immobilization, remove glyoxal from RNA by washing the filter for 15 min at 65°C in 20 mM Tris-Cl, pH 8. Alternatively, the glyoxal will be removed during incubation with hybridization buffer in the pre-hybridization step.

3.3. Denaturing Polyacrylamide Gel

The example is a denaturing (urea) 5% polyacrylamide gel. The effective separation range is 50–500 nt and the migration of the marker dyes are approximately 130 nt (xylene cyanol FF) and 35 nt (bromophenol blue), respectively. A Semi-Phor Blotter (Hoefer Instruments) was used for semi-dry transfer of the RNA. This type experiment is used for analysis of small RNA molecules, such as snRNA or snoRNA.

1. Cast the gel. For a 5% UPAG pour 35 mL of gel-mix into a beaker, add 135 μL of 10% ammonium persulfate and 70 μL of TEMED. Cast the gel in the pre-assembled sandwich of glass-plates. Allow to polymerize for 45 min.
2. Remove the slot former. Mount the gel in the electrophoresis unit. Pre-run the gel for 10–30 min.
3. Add one vol of loading buffer to the sample RNA. Heat-denature the RNA at 70°C for 2 min. Meanwhile flush

the slots with 1× TBE running buffer. Load and run the gel.

4. After gel electrophoresis: stain the gel with ethidium bromide and place it on a quartz plate. Take a photo and cut the gel to appropriate size. Measure its size with a ruler, and note the position of the wells. Do not remove the gel from the quartz plate, but cover it with plastic wrap to ensure that it does not dry out.
5. Cut out a piece of nylon membrane and three pieces of Whatman 3MM paper to the size of the gel to be blotted (*see* **Note 9**).
6. Pour a small amount of 1× TBE in a wash tray and wet the membrane (cut to the exact size of the gel) by floating it carefully on the buffer surface before immersing it. This will avoid trapping of air bubbles in the pores of the membrane. Do not allow the membrane to dry at any time prior to the completion of the transfer procedure.
7. Remove the plastic foil from the gel and soak the surface of it with 1× TBE. Lay the wet membrane carefully on the gel. Align it with two diagonal corners, and then gently roll the membrane down onto the gel. Remove any trapped bubbles by gently pushing them to the side or rolling them out with a pipette. Flood the surface of the membrane with additional 1× TBE.
8. Wet one of the Whatman 3MM papers (cut to exact size of the gel) with 1× TBE and place it over the membrane. Flood the lower grating of the electroblotter with 1× TBE. Now take the quartz plate with the gel, membrane, and Whatman paper between your hands and invert it. When the gel has slipped off the quartz plate, lay the sandwich, paper down, on the grating. Flood the surface of the gel with 1× TBE.
9. Wet the two remaining pieces of 3MM paper in 1× TBE and place them, one at a time, precisely on top of the gel. Carefully remove all air bubbles. Remove buffer around the gel with paper towels and align pieces of parafilm along the edges of the gel. The buffer must not bypass the gel, as it will result in inefficient and uneven transfer of the RNA.
10. Flood the surface of the 3MM paper with 1× TBE and remove any excess of buffer from the parafilm. Place the lid of the electroblotter on top of the paper and blot for 30 min at 150 mA.
11. After transfer is complete, dismantle the electroblotter. Be sure that you are wearing gloves. Take out the sandwich of paper, membrane and gel and place it inverted (membrane

over the gel) on a piece of plastic foil. Remove the upper paper and cut off the lower left corner of the membrane (this will correspond to the lower right corner of the gel). Mark the membrane with a pencil to allow identification of tracks.

12. Lay the membrane, RNA side up, on plastic wrap (cut-off corner at lower right). If an effective RNA transfer has occurred, this will show up as ethidium bromide fluorescence under a UV lamp. Fix the RNA covalently to the membrane by UV irradiation in the UV crosslinker. The membrane is now ready for hybridization analysis (*see* **Note 12**).

3.4. Hybridization Analysis

The example describes a standard hybridization aimed at detection of mRNA in whole cell RNA and using a formamide containing hybridization buffer to avoid extended incubations at high temperatures that leads to RNA degradation.

1. Place the hybridization membrane in a hybridization bottle. Prehybridize the membrane for 1–2 h in 10 mL of hybridization solution at 42°C.
2. 5×10^6 cpm of probe/mL of hybridization buffer is adjusted to 500 μL with hybridization solution and denatured by boiling in a boiling water bath for 2 min. The probe is added to the hybridization solution in the bottle without spotting it directly on the membrane.
3. Hybridize over night at 45°C.
4. Remove the membrane from the bottle using forceps and submerge immediately in low stringency washing buffer.
5. Wash the membrane 2 × 10 min in low stringency washing buffer at RT. Follow the progress of the washing steps using a Geiger-Müller tube monitor.
6. Wash the membrane 2 × 20 min in medium stringency washing buffer at 65°C.
7. (Optional). Wash the membrane 2 × 20 min in high stringency washing buffer at 65°C.
8. Autoradiography or phosphorimager analysis of the membrane.
9. (Optional). The filter can be stripped of the probe by incubation for 1 h at 65°C in 50% formamide, 2× SSPE. This is followed by a brief rinsing in 0.1× SSPE and blotting of excess liquid. The filter should be kept damp until further hybridization steps. It is extremely difficult to completely remove probe from a filter that has been allowed to dry out and to re-hybridize a filter that been dried in the presence of SDS.

4. Notes

1. The MOPS-buffer turns yellow with age if exposed by light or by autoclaving. Straw-coloured buffers can be used, but darker coloured buffers should be discarded.
2. Formaldehyde is supplied as a 37% (12.3 M) solution, containing 10–15% methanol. It should be stored in tight bottles out of direct sunlight. The pH should be checked before use (should be grater than 4.0). Formaldehyde is carcinogenic.
3. Many batches of formamide are sufficiently pure to be used directly. If a yellow colour is present, the formamide must be deionized by batch treatment with a mixed-bed resin (e.g. Dowex XG8) prior to use. The formamide is stirred with the resin for 30–60 min and then filtered through a filter paper. Aliquots are stored at –70°C under nitrogen. Formamide is a suspected teratogen and should not be handled by expectant mothers.
4. Glyoxal is obtained as a 6 M (40%) solution. It is readily oxidized in air and the oxidation products (glyoxylic acid) will cause fragmentation of the RNA sample. For this reason, glyoxal must be deionized before use. This can be done by passage through a mixed-bed resin (Bio-Rad AG 501-X8) until its pH is greater than 5.0. The deionized glyoxal is stored in small aliquots at –20°C in tightly capped tubes. Each aliquot is only use once. Glyoxal is a mutagen.
5. Acrylamide is a very potent neurotoxin that is readily absorbed through the skin. A mask and gloves should be worn when handling unpolymerized acrylamide and gloves should also be used when handling gels. Handling of acrylamide gels, e.g. for staining is difficult, and it is advisable to seek assistance from more experienced persons.
6. If the formamide is left out of the hybridization buffer, standard hybridization is performed at 68°C.
7. Many different kind of probes can be used, including double-stranded DNA labelled according to the random priming labelling method, radioactive in vitro transcripts, and asymmetric PCR-product. The two latter are preferred because of the single-stranded nature of the probe. The probes should be labelled to a specific activity exceeding 10^9 cpm/μg.
8. The efficiency of transfer can also be improved by slight alkaline treatment of the gel prior to transfer. The gel is placed in a tray with 50 mM NaOH, 0.1 M NaCl for

20 min with gentle shaking. Extensive alkaline treatment will fragment the RNA to the extent that it is no longer hybridization competent. Before transfer, the gel is neutralized in 0.1 M Tris-HCl, pH 7.6, for 10–15 min.

9. Membranes should be handled carefully. Finger grease on the membrane will reduce its performance and contamination with finger RNases are detrimental. Membranes are supplied in a sandwich between two sheets of protective paper. It is a good idea to keep the protective paper in place while cutting out a gel-sized piece of membrane.
10. A wetted membrane appears gray. Patches of white indicate areas that have been damaged. If these are in critical parts, the membrane should be discarded.
11. RNA on nylon membranes can be stained by the non-toxic methylene blue. The staining solution is 0.02% methylene blue in 0.3 M sodium acetate, pH 5.5. The RNA will stain in a matter of few minutes. De-staining is by incubation in 1× SSPE (10 mM phosphate buffer, pH 7.4, containing 150 mM NaCl and 1 mM EDTA).
12. Filters that have not been used for hybridization experiments are stored dry between sheets of Whatman 3MM paper. Filters that are stored after a hybridization experiments are stored damp in vacuo or frozen to avoid opportunistic growth. Storage can be for several months.
13. In formaldehyde gels, RNases are inhibited due to the presence of formaldehyde in the gel. In glyoxal gels, inhibition of RNases can be achieved by addition of solid sodium iodoacetate to 10 mM to the melted agarose.

References

1. Alwine, J. C., Kemp, D. J., Stark, G. R. (1977) Method for detection of specific RNAs in agarose gels by transfer to diazobenzyloxymethyl-paper and hybridization with DNA probes. *Proc Natl Acad Sci USA* 74, 5350–5354.
2. Alwine, J. C., Kemp, D. J., Parker, B. A., Reiser, J., Renart, J., Stark, G. R., Wahl, G. M. (1979) Detection of specific RNAs or specific fragments of DNA by fractionation in gels and transfer to diazobenzyloxymethyl paper. *Methods Enzymol* 68, 220–242.
3. Southern, E. M. (1975) Detection of specific sequences among DNA fragments separated by gel electrophoresis. *J Mol Biol* 98, 503–517.
4. Lamond, A. I., Sproat, B. S. (1994) Isolation and characterization of Ribonucleoprotein complexes, in (Higgins, S. J. and Hames, B. D., eds.) *RNA Processing. A Practical Approach*. IRL Press, Oxford, Vol. 1, pp. 103–140.
5. Sambrook, J., Fritsch, E. F., Maniatis, T. (1989) *Molecular Cloning: A Laboratory Manual*. Cold Spring Harbor Laboratory Press, Cold Spring Harbor, NY.
6. Darling, D. C., Brickell, P. M. (1994) *Nucleic Acid Blotting. The Basics.*. IRL Press, Oxford.
7. Farrell, R. E.,Jr. (1993) *RNA Methodologies. A Laboratory Guide for Isolation and Characterization*. Academic, San Diego, CA.
8. Chomczynski, P., Sacchi, N. (1987) Single-step method of RNA isolation by acid guanidinium thiocyanate-phenol-chloroform extraction. *Anal Biochem* 162, 156–159.
9. Aviv, H., Leder, P. (1972) Purification of biologically active globin messenger

RNA by chromatography on oligothymidylic acid-cellulose. *Proc Natl Acad Sci USA* 69, 1408–1412.

10. Lehrach, H., Diamond, D., Wozney, J. M., Boedtker, H. (1977) RNA molecular weight determinations by gel electrophoresis under denaturing conditions, a critical reexamination. *Biochemistry* 16, 4743–4751.
11. McMaster, G. K., Carmichael, G. G. (1977) Analysis of single- and double-stranded nucleic acids on polyacrylamide and agarose gels by using glyoxal and acridine orange. *Proc Natl Acad Sci USA* 74, 4835–4838.
12. Reijnders, L., Sloof, P., Sival, J., Borst, P. (1973) Gel electrophoresis of RNA under denaturing conditions. *Biochim Biophys Acta* 324, 320–333.
13. Chomczynski, P., Mackey, K. (1994) One-hour downward capillary blotting of RNA at neutral pH. *Anal Biochem* 221, 303–305.
14. Bittner, M., Kupferer, P., Morris, C. F. (1980) Electrophoretic transfer of proteins and nucleic acids from slab gels to diazobenzyloxymethyl cellulose or nitrocellulose sheets. *Anal Biochem* 102, 459–471.
15. Herrin, D. L., Schmidt, G. W. (1988) Rapid, reversible staining of northern blots prior to hybridization. *Biotechniques* 6, 196–200.
16. Church, G. M., Gilbert, W. (1984) Genomic sequencing. *Proc Natl Acad Sci USA* 81, 1991–1995.
17. Saito, I., Sugiyama, H., Furukawa, N., Matsuura, T.. (1981) Photoreaction of thymidine with primary amines. Application to specific modification of DNA. *Nucleic Acids Symp Ser* 10, 61–64.
18. Feinberg, A. P., Vogelstein, B.. (1983) A technique for radiolabeling DNA restriction endonuclease fragments to high specific activity. *Anal Biochem* 132, 6–13.
19. Casey, J., Davidson, N. (1977) Rates of formation and thermal stabilities of RNA:DNA and DNA:DNA duplexes at high concentrations of formamide. *Nucleic Acids Res* 4, 1539–1552.
20. Bonner, J., Kung, G., Bekhor, I. (1967) A method for the hybridization of nucleic acid molecules at low temperature. *Biochemistry* 6, 3650–3653.

Chapter 8

Rapid Amplification of cDNA Ends (RACE)

Oladapo Yeku and Michael A. Frohman

Abstract

Rapid Amplification of cDNA ends (RACE) provides an inexpensive and powerful tool to quickly obtain full-length cDNA when the sequence is only partially known. Starting with an mRNA mixture, gene-specific primers generated from the known regions of the gene and non-specific anchors, full-length sequences can be identified in as little as 3 days. RACE can also be used to identify alternative transcripts of a gene when the partial or complete sequence of only one transcript is known. In the following sections, we outline details for rapid amplification of 5′ and 3′ cDNA ends using the "new RACE" technique.

Key words: RACE, new RACE, alternative transcripts.

1. Introduction

The advent of microarrays and powerful bioinformatics analysis has not only led to the discovery of new genes but has also provided tools for analysis of existing genes (1–3). Full or partial sequences of newly discovered genes can be aligned against entire organism genomes in search of homology or any other clue that can yield insight into the identity or function of the gene. There are instances, however, where bioinformatics data is unavailable or incomplete. In these instances, rapid amplification of cDNA ends (RACE) can be used to identify full-length sequences of a gene if only part of its sequence is known (4–6). RACE can be used to amplify both the 5′ and the 3′ ends of genes yielding valuable information such as the location of transcription initiation sites, *cis*-acting elements, and the localization and stability of the transcript. By using 3′ and 5′ RACE, full-length cDNA sequences can be cloned from partial sequences in as little as 3 days.

H. Nielsen (ed.), *RNA*, Methods in Molecular Biology 703,
DOI 10.1007/978-1-59745-248-9_8,

Since the initial description of RACE (7), many labs and commercial companies have adapted and modified the protocol to increase specificity and user friendliness (8–18) (*see* **Note 1**). Amplification of the 5′ end can generally be divided into two classifications; classic RACE and "new RACE." The same principles underlie all RACE protocols and they differ only in terms of specificity, ease of use and cost. While the principles for classic RACE will be touched on, it has been thoroughly described elsewhere for both 5′ (19) and 3′ (20) end amplification. This protocol will address the identification of alternative transcripts using the "new RACE" protocol (21).

RACE primarily utilizes RT-PCR (Reverse Transcriptase-Polymerase Chain Reaction) and PCR to amplify the ends of transcripts starting with mRNA and cDNA, respectively (**Fig. 8.1**). To perform RACE, the partial or a complete sequence of the mRNA of interest has to be known. This is required to generate the Gene Specific Primers (GSPs) that will be used for the amplifications. A second set of primers that extend from the unknown end of the of the message back to the known region of the 3′ end is provided by the poly (A) tail (or an appended homopolymer), while an appended homopolymer tail is used for the 5′ end. Herein lies the first and perhaps most important difference between "classic" and "new" RACE (22, 23). In classic RACE (19), the homopolymer is appended after the mRNA is reverse transcribed, whereas in "new" RACE, the homopolymer is appended before the reverse transcriptase reaction (*see* **Fig. 8.2**). This simple difference eliminates the amplification of non-full length products, which greatly improves the ability to identify transcription start sites. After the mRNA is reverse transcribed to cDNA, the 5′ and 3′ ends are amplified using two nested PCR reactions using gene-specific primers and primers derived from the sequence of the RNA oligonucleotide (appended homopolymer). The use of sequential nested PCR reactions is important to reduce the amplification of unwanted products, since in each reaction, only one of the primers is specific for the gene of interest and the other binds to all cDNAs present in the starting mixture (by comparison, standard PCR reactions employ two gene-specific primers).

The starting mRNA mixture is dephosphorylated with shrimp alkaline phosphatase (SAP). This dephosphorylates degraded and uncapped (non full-length) mRNAs, leaving the full length mRNAs with methylated "G" caps intact (24). The methylated "G" cap is then removed with tobacco acid pyrophosphatase (TAP). Treatment with TAP exposes the phosphorylated 5′ end of the mRNA and prepares it for ligation to the linker or homopolymer. A short synthetic RNA, prepared by in vitro transcription of a linearized plasmid (25), is ligated to the uncapped 5′ and 3′ end using T4 RNA ligase (*see* **Fig. 8.3**). Degraded or other

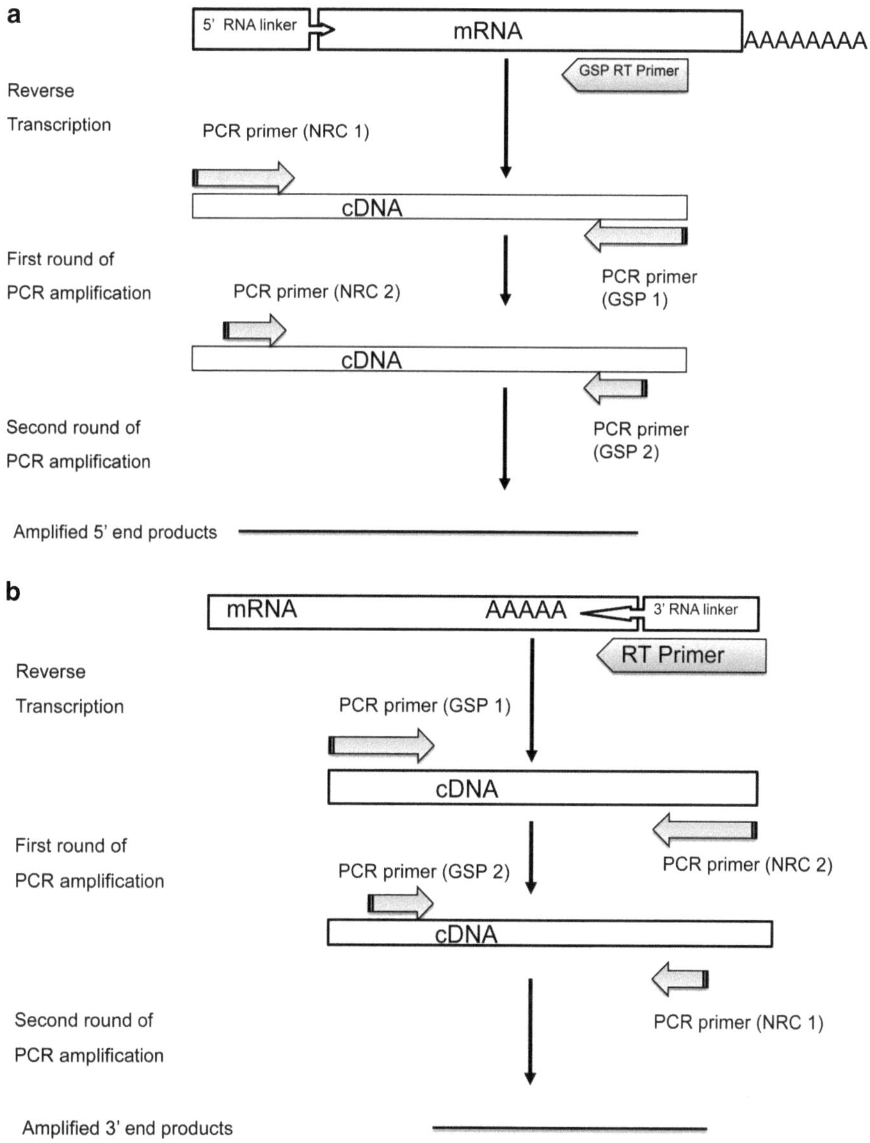

Fig. 8.1. Figure showing the general schematic for Rapid Amplification of cDNA Ends (RACE) starting with an mRNA pool. **a** Amplification of the 5′ end. See text for full details. **b** Amplification of the 3′ end. Note that the NRC 1, NRC 2 and NRC 3 primers are the reverse compliments of sequences described in text. RT: Reverse Transcriptase. GSP: Gene Specific Primer. NRC: New RACE Complement.

non-full length mRNA will not be ligated with the synthetic oligonucleotide, since they are dephosphorylated. The mRNA-RNA oligonucleotide hybrids are then reverse-transcribed using a GSP. Full-length cDNA generated from the previous step then contains specific sequences at the 5′ and 3′ ends that can be used to amplify the full-length cDNA ends. Two nested PCR reactions are used to amplify the full-length cDNA product with high specificity (*see* **Fig. 8.3**).

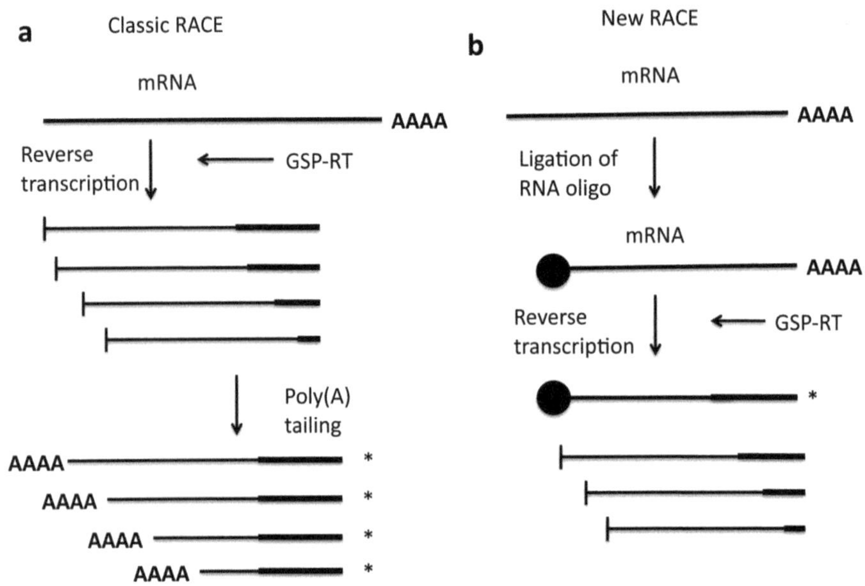

Fig. 8.2. Comparison between "classic RACE" and "new RACE." **a** In "classic RACE," the homopolymer (PolyA tail) is added after the reverse transcription process. This anchor provides specificity for the amplifications downstream of the GSP-reverse transcription step. **b** In "new RACE," the homopolymer is appended before the reverse transcription takes place. This ensures full length products from the onset. Figure reproduced from citation (21). **http://www.nature.com/nprot/journal/v1/n6/images/nprot.2006.480-F3.jpg**.

2. Materials

2.1. Dephosphorylation of Degraded RNA

1. RNA sample (TAP-treated and untreated). All reagents must be RNase free.
2. Phosphatase buffer (10×): 0.1 M Tris-HCl (pH 7.5 at 37°C), 0.1 M $MgCl_2$, and 1 mg/mL BSA.
3. DTT (0.1 M).
4. RNasin (40 U/μL).
5. Shrimp Alkaline Phosphatase SAP (1 U/μL; Fermentas).

2.2. Decapping of Intact RNA

1. Tobacco Acid Pyrophosphatase (TAP) buffer (10×): 0.5 M sodium acetate (pH 6.0), 10 mM EDTA, 1% β-mercaptoethanol, and 0.1% Triton X-100
2. TAP (5 U/μl; Epicentre)
3. TE buffer: 10 mM Tris-HCl, pH 7.5, and 1 mM EDTA, pH 8.0
4. RNA spin column (Qiagen)
5. DTT (0.1 M)
6. RNasin (40 U/μL).

2.3. Preparation of RNA Oligonucleotide

1. TE buffer: 10 mM Tris-HCl, pH 7.5, and 1 mM EDTA, pH 8.0
2. Plasmid template DNA for transcribing RNA oligonucleotide (*see* **Note 2**)
3. Restriction enzymes and buffers
4. Proteinase K (5 mg/mL stock solution (100×))
5. H_2O treated with diethylpyrocarbonate (DEPC)
6. Transcription buffer (5×): Provided with the enzyme *or* 200 mM Tris-HCl (pH 8.0), 40 mM $MgCl_2$, 10 mM spermidine, 250 mM NaCl
7. rUTP solution (10 mM)
8. rATP solution (10 mM)
9. rCTP solution (10 mM)
10. rGTP solution (10 mM)
11. DNA-dependent RNA polymerase (20 U/μL)
12. RNase-free DNase I (0.5 Kunitz units)
13. DTT (0.1 M)
14. RNasin (40 U/μL).

2.4. RNA Oligonucleotide–Cellular RNA Ligation

1. RNA sample (TAP-treated and untreated)
2. Ligation buffer (10×): 500 mM Tris-HCl, 100 mM $MgCl_2$, 100 mM DTT, 10 mM ATP (pH 7.8 at 25°C) (*see* **Note 3)**
3. RNA oligonucleotide (*see* **Note 4**)
4. TP (2 mM)
5. T4 RNA ligase (New England BioLabs) (20 U/μL).

2.5. Reverse Transcription

1. SuperScript II reverse transcriptase (Invitrogen) (200 U/μL). Reverse-transcription buffer (5×) is supplied with the transcriptase.
2. dNTP solution (containing all four dNTPs, each at 10 mM)
3. Gene-specific antisense primer for 5′ end RACE or reverse complement of NRC3 (new Race Compliment primer) for 3′ end RACE (20 ng/μL)
4. RNase H (2 U/μL)
5. DTT (0.1 M)
6. RNasin (40 U/μL)
7. TE buffer: 10 mM Tris-HCl, 1 mM EDTA, pH 8.0.

2.6. PCR Amplifications

1. Hercules Hot-Start polymerase buffer (10×; Stratagene). Do not add additional nucleotides if they are already contained in the buffer.

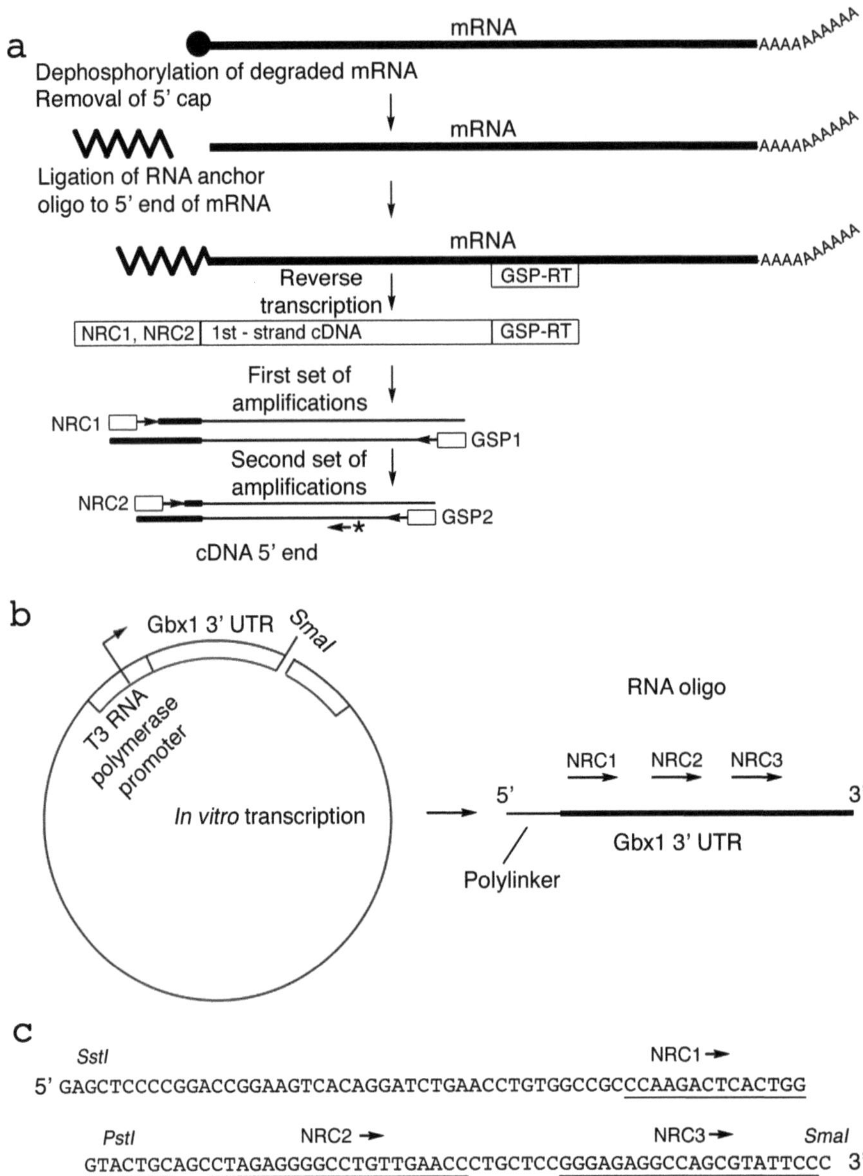

Fig. 8.3. New RACE overview. **a** Outline of steps involved in the amplification of cDNA 5′ ends. See text for detailed explanation. **b** Plasmid map and primers used for the example in the text. **c** In vitro transcription of the T3 plasmid in (**b**) produces a 132-nt product. The sequences for NRC1, NRC2 and NRC3 used for the generation of corresponding primers are underlined. Figure reproduced from citation (21). **http://www.nature.com/nprot/journal/v1/n6/fig_tab/nprot.2006.479_F1.html**.

2. Hercules Hot-Start polymerase (Stratagene). Hot-start protocol is strongly recommended (*see* **Note 5**).
3. User-defined gene-specific oligonucleotide primers GSP1, GSP2, NRC1, and NRC2 to clone the 5′ end and the reverse compliment of GSP1, GSP2, NRC1, NRC2, and NRC3 to

clone the 3′ end (*see* **Note 6** for primer design considerations and **Fig. 8.3** for details of primers NRC1 and NRC2).

2.7. DNA/RNA Purification and Electrophoresis

1. Agarose gel (1%) in TAE buffer
2. Ethidium bromide staining bath (0.5 μg/mL ethidium bromide in electrophoresis buffer; prepared from a stock solution of 10 mg/mL in water)
3. Sodium acetate: 3 M, pH 5.2
4. Phenol/chloroform (1:1 (vol/vol))
5. Chloroform
6. Ethanol
7. Microcon spin filters (Millipore).

3. Methods

If large amounts of RNA are used, it is strongly recommended that the quality control steps be performed. Small samples of the RNA can be run on an agarose gel to check for degradation. Aliquots can then be stored at –80°C for future experiments. The quantities presented in this protocol are the ideal amounts of RNA to be used. Reactions can be scaled down in the event that RNA quantities are limited.

3.1. Dephosphorylation of Degraded RNA

1. Combine reagents listed in **Table 8.1** in a sterile microfuge tube. Follow the manufacturers guidelines regarding the use of SAP.
2. The reaction should be incubated for 1 h at 37°C. Dephosphorylation of uncapped mRNA prevents degraded fragments from participating in the subsequent ligation step.

Table 8.1
Dephosphorylation of degraded RNA

Component	Amount	Final
RNA	50 μg	50 μg
10X phosphatase buffer	5 μL	1×
DTT (0.1 M)	0.5 μL	1 mM
RNasin (40 U/μL)	1.25 μL	50 U
SAP (1 U/μL)	3.5 μL	3.5 U
H_2O	to 50 μL	–

3. Incubate the reaction for 15 min at 65°C to inactivate SAP, then spin briefly in a microcentrifuge. At this point, the products can be stored at –80°C.
4. To visually confirm that the RNA remained intact during the dephosphorylation step, 2 μg (2 μL) of RNA should be analyzed by electrophoresis through a 1% agarose gel (TAE buffer) adjacent to a lane containing 2 μg of the original RNA preparation. Upon ethidium bromide staining, degraded RNA will be present as a low molecular weight smear on the gel rather than a discrete band of high molecular weight. If evidence of RNA degradation is observed, then fresh reagents should be prepared and the step repeated.

3.2. Decapping of Intact RNA

1. Combine and mix the reagents listed in **Table 8.2** in a sterile microfuge tube. Note that the quantity of TAP proposed is sufficient for the reaction.
2. Incubate the reaction for 1 h at 37°C; then add 200 μL TE buffer to stop the reaction.
3. Spin columns provided in any RNA extraction kit or from varied companies can be used for separation of the RNA from the reaction components. Resuspend the RNA in 40 μL H_2O. It is permissible to store the products at this point at –80°C.
4. 2 μg of RNA should be analyzed by electrophoresis (through a 1% agarose gel in TAE buffer) adjacent to a lane containing 2 μg of the original RNA preparation; stain the gel with ethidium bromide and visually confirm that the RNA remained intact during the decapping step. A low molecular weight smear on the gel indicates the presence of RNA fragments, whereas a discrete band of high molecular weight corresponds to the desired product.

Table 8.2
Decapping of intact RNA

Component	Amount	Final
RNA (obtained in **Section 3.1, Step 4**)	48 μg	48 μg
TAP buffer (10×)	5 μL	1×
DTT (0.1 M)	0.5 μL	1 mM
RNasin (40 U/μL)	1.25 μL	50 U
TAP (5 U/μL)	1 μL	5 U
H_2O	to 50 μL	–

3.3. Preparation of RNA Oligonucleotide

1. Using digestion by appropriate restriction enzymes and buffers, linearize 25 μg of the plasmid to be transcribed. Ensure that the plasmid is reasonably RNase-free.
2. Eliminate any residual RNase activity in the reaction by treating the digestion for 30 min at 37°C with 50 μg/mL proteinase K. Extract twice with phenol–chloroform and once with chloroform, and collect the DNA by standard ethanol precipitation.
3. Redissolve the template DNA in 25 μL TE buffer, pH 8.0. This yields a final concentration of approximately 1 μg/μL.
4. Mix the transcription reagents in the order listed in **Table 8.3** in a sterile microfuge tube at room temperature (25°C).
5. Incubate for 1 h at 37°C to allow transcription to occur.
6. Remove the DNA template by adding 0.5 Kunitz units of DNase I (RNase-free) for every 20 μL of reaction volume and incubate for 10 min at 37°C.
7. Analyze 5 μL of the test or preparative reaction by electrophoresis through a 1% agarose gel (TAE buffer). Expect to see a diffuse band of about the right size for the expected product (or a bit smaller) in addition to some smearing up and down the gel.
8. Purify the oligonucleotide by extracting the reaction mixture with phenol–chloroform and then chloroform (keeping

Table 8.3
Preparation of RNA oligonucleotide

Component	Amount (test scale)	Final	Amount (preparative scale)	Final
DEPC-treated H_2O	4 μL	–	80 μL	–
Transcription buffer (5×)	2 μL	1×	40 μL	1×
DTT (0.1 M)	1 μL	10 mM	20 μL	10 mM
rUTP (10 mM)	0.5 μL	0.5 mM	10 μL	0.5 mM
rATP (10 mM)	0.5 μL	0.5 mM	10 μL	0.5 mM
rCTP (10 mM)	0.5 μL	0.5 mM	10 μL	0.5 mM
Rgtp (10 mM)	0.5 μL	0.5 mM	10 μL	0.5 mM
Linearized DNA from **Section 3.3 Step 3** (1 μg/μL)	0.5 μL	0.5 μg	10 μL	10.0 μg
RNasin (40 U/μL)	0.25 μL	10 U	5 μL	200 U
RNA polymerase (20 U/μL)	0.25 μL	5 U	5 μL	100 U
Total	10 μL	–	200 μL	–

the aqueous phase each time). Perform buffer exchange and clean-up of the RNA oligonucleotide by diluting it to 1 mL with H_2O and then concentrating it using a Microcon spin filter (pre-rinse the spin filter with H_2O before addition of sample). Rinse twice more with 1 mL of H_2O as above. Note that Microcon 10 spin filters have a cutoff size of 20 nt and Microcon 30 spin filters have a cutoff size of 60 nt. Be sure to use Microcon 10 filters for oligonucleotides smaller than 100 nt, and Microcon 30 spin filters for anything larger.

9. To check for the concentration and integrity of the sample, analyze a second aliquot of the oligonucleotide by electrophoresis through a 1% agarose gel in TAE buffer. The oligonucleotide distribution pattern should look like that in **Step** 7 above. A much smaller band likely indicates that degradation has occurred and the procedure should be repeated with fresh material. At this juncture, the product can be stored at –80°C indefinitely.

3.4. RNA Oligonucleotide–Cellular RNA Ligation

1. Set up two sterile microfuge tubes, one with TAP-treated cellular RNA and another with untreated cellular RNA, and add the remaining components for the ligation reaction (*see* **Table 8.4**). The tube with the untreated cellular RNA serves as a negative control.
2. Incubate for 16 h at 17°C. Overnight is permissible.
3. Purify and concentrate the ligation product by spin filtration with a Microcon 100 spin filter (rinsing product three times with RNase-free H_2O; pre-rinse filter with H_2O). The recovered volume after the Microcon filtration should not exceed 20 μL. Products can be stored at –80°C.
4. Analyze one third of the product by electrophoresis through a 1% agarose gel (TAE buffer) to verify the integrity of the

Table 8.4
RNA oligonucleotide–cellular RNA ligation

Component	Amount	Final
Ligation buffer (10×)	3 μL	1×
RNasin (40 U/μL)	0.75 μL	30 U
RNA oligonucleotide[a]	4 μg	4 μg
TAP-treated or untreated RNA	10 μg	10 μg
ATP (2 mM)	1.5 μL	0.1 mM
T4 RNA ligase (20 U/μL)	1.5 μL	30 U
H_2O	to 30 μL	–

[a] RNA oligonucleotides are at a molar excess of 3–6 over target cellular RNA.

ligation. It should look like the previous samples, as discussed (*see* **Section 3.3**, **Step** 7).

3.5. Reverse Transcription

1. For each experimental and control sample from **Section 3.4**, **Step 3**, assemble the transcription components listed in **Table 8.5** on ice in a sterile microfuge tube.

Table 8.5
Reverse transcription

Component	Amount	Final
Reverse-transcription buffer (5×)	4 μL	1×
dNTP mixture (containing all four dNTPs, each at 10 mM)	1 μL	471 μM
DTT (0.1 M)	2 μL	9.41 mM
RNasin (40 U/μL)	0.25 μL	10 U
Total	7.25 μL	–

2. In a separate tube, add 20 ng of a gene-specific antisense primer (e.g., GSP-RT to generate the 5′ end and the reverse compliment of NRC3 to generate the 3′ end) to the remaining RNA (about 6.7 μg) from **Section 3.4**, **Step 3** in 13 μL of H_2O. Incubate for 3 min at 80°C, cool rapidly on ice and centrifuge for 5 s. To run an optional control reaction, set up an identical tube in parallel to which no reverse transcriptase is added in **Step 3**.
3. Add the RNA-primer mix to the reverse-transcription components, subsequently add 1 μL (200 U) of SuperScript II reverse transcriptase, and incubate for 1 h at 42°C and for 10 min at 50°C.
4. Inactivate the reverse transcriptase by incubating at 70°C for 15 min then centrifuge for 5 s.
5. Destroy the RNA template by adding 0.75 μL (1.5 U) of RNase H to the reaction mixture and incubating it for 20 min at 37°C.
6. Dilute the reaction mixture to 100 μL with TE buffer and store at 4°C. Note that the 5′ end oligonucleotide–cDNA pool can be stored indefinitely at 4°C. Avoid storing at –20°C, which could snap some of the cDNA strands with repeated freeze–thawings.

3.6. First-Round Amplification

1. In a sterile 0.2-mL microfuge tube, mix the reagents listed in **Table 8.6** for each experimental and control sample from **Section 3.5**, **Step 6**.

Table 8.6
First-round amplification (reaction pre-mix)

Component	Amount	Final
Hercules Hot-Start polymerase buffer (10×)	5 μL	1×
dNTP solution (10 mM)	1.0 μL	200 μM
Hercules Hot-Start polymerase	2.5 U	2.5 U
H_2O	to 50 μL	–

2. Add a 1 μL aliquot of the 5′ end oligonucleotide–cDNA pool and 25 pmol each of primers GSP1 and NRC1 (or the reverse compliment of GSP1 and NRC1 if you are cloning the 3′ end). A "no-template" control should also be prepared.
3. Heat in a DNA thermal cycler for 5 min at 98°C to denature the first-strand (and to activate the polymerase). Cool for 2 min to the appropriate annealing temperature (56–68°C), followed by cDNA extension for 40 min at 72°C.
4. Carry out 35 cycles of amplification with a "step" program as listed in **Table 8.7**. Products can be stored at 4°C.

Table 8.7
First-round amplification (heat-block settings)

Cycle number	Denaturation	Annealing	Polymerization-extension
1–30	10 s at 94°C	10 s at 52–60°C	3 min at 72°C
31–35	10 s at 94°C	10 s at 52–60°C	15 min at 72°C; then cool to room temperature

3.7. Second-Round Amplification

1. Dilute a portion of the amplification products from the first round amplification 1:20 in TE buffer. Two rounds of amplification ensure the specificity of the yield by quenching the background amplification. The first round generates a lot of background because only one gene-specific primer is utilized. By using a second gene-specific primer, the specificity of the yield is significantly increased.
2. In a sterile 0.2 mL microfuge tube, the reagents listed in **Table 8.8** should be mixed (on ice).
3. Add a 1-mL aliquot of the diluted first-round amplification products (obtained in **Step 1** above) and 25 pmol each of

Table 8.8
Second-round amplification (reaction pre-mix)

Component	Amount	Final
Hercules Hot-Start polymerase buffer (10×)	5 μL	1×
dNTP solution (10 mM)	1.0 μL	200 μM
Hercules Hot-Start polymerase	2.5 U	2.5 U
H_2O	to 50 μL	–

primers GSP2 and NRC2 (or their reverse compliments for amplifying the 3′ end). Set up a "no-template" control as well.

4. Mix and heat in a DNA thermal cycler for 5 min at 98°C to denature the first-strand products and to activate the polymerase.
5. Carry out 30 cycles of amplification with a "step" program as described in **Table 8.9**. The product can be stored at 4°C.
6. Separate 20% of the products of first- and second-round amplification by electrophoresis through a 1% agarose gel. Specific partial cDNAs can be checked by Southern blot analysis. Information gained from this analysis can be used to optimize the RACE procedure.

Table 8.9
Second-round amplification (heat-block settings)

Cycle number	Denaturation	Annealing	Polymerization-extension
1–29	10 s at 94°C	10 s at 52–68°C	3 min at 72°C
30	10 s at 94°C	10 s at 52–68°C	15 min at 72°C; then cool to room temperature

3.8. Anticipated Results

Depending on the stringency of the amplification cycles, the yield of the desired product could range from less than 1 to 100% of the amplified product. After the second set of amplifications, the expected yield of specific products should be about 100%.

If there are alternative transcripts in the starting pool of mRNA, then several different specific products should be seen after the second round of amplifications. These products are identified as distinct bands when visualized on an ethidium bromide stained gel. Upon sequencing, the presence of TATA or CCAAT, or other promoter elements, and initiator sites around your gene should identify these products as genuine alternative transcripts.

4. Notes

1. RACE kits are available from Invitrogen, Clontech, and Ambion. These ready-made systems are user friendly and powerful. However they are not suited for every purpose. The investigator might have to optimize the kits using principles outlined in this protocol, e.g., using your own gene-specific primers (GSP) and/or varying the annealing temperature. The entire procedure can be completed in 3 days but will take up to 5 days if the procedure is stopped at the storage points.
2. Adenosine residues are the best "acceptors" for the ligation of the 3′ end of the RNA oligonucleotide to the 5′ end of its target, meaning that the restriction enzyme for linearization should be chosen accordingly (for example, *Nde*I, *Sal*I, or *Xho*I, which terminate in an A at the end of the transcription product). A test transcription should be performed to ensure that all reactions are working properly, and then the reaction can be further "scaled up." The oligonucleotide can be stored indefinitely at –80°C for future experiments. It is also important to synthesize sufficient oligonucleotide to allow for losses due to purification and "spot-checks" of the oligonucleotide during use. Alternatively, the DNA or RNA oligonucleotide can be ordered from a commercial source.
3. 10× T4 RNA ligase buffers supplied by some manufacturers contain too much ATP (25). If the composition contains more than 1 mM ATP, it is strongly recommended that you make your own. You can do this by regenerating the components used in the supplied buffer and adjusting the ATP concentration to 1 mM (final 1× concentration should be 0.1 mM). Alternatively, the 10× buffer described in ref (25) can be used.
4. For the synthesis of the RNA oligonucleotide, it is important to select a plasmid that can be linearized at a site about 100 bp downstream from a T7 or T3 RNA polymerase site. A plasmid containing an insert cloned into the first polylinker site should be used because primers made from palindromic polylinker DNA do not amplify well during PCR. In the scenario presented here, pBS-SK-GBX-1-3′UTR contains the 3′ untranslated region of the mouse Gbx1 gene7 cloned into the SstI restriction site of the

plasmid pBS-SK (Stratagene). This plasmid can be linearized with the restriction enzyme SmaI and transcribed with T3 RNA polymerase to produce a 132-nt RNA oligonucleotide, although as suggested in **Note 2**, using a restriction enzyme that generates a cleavage site that terminates in an A residue would be preferable. Any other readily amplifiable sequence can be substituted, but avoid ones derived from abundant endogenous genes (such as actin) to prevent spurious nonspecific amplification.

5. The presence of nonspecific amplification products could be the result of inappropriate annealing temperature. Using "Touch down PCR" is an effective approach to optimize the annealing temperature. Alternatively, the same optimization can be attained by gradually increasing the annealing temperature in 2°C intervals at each relevant step of the protocol.
6. All the primers used for amplification are derived from the RNA oligonucleotide. It is recommended that primer design is software assisted in order to ensure similar melting temperatures. For instance, Primer3 from the Massachusetts Institute of Technology (**http://frodo.wi.mit.edu/cgi-bin/primer3/primer3_www.cgi**) can be used to generate suitable candidates.

Acknowledgments

This work was supported by NIH awards GM071520 and GM084251 to MAF and an NIH T32 Medical Scientist Training Grant, an F31 National Research Service Award from the National Institute of Diabetes and Digestive and Kidney Diseases, and the Turner Foundation award to OY.

References

1. Ferreira, E. N., Galante, P. A., Carraro, D. M., de Souza, S. J. (2007) Alternative splicing: a bioinformatics perspective. *Mol Biosyst* 3, 473–477.
2. Kashyap, L., Tabish, M., Ganesh, S., Dubey, D. (2007) Identification and comparative analysis of novel alternatively spliced transcripts of RhoGEF domain encoding gene in C. elegans and C. briggsae. *Bioinformation* 2, 43–49.
3. Seim, I., Collet, C., Herington, A. C., Chopin, L. K. (2007) Revised genomic structure of the human ghrelin gene and identification of novel exons, alternative splice variants and natural antisense transcripts. *BMC Genomics* 8, 298.
4. Allen, R. D., 3rd, Dickerson, S., Speck, S. H. (2006) Identification of spliced gammaherpesvirus 68 LANA and v-cyclin transcripts and analysis of their expression in

vivo during latent infection. *J Virol* 80, 2055–2062.
5. Frohman, M. A. (1994) On beyond classic RACE (rapid amplification of cDNA ends. *PCR Methods Appl* 4, S40–S58.
6. Jarosinski, K. W., Schat, K. A. (2007) Multiple alternative splicing to exons II and III of viral interleukin-8 (vIL-8) in the Marek's disease virus genome: the importance of vIL-8 exon I. *Virus Genes* 34, 9–22.
7. Frohman, M. A., Dush, M. K., Martin, G. R. (1988) Rapid production of full-length cDNAs from rare transcripts: amplification using a single gene-specific oligonucleotide primer. *Proc Natl Acad Sci USA* 85, 8998–9002.
8. Bertling, W. M., Beier, F., Reichenberger, E. (1993) Determination of 5′ ends of specific mRNAs by DNA ligase-dependent amplification. *PCR Methods Appl* 3, 95–99.
9. Borson, N. D., Salo, W. L., Drewes, L. R. (1992) A lock-docking oligo(dT) primer for 5′ and 3′ RACE PCR. *PCR Methods Appl* 2, 144–148.
10. Edwards, J. B., Delort, J., Mallet, J. (1991) Oligodeoxyribonucleotide ligation to single-stranded cDNAs: a new tool for cloning 5′ ends of mRNAs and for constructing cDNA libraries by in vitro amplification. *Nucleic Acids Res* 19, 5227–5232.
11. Fritz, J. D., Greaser, M. L., Wolff, J. A. (1991) A novel 3′ extension technique using random primers in RNA-PCR. *Nucleic Acids Res* 19, 3747.
12. Frohman, M. A. (1989) Creating full-length cDNAs from small fragments of genes: Amlification of rare transcripts using a single gene-specific oligonucleotide primer, in (Innis, M., Gelfand, D., Sninsky, J. and White T. C., Eds.), *PCR Protocols and Applications: A Laboratory Manual.* Academic, New York, NY, pp. 28–38.
13. Frohman, M. A. (1993) Rapid amplification of cDNA for generation of full-length cDNA ends: thermal RACE. *Methods Enzymol* 218, 340–356.
14. Frohman, M. A. M., Martin, G. R. (1989) Rapid amplification of cDNA ends using nested primers. *Tech* 1, 165–173.
15. Jain, R., Gomer, R. H., Murtagh, J. J.,Jr. (1992) Increasing specificity from the PCR-RACE technique. *Biotechniques* 12, 58–59.
16. Monstein, H. J., Thorup, J. U., Folkesson, R., Johnsen, A. H., Rehfeld, J. F. (1993) cDNA deduced procionin. Structure and expression in protochordates resemble that of procholecystokinin in mammals. *FEBS Lett* 331, 60–64.
17. Rashtchian, A., Buchman, G. W., Schuster, D. M., Berninger, M. S. (1992) Uracil DNA glycosylase-mediated cloning of polymerase chain reaction-amplified DNA: application to genomic and cDNA cloning. *Anal Biochem* 206, 91–97.
18. Templeton, N. S., Urcelay, E., Safer, B. (1993) Reducing artifact and increasing the yield of specific DNA target fragments during PCR-RACE or anchor PCR. *Biotechniques* 15, 48–50, 52.
19. Scotto-Lavino, E., Du, G., Frohman, M. A. (2006) 5′ end cDNA amplification using classic RACE. *Nat Protoc* 1, 2555–2562.
20. Scotto-Lavino, E., Du, G., Frohman, M. A. (2006) 3′ end cDNA amplification using classic RACE. *Nat Protoc* 1, 2742–2745.
21. Scotto-Lavino, E., Du, G., Frohman, M. A. (2006) Amplification of 5′ end cDNA with 'new RACE'. *Nat Protoc* 1, 3056–3061.
22. Fromont-Racine, M., Bertrand, E., Pictet, R., Grange, T. (1993) A highly sensitive method for mapping the 5′ termini of mRNAs. *Nucleic Acids Res* 21, 1683–1684.
23. Liu, X., Gorovsky, M. A. (1993) Mapping the 5′ and 3′ ends of Tetrahymena thermophila mRNAs using RNA ligase mediated amplification of cDNA ends (RLM-RACE). *Nucleic Acids Res* 21, 4954–4960.
24. Volloch, V., Schweitzer, B., Zhang, X., Rits, S. (1991) Identification of negative-strand complements to cytochrome oxidase subunit III RNA in Trypanosoma brucei. *Proc Natl Acad Sci USA* 88, 10671–10675.
25. Tessier, D. C., Brousseau, R., Vernet, T. (1986) Ligation of single-stranded oligodeoxyribonucleotides by T4 RNA ligase. *Anal Biochem* 158, 171–178.

Chapter 9

Fractionation of mRNA Based on the Length of the Poly(A) Tail

Hedda A. Meijer and Cornelia H. de Moor

Abstract

Poly(A) tail length plays an important role in mRNA stability and translational control. Poly(A) fractionation is a very powerful technique to separate mRNAs according to the length of the poly(A) tail. Poly(A) fractionation can be used to detect small changes in poly(A) tail length or to prepare samples for microarray analysis. RNA or crude lysate is mixed with biotinylated oligo(dT), which is then bound to paramagnetic streptavidin beads. Oligoadenylated mRNA is eluted first with a high salt buffer, followed by a low salt elution for polyadenylated mRNA. Elution of the RNA in two fractions can be used as a preparation of samples for microarray analysis while elution of the mRNA in several fractions can be used to analyse (changes in) poly(A) tail length. This method allows for accurate quantification of the amount of oligoadenylated/polyadenylated RNA in each fraction because it is not dependent on visualising the smears representing the variations in poly(A) tail length. The method is technically easy, fast, highly reproducible and can be performed on almost any sample containing RNA.

Key words: Poly(A) tail length, fractionation, translation, mRNA stability, polysome profiling.

1. Introduction

The poly(A) tail of an mRNA plays an important role in both the translational activity and the stability of the mRNA. Most mRNAs obtain a long poly(A) tail (100–250 nt) in the nucleus, which gets gradually shortened after exportation to the cytoplasm (1). However, for some mRNAs the length of the poly(A) tail is strongly regulated and can therefore be altered dramatically. These mRNAs can be polyadenylated and/or deadenylated in the cytoplasm. mRNA deadenylation is thought to precede mRNA degradation in most cases (2). However, an mRNA with a short

H. Nielsen (ed.), *RNA*, Methods in Molecular Biology 703,
DOI 10.1007/978-1-59745-248-9_9,

poly(A) tail (oligoadenylated RNA) is not necessarily unstable (3). The length of the poly(A) tail is also thought to be linked to the translational activity of that mRNA. A long poly(A) tail seems to coincide with active translation while a short poly(A) tail correlates with translational repression (4).

One of the most challenging parts of studying poly(A) tail length is that all methods that are dependent on visualisation of the size of the poly(A) tail by gel electrophoresis tend to underestimate the length of the poly(A) tail because the signal is spread out over a large area which weakens the signal. Most traditional techniques are based on either northern blotting (for long mRNAs in combination with RNase H cleavage) or PCR. The northern blotting approach can only be used for relatively abundant mRNAs (5). Several PCR-based protocols (6–8) have improved detection limits but still suffer from an underestimation of long poly(A) tails. Both northern blotting and the PCR-based protocols require the preparation of a sample treated with RNase H in the presence of oligo(dT) to remove the poly(A) tail as a marker (5). For some systems, such as *Xenopus* oocytes, it is possible to inject radiolabelled reporter mRNAs which can then be analysed by gel electrophoresis (9). The use of radiation and the possibility to use very short probes limits the underestimation of the poly(A) tail length but unfortunately cannot be used in many experimental systems.

When studying poly(A) tail changes on a global scale, mRNAs need to be separated on the basis of the length of their poly(A) tail. The resulting fractions can then be used for microarray analysis. mRNA can be bound to poly(U) agarose and then eluted at different temperatures (10). While this technique can be used to separate the RNA into fractions dependent on poly(A) tail length, it is labour intensive and difficult to get this technique working reproducibly.

The protocol detailed here can be used for both the analysis of poly(A) tail length, especially for changes in poly(A) tail length, as well as for the fractionation of mRNA based on the length of the poly(A) tail as a preparation of samples for microarray analysis (*see* **Fig. 9.1**). Total RNA or crude lysate is mixed with a radiolabelled and polyadenylated probe (*see* **Fig. 9.2**), and biotinylated oligo(dT). The probe allows for visualisation of the fractionation and is furthermore used to estimate the poly(A) tail length in the sample RNA. The poly(A+) mRNA will bind to the biotinylated oligo(dT), which in turn is bound to paramagnetic streptavidin beads. The amount of biotinylated oligo(dT) needs to be optimised for each sample type (*see* **Fig. 9.3**). A magnet is used to separate the bound mRNA from the unbound RNA without a poly(A) tail or a poly(A) shorter than 15 nt. The poly(A+) mRNA is then eluted from the beads by decreasing the salt concentration, which results in elution of mRNAs with increasing

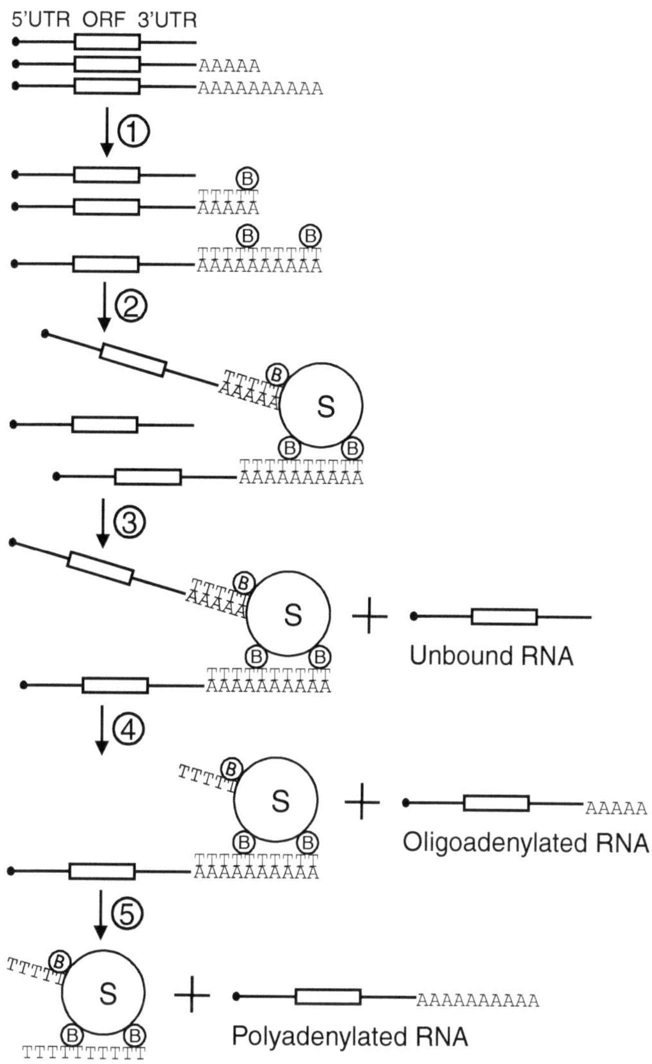

Fig. 9.1. Schematic representation of the poly(A) fractionation protocol. Total RNA is mixed with biotinylated oligo(dT) in a GTC buffer (**Step 1**). The biotin is then allowed to bind to the streptavidin paramagnetic beads (**Step 2**). The unbound fraction is collected (**Step 3)** and the beads are washed in 0.5× SSC. The oligoadenylated (**Step 4**) and polyadenylated RNA (**Step 5)** are then eluted in two or more steps by decreasing the SSC concentration in the elution buffer. A radiolabelled and polyadenylated probe is added to the RNA sample prior to fractionation. This probe is co-fractionated and allows for analysis of the fractionation and estimation of the poly(A) tail length of the RNA in the different fractions. UTR = untranslated region, ORF = open reading frame, B = biotin, S = paramagnetic streptavidin bead.

poly(A) tail length. For the analysis of (changes in) poly(A) tail length the mRNA is eluted in 6 fractions (*see* **Fig. 9.4**). For the preparation of samples for microarray analysis the mRNA is eluted in 2 fractions, (oligoadenylated and polyadenylated mRNA) (*see*

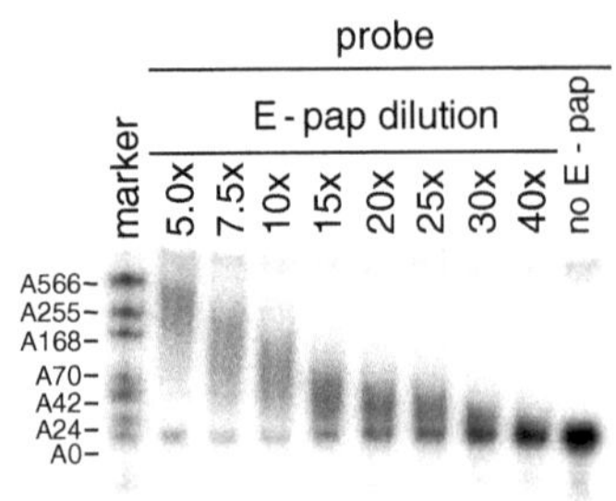

Fig. 9.2. Polyadenylation of radiolabelled probe. A radiolabelled probe was incubated with several dilutions of E-PAP. 0.5 μL of marker, 0.5 μL of probe and 1 μL of each polyadenylation reaction was analysed on a 5% acrylamide/urea/TBE gel. The length of the marker bands is indicated as the number of nucleotides in excess over the unadenylated probe (indicated with A0, actual length 79 nt).

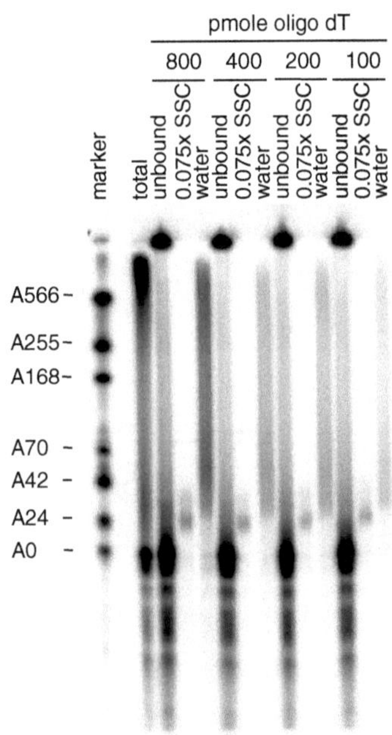

Fig. 9.3. Poly(A) fractionation into two fractions with different amounts of biotinylated oligo(dT). 80 μg of total RNA from NIH3T3 cells was mixed with polyadenylated radiolabelled probe and different amounts of biotinylated oligo (dT) and separated using the poly(A) fractionation protocol. RNA was eluted with 0.075× SSC (oligoadylated RNA) and water (polyadenylated RNA). This shows that 800 pmole of oligo dT is required for efficient fractionation of 80 μg of RNA from NIH3T3 cells.

Fig. 9.3). Since poly(A) tail length and translational activity are thought to be correlated for many mRNAs, poly(A) fractionation can be used as an alternative for other posttranscriptional profiling methods. Poly(A) fractionation is a fast and reproducible method which can be used for a wide variety of samples such as tissue culture cells, oocytes, embryos, clinical samples and total RNA (3).

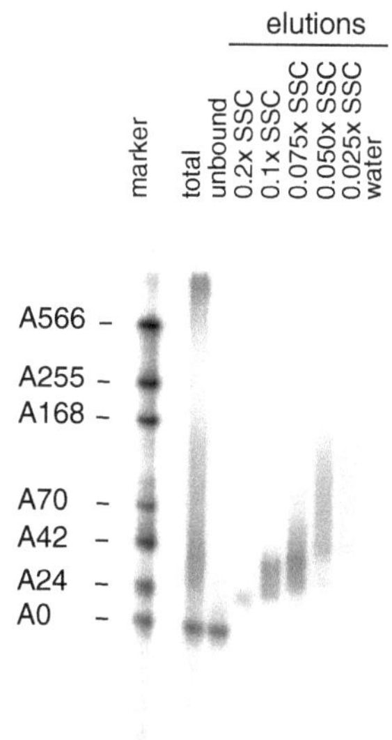

Fig. 9.4. Poly(A) fractionation into 6 fractions. 80 μg of total RNA from NIH3T3 cells was mixed with polyadenylated radiolabelled probe and 750 pmole of biotinylated oligo(dT) and separated using the poly(A) fractionation protocol. The marker can be used to estimate the range of poly(A) tail length in each fraction. Note that the signal in the water lane is weak due to the lack of probe with a very long poly(A) tail.

2. Materials

2.1. Molecular Weight Marker

1. Linearised plasmid DNA. The marker is a mixture of radioactively labelled in vitro transcripts of known lengths. In order to make this, you will need several plasmids which differ slightly in size when digested with the same restriction enzyme. Alternatively, you can use one plasmid which can be digested by several different restriction enzymes (one at a time).
2. The same materials as for the purification of the template (*see* **Section 2.2**) and for the probe synthesis (*see* **Section 2.3**).

2.2. Template Purification

1. Linearised plasmid DNA (this plasmid will also generate the lowest band of the molecular weight marker).
2. Water-saturated phenol.
3. Chloroform.
4. 3 M NaAc, pH 5.2.
5. 100% isopropanol.
6. 70% ethanol.
7. RNase-free water.

2.3. Probe Synthesis

1. RNA polymerase (including 5× transcription buffer (TSC) and 100 mM dithiothreitol (DTT)) (Promega).
2. NTP mix: 5 mM ATP, 5 mM CTP, 0.1 mM UTP, 0.5 mM GTP.
3. 40 U/μL RNasin.
4. 5 mM cap analogue (m^7GpppG) (Ambion).
5. 1 μg/μL template in RNase-free water.
6. 800 Ci/mmole, 20 μCi/μL [α-^{32}P]UTP.
7. Water-saturated phenol.
8. Chloroform.
9. G-50 Sephadex in water, autoclaved: store at 4°C.
10. 1-mL syringes.
11. Sterilised and silanised glass wool.

2.4. Polyadenylation

1. Poly(A) Tailing Kit (containing 2 U/μL *Escherichia coli* poly(A) polymerase (E-PAP), 5× E-PAP buffer, 10 mM ATP, 25 mM $MnCl_2$, nuclease free water) (Ambion).
2. RNase-free TE: 10 mM Tris-HCl, 1 mM ethylenediamine tetraacetic acid (EDTA) (pH 8.0), autoclaved.
3. Water-saturated phenol.
4. Chloroform.

2.5. Analysis of Probes

1. Urea.
2. RNase-free 10× TBE: 500 mM Tris-base, 500 mM boric acid, 1 mM EDTA, pH 8.0; autoclaved.
3. 30% acrylamide/bisacrylamide (37.5:1).
4. 10% ammonium persulfate (APS); freshly prepared or aliquoted and stored at –20°C.
5. N,N,N,N'-tetramethyl-ehylenediamine (TEMED).
6. Gel loading buffer: 95% formamide, 0.025% xylene cyanol, 0.025% bromophenol blue, 18 mM EDTA, 0.025% sodium dodecyl sulfate (SDS).

2.6. Oligo(dT)-Based Fractionation of RNA

1. 80 μg of total RNA in RNase-free water.
2. PolyA Tract System 1000, containing paramagnetic streptavidin beads, guanidinium thiocyanate (GTC) extraction buffer, dilution buffer, biotinylated oligo(dT), β-mercaptoethanol (BME), 0.5× SSC and magnetic separation stand (Promega).
3. Extra 50 pmol/μL biotinylated oligo(dT) (Promega).
4. RNase-free water.

5. 3 M NaAc, pH 5.2.
6. 100% isopropanol.
7. 10 μg/μL yeast tRNA.
8. 70% ethanol.

2.7. Analysis of Fractionation

1. The same materials as for the analysis of the probes (*see* **Section 2.5**)

3. Methods (see Note 1)

3.1. Molecular Weight Marker

Linearise the plasmids as described for the purification of the template (*see* **Section 3.2**) and then mix them and use for the synthesis of the molecular weight marker as described for the probe synthesis (*see* **Section 3.3**) (*see* **Note 2**).

3.2. Template Purification by Phenol Extraction and Ethanol Precipitation

1. Linearise the plasmid DNA.
2. Add 0.5 volume of phenol and vortex for 10 s (*see* **Note 3**).
3. Add 0.5 volume of chloroform and vortex for 30 s.
4. Spin at 20,000×*g* for 10 min at room temperature.
5. Transfer the supernatant to a new tube. Add 1 volume of chloroform and vortex for 30 s.
6. Spin at 20,000×*g* for 10 min at room temperature.
7. Transfer the supernatant to a new tube. Add 0.1 volume of 3 M NaAc, pH 5.2, and vortex.
8. Add 1 volume of isopropanol and vortex.
9. Incubate on ice for 30 min.
10. Spin at 20,000×*g* for 10 min at 4°C.
11. Aspirate the supernatant. Wash the pellet with 2 volumes of 70% ethanol.
12. Spin at 20,000×*g* for 5 min at 4°C and remove residual ethanol.
13. Dry the pellet and dissolve it in 10 μL of H_2O.
14. Check an aliquot of the template on a 1.5% agarose gel.
15. Quantify the template and dilute it to 1 μg/μL.

3.3. Probe Synthesis by In Vitro Transcription

1. Mix:1 μL 5×TSC
   0.5 μL 0.1 M DTT
   0.5 μL NTP mix
   0.5 μL 40 U/μL RNasin

0.5 μL 5 mM cap analogue

0.5 μL template (1 μg/μL)

1.5 μL 800 Ci/mmole, 20 μCi/μL [α-^{32}P]UTP (*see* **Note 4**)

0.25 μL RNA polymerase (*see* **Note 5**).

2. Incubate for 1 h at 37°C.
3. In the meantime prepare a G-50 Sephadex column per sample. Remove the plunger from a 1-mL syringe and put a piece of glass wool in the syringe. Push it down with the plunger to the bottom of the syringe and then remove the plunger. Swirl the Sephadex to mix it and pipette it into the syringe, avoiding air bubbles. Completely fill the syringe. Place the syringe in a 15-mL Falcon tube. Spin for 1 min at 1,300×*g* at 4°C. Discard the flow through. Keep adding and spinning until the syringe contains 1 mL of Sephadex. Spin for 5 min at 1,300×*g* at 4°C. Discard the flow through and move the syringe to a new 15-mL Falcon tube. The G-50 Sephadex column is now ready for the purification of the probe.
4. After the incubation add 45 μL of H_2O to the probe.
5. Add 25 μL of phenol and vortex for 10 s.
6. Add 25 μL of chloroform and vortex for 30 s .
7. Spin at 20,000×*g* for 10 min at 4°C.
8. Load the supernatant on the G-50 Sephadex column.
9. Spin for 1 min at 1,300×*g* at 4°C.
10. Transfer the eluate to a 1.5-mL tube.
11. Load 50 μL of RNase-free H_2O onto the Sephadex column.
12. Spin for 1 min at 1,300×*g* at 4°C.
13. Transfer the eluate to the tube with the first eluate. The combined eluates is the unadenylated probe.

3.4. Polyadenylation

Assemble on ice! (*see* **Note 6**)

1. Prepare E-PAP dilutions (7.5×, 15×, 30× in 1× E-PAP buffer).
2. For each E-PAP dilution mix:

   5 μL 5× E-PAP buffer

   2.5 μL 25 mM MnCl2

   2.5 μL 10 mM ATP

   5 μL diluted E-PAP

   10 μL unadenylated probe (*see* **Section 3.3**)
3. Incubate for exactly 30 min at 37°C.
4. Immediately put samples on ice.

5. Add 25 μL of TE and mix.
6. Add 25 μL of phenol and vortex for 10 s.
7. Add 25 μL of chloroform and vortex for 30 s.
8. Spin at 20,000×*g* for 10 min at 4°C.
9. Transfer the supernatant to a new tube. The polyadenylated probe is used without further purification steps or concentration by ethanol precipitation.

3.5. Analysis of Probes on a 5% Acrylamide/Urea/TBE Mini Gel

1. Mix:
   6 g urea
   1.2 mL 10×TBE
   2 mL 30% acrylamide/bisacrylamide (37.5:1) (*see* **Note 7**)
   4 mL H_2O.
2. Microwave briefly to dissolve urea. Do not allow the solution to get warmer than 60°C.
3. Let the solution cool down to room temperature.
4. Add 24 μL of APS and 24 μL of TEMED (*see* **Note 8**).
5. Pour the gel and allow it to set.
6. Pre-run the gel at 100 V for 15 min.
7. Prepare samples:
   0.5 μL marker in gel loading buffer
   0.5 μL probe w/o poly(A) tail in gel loading buffer
   1 μL polyadenylated probe (30× diluted E-PAP) in gel loading buffer
   1 μL polyadenylated probe (15× diluted E-PAP) in gel loading buffer
   1 μL polyadenylated probe (7.5× diluted E-PAP) in gel loading buffer.
8. Denature the samples at 95°C for 5 min.
9. Load the samples and run at 100 V (until bromophenol blue is almost at the bottom of the gel).
10. Wrap the gel in Saran wrap and expose it to a PhosphoImager screen or film for 5–30 min.
11. Mix the polyadenylated samples to produce a range from A0 to A400, including a small amount of probe without poly(A) tail. This constitutes the probe mix (*see* **Fig. 9.2**).

3.6. Oligo(dT)-Based Fractionation of RNA

All steps are at room temperature unless indicated otherwise.

1. Allow the GTC, BME, biotinylated oligo(dT), 0.5×SSC and H_2O to reach room temperature.
2. Add 41 μL of BME per mL of GTC (GTC/BME).
3. Add 20.5 μL of BME per mL of dilution buffer (DIL/BME). Preheat the DIL/BME to 70°C.

4. Make the required SSC dilutions (0.200×, 0.100×, 0.075×, 0.050×, 0.025×).
5. Mix a maximum of 40 μL total RNA (max 80 μg) with 400 μL GTC/BME in a 2-mL tube (*see* **Note 9**).
6. Add 5 μL of polyadenylated probe mix (*see* **Section 3.5**), 15 μL biotinylated oligo(dT) and 816 μL DIL/BME (*see* **Note 10**, and **Fig. 9.3**).
7. Incubate at 70°C for 5 min.
8. Spin at 12,000×*g* for 10 min at room temperature.
9. In the meantime prepare the paramagnetic streptavidin beads. Completely resuspend the beads by gently rocking the bottle. Transfer 600 μL of resuspended beads to a 2-mL tube for each sample. Place the tube on the magnetic stand and slowly move the stand to a horizontal position until the beads are collected at the tube side. Carefully pour off the storage buffer by tilting the tube so that the solution runs over the captured beads. Resuspend the beads in 600 μL 0.5×SSC and capture the beads again using the stand. Repeat this wash step twice. Resuspend the beads in 600 μL 0.5×SSC.
10. After the spin add the supernatant to the beads. Allow the biotinylated oligo(dT) to bind to the beads by rotation at room temperature for 15 min.
11. After the biotinylated oligo(dT) has bound to the beads capture the beads using the magnetic stand. Transfer the supernatant to a new tube and keep on ice (unbound fraction).
12. Wash the beads three times as described above (*See* **Step 9**). Rotate for at least 5 min between each wash.
13. Resuspend the beads in 400 μL 0.200× SSC. Rotate for at least 5 min at room temperature. Capture the beads using the magnetic stand. Transfer the eluate (400 μL) to a new tube. Be careful not to disturb the beads when doing so. Keep the 400 μL in a tube on ice (eluate 1) (*see* **Note 11**).
14. Repeat the elution (**Step 13**) with 400 μL 0.100× SSC, 0.075× SSC, 0.050× SSC, 0.025× SSC and H_2O (eluates 2–6) (*see* **Note 11**).
15. Spin all collected fractions (unbound and eluates 1–6) at 20,000×*g* for 5 min at 4°C to remove any transferred beads.
16. Transfer 2× 650 μL of the unbound supernatant to two new tubes. Add 650 μL of isopropanol to each. Mix and precipitate overnight at –20°C.

17. Transfer 360 μL of each eluate supernatant to a new tube. Add 2 μL yeast tRNA, 36 μL 3 M NaAc pH 5.2 and 360 μL isopropanol. Mix and precipitate overnight at –20°C.
18. Spin all samples at 20,000×*g* for 30 min at 4°C.
19. Remove the supernatant and wash with 800 μL 70% ethanol.
20. Spin at 20,000×*g* for 10 min at 4°C.
21. Completely remove the supernatant with a drawn out glass pipette and allow the pellets to dry.
22. Resuspend the pellets from the eluates in 10 μL of H_2O. Resuspend the pellets from the unbound fraction in 5 μL of H_2O per tube and pool the two samples.

3.7. Analysis of the Fractionation on a 5% Acrylamide/UREA/TBE Maxi Gel

1. Mix per gel:
   30 g urea
   6 mL 10×TBE
   10 mL 30% acrylamide/bisacrylamide (37.5:1)
   20 mL H_2O.
2. Microwave briefly to dissolve urea. Do not allow the solution to get warmer than 60°C.
3. Let the solution cool down to room temperature.
4. Add 120 μL of APS and 120 μL of TEMED.
5. Pour the gel and allow to set.
6. Pre-run the gel at 200 V for 15 min.
7. Prepare samples:
   1 μL of molecular weight marker 20× diluted in gel loading buffer.
   0.9 μL of probe mix in gel loading buffer.
   1 μL of unbound 5× diluted in gel loading buffer.
   1 μL for each eluate in gel loading buffer (*see* **Note 12**).
8. Denature the samples at 95°C for 5 min.
9. Load the samples and run at 200 V (until the dark blue colour of the dye indicator is almost at the bottom of the gel).
10. Wrap the gel in Saran wrap and expose it to a PhosphoImager screen or film for 24 h at –20°C. Allow the cassette to reach room temperature before opening (*see* **Fig. 9.4**).

3.8. Further Analysis

The remainder of the samples can be used for subsequent analysis by RT-qPCR and/or northern blotting to show in which fractions endogenous and/or reporter RNAs are (*see* **Note 13**).

4. Notes

1. This protocol requires RNase-free handling. Gloves must be worn at all times. Glassware, tubes and tips should be sterile. All solutions have to be RNase-free. DEPC treatment might be required.
2. The linearised plasmids should when transcribed result in products slightly longer (up to about 400 nt extra) than the probe used for the polyadenylation. Ideally they all are transcribed using the same polymerase. Mix the templates, including the template used for the probe for the polyadenylation and synthesise a mixed set of radiolabelled transcripts which then can be used as a molecular weight marker. If not all plasmids are transcribed using the same polymerase, it is not possible to mix the templates and the transcription reactions will have to be performed separately.
3. Handle phenol and chloroform in a fume hood.
4. ^{32}P is a highly energetic radioactive isotope. Take precautions not to expose yourself or others. If you have not handled ^{32}P before, ask for instructions from someone who has experience in handling ^{32}P.
5. The choice of RNA polymerase is dependent on the plasmid used.
6. The polyadenylation reaction is very sensitive to temperature and timing. If the polyadenylation reaction does not result in a range from A0 to A400 (length in excess over unadenylated probe), then it is necessary to try different dilutions of E-PAP until a full range is obtained. Alternatively, it is possible to vary the incubation times.
7. Acrylamide is poisonous and needs to be handled with care.
8. Pour the gel immediately after adding the TEMED.
9. This protocol is optimised for the use of total RNA. However, the protocol has also been tested on crude lysates of tissue culture cells (NIH3T3) and *Xenopus laevis* oocytes. For tissue culture cells use a 10 cm diameter plate with 4 $\times$ 10^6 cells. Remove medium, wash cells in ice cold PBS and remove as much of the PBS as possible. Lyse cells in 500 μL GTC/BME per plate, scrape cells and transfer 440 μL of lysate to a 2-mL tube. For *X. laevis* oocytes use ten oocytes per sample. Collect oocytes in tubes and freeze on dry ice. Lyse the frozen oocytes in 416 μL GTC/BME. These amounts are sufficient for subsequent microarray analysis. For detection of a specific mRNA by RT-qPCR the

procedure can be performed on smaller amounts of starting material.

10. The amount of biotinylated oligo(dT) will have to be optimised for each sample type to ensure efficient binding and elution. We tested biotinylated oligo(dT) from different suppliers but the protocol only works efficiently with biotinylated oligo(dT) from Promega.
11. This protocol can be adjusted to elute in fewer fractions. To prepare samples for microarray analysis elute with 0.075× SSC (oligoadenylated RNA) and water only (polyadenylated RNA).
12. The amount loaded for total RNA and unbound is 20% of starting material.
13. For examples of further analysis *see* Meijer et al. (3).

References

1. Piccioni, F., Zappavigna, V., Verrotti, A. C. (2005) Translational regulation during oogenesis and early development: the cap-poly(A) tail relationship. *C.R. Biol* 328, 863–881.
2. Meyer, S., Temme, C. and Wahle, E. (2004) Messenger RNA turnover in eukaryotes: pathways and enzymes. *Crit Rev Biochem Mol Biolv* 39, 197–216.
3. Meijer, H.A., Bushell, M., Hill, K., Gant, T.W., Willis, A.E., Jones, P. and De Moor, C.H. (2007) A novel method for poly(A) fractionation reveals a large population of mRNAs with a short poly(A) tail in mammalian cells. *Nucleic Acids Res* 35, e132.
4. Gorgoni, B., Gray, N.K. (2004) The roles of cytoplasmic poly(A)-binding proteins in regulating gene expression: a developmental perspective. *Brief Funct Genomics Proteomics* 3, 125–141.
5. Zangar, R.C., Hernandez, M., Kocarek, T.A. and Novak, R.F. (1995) Determination of the poly(A) tail lengths of a single mRNA species in total hepatic RNA. *Biotechniques* 18, 465–469.
6. Couttet, P., Fromont-Racine, M., Steel, D., Pictet, R. and Grange, T. (1997) Messenger RNA deadenylation precedes decapping in mammalian cells. *Proc Natl Acad Sci USA* 94, 5628–5633.
7. Sallés, F.J., Richards, W.G. and Strickland, S. (1999) Assaying the polyadenylation state of mRNAs. *Methods* 17, 38–45.
8. Rassa, J.C., Wilson, G.M., Brewer, G.A. and Parks, G.D. (2000) Spacing constraints on reinitiation of paramyxovirus transcription: the gene end U tract acts as a spacer to separate gene end from gene start sites. *Virology* 274, 438–449.
9. Piqué, M., López, M. and Méndez, R. (2006) Cytoplasmic mRNA polyadenylation and translation assays. *Methods Mol Biol* 322, 183–198.
10. Belloc, E., Méndez, R. (2008) A deadenylation negative feedback mechanism governs meiotic metaphase arrest. *Nature* 452, 1017–1022.

Chapter 10

Analysis of Mutations that Influence Pre-mRNA Splicing

Zhaiyi Zhang and Stefan Stamm

Abstract

A rapidly increasing number of human diseases are now recognized as being caused by the selection of wrong splice sites. In most cases, these changes in alternative splice site selection are due to single nucleotide exchanges in splicing regulatory elements. This chapter describes the use of bioinformatics tools to predict the influence of a mutation on alternative pre-mRNA splicing and the experimental testing of these predictions. The bioinformatic analysis determines the influence of a mutation on splicing enhancers and silencers, splice sites and RNA secondary structures. This approach generates hypotheses that are tested using splicing reporter constructs, which are then analyzed in transfection assays. We describe a recombination-based system that allows for the generation of splicing reporter constructs in the first week and their subsequent analysis in the second week.

Key words: Alternative splicing, mutation, splicing enhancer, splicing silencer, minigene analysis, single nucleotide polymorphisms.

1. Introduction

In the flow of genetic information, pre-messenger RNAs (pre-mRNAs) transcribed from DNA templates carry the information from genomic sequences to protein synthesis. The pre-mRNA is processed, and sequences known as *exons* are incorporated into the mRNA and exported into the cytosol. The remaining intervening sequences, known as *introns,* stay in the nucleus where they are eventually degraded. A pre-mRNA sequence can be recognized as an intron or an exon, depending on the cellular environment. This process is called alternative splicing. Almost all (95%) of human multi-exon genes undergo alternative splicing (1, 2). Unlike promoter activity that predominantly regulates the

H. Nielsen (ed.), *RNA*, Methods in Molecular Biology 703,
DOI 10.1007/978-1-59745-248-9_10, © Springer Science+Business Media, LLC 2011

abundance of transcripts, alternative splicing influences the structure of the mRNAs and their encoded proteins. As a result, it influences binding properties, intracellular localization, enzymatic activity, protein stability, and post-translational modification of numerous gene products (reviewed in (3)).

1.1. Patterns of Alternative Splicing

Alternative splicing events can be subdivided into five basic patterns, as shown in **Fig. 10.1**. Exons can be skipped or included, extended or shortened, or included in a mutually exclusive manner; introns can be either removed or retained. Cassette exons (or exon skipping) account for the majority of alternative splicing events conserved between human and mouse genomes. Less frequent are alternatively used 3′ and 5′ splice sites. Intron retention is the least frequently used pattern and is responsible for less than 3% of the alternative splice events conserved between human and mouse genomes. There are more complex events, such as mutually exclusive events, alternative transcription start sites, and multiple polyadenylation sites. In addition, these basic patterns can be combined resulting in highly complex transcripts (4). An estimated 75% of all alternative splicing patterns change the coding sequence (5), indicating that alternative splicing is a major mechanism for enhancing protein diversity (reviewed by (3)).

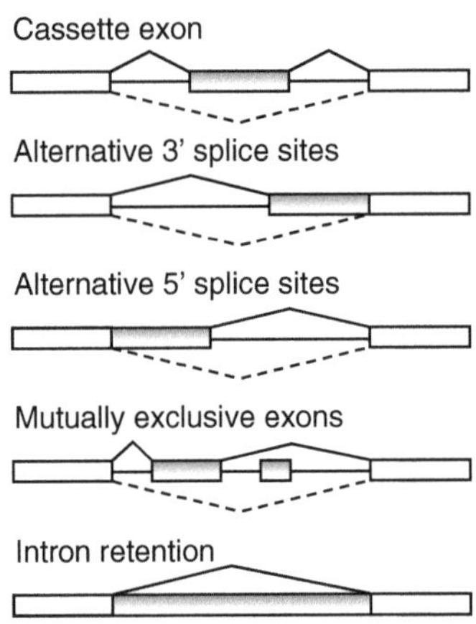

Fig. 10.1. Types of alternative splicing. Flanking constitutive exons are indicated as *white boxes*, and alternative spliced regions are shown in *grey gradient*; introns are shown by *solid lines*. The splicing patterns are indicated.

1.2. Mechanism of Splice Site Selection

The introns that are removed from the pre-mRNA are defined by the 5′ and 3′ splice sites located at their ends and by the branch point upstream of the 3′ splice site (*see* **Fig. 10.2**). The molecular mechanism of the splicing reaction that connects these sites has been determined in great detail (6, 7). In contrast, it is not

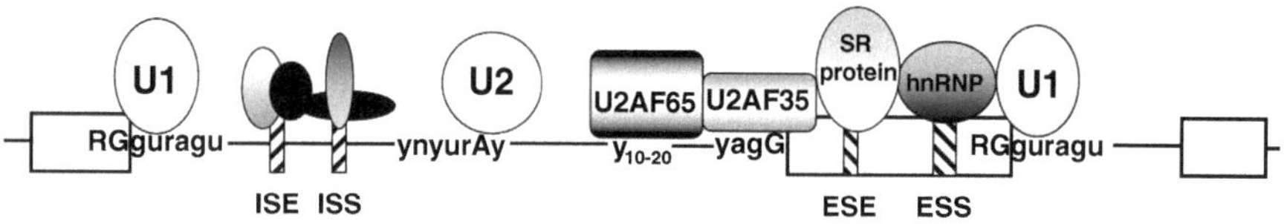

Fig. 10.2. Splicing elements and splicing factors. Exons are indicated as *boxes* and introns as *thick lines*. Splicing regulator elements (enhancers or silencers) are shown as boxes labeled as ESE/ESS in exons or as ISE/ISS in introns. The 5′ splice site (RGguragu), 3′ splice site $(y)_{10}$ncagG, and the branch point (ynyurAy) are indicated (r= a or g, y=c or u, n=a, t, c, or g). Two major groups of proteins, hnRNPs and SR or SR-related proteins bind to splicing regulator elements. ISE: intronic splicing enhancer, ISS: intronic splicing silencer, ESE: exonic splicing enhancer, ESS: exonic splicing silencer.

clear how splice sites are recognized in the large background of the pre-mRNA sequences. One major difficulty is the degeneracy of 5′ and 3′ splice sites and branch point. To precisely identify the relatively small exons and excise large introns, additional regulatory elements play an important role in splice site recognition. They are classified according to their location and their functional effect on splicing as exonic splicing enhancer (ESE), exonic splicing silencer (ESS), intronic splicing enhancer (ISE), or intronic splicing silencer (ISS). These elements are again characterized by the loose consensus sequences (8). This sequence degeneracy prevents splicing regulatory elements from interfering with the coding capacity of the exons (9). As a result, the accurate splice site selection in vivo is achieved through a combinatorial regulatory mechanism by which the exonic or intronic auxiliary elements aid exon recognition by binding to regulatory proteins (10). Proteins binding to regulatory sequence elements can be classified into two groups: serine/arginine-rich (SR) proteins and heterogeneous nuclear ribonucleoproteins (hnRNPs). Generally, these proteins not only bind to RNA, but also to other regulatory proteins. The interaction between individual splicing factors and the regulatory sequence is weak, which allows easy dislodging of the proteins from processed RNA after the splicing reaction. As the protein:RNA interaction is weak, different SR and SR-like proteins can act through the same regulatory elements and influence the same splice sites. Higher specificity is achieved by protein–protein interactions that allow simultaneous binding of multiple proteins to RNA. Several SR and SR-like proteins bind to the catalytic component of the spliceosome, e.g., to the U1 and U2 snRNP. Therefore, the transient formation of these protein complexes facilitates splicing complex assembly (11). In addition, SR and SR-like proteins can bridge the introns by interacting with themselves and the classical spliceosomal components (9), (*see* **Fig. 10.2**). In summary, numerous factors binding to RNA elements influence splice site selection. This degeneracy makes it difficult to accurately predict splice site usage.

1.3. Single Nucleotide Changes Can Influence Splicing Pattern

Single nucleotide changes in the pre-mRNA can disturb the fragile balance of multiple weak interactions governing exon recognition and alter pre-mRNA splicing pattern. Mutations of *cis*-elements can be classified into four categories according to their location and effect. Type I and type II mutations occur in the splice sites. They either destroy known splice sites or create novel ones, which leads to exon skipping or inclusion. Type III and type IV mutations take place in exons or introns, respectively, and alter exon usage. From all the mutations that are annotated in the human genome, about 10% of the 80,000 reported affect canonical splice site sequences (12). This number is most likely an underestimation since most mutations that affect the intronic and exonic splicing elements are not included in the statistic. Exonic nucleotide exchanges that influence splice site selection are often synonymous (13) and polymorphisms located in introns can influence splice site selection (14). Because these single nucleotide polymorphisms (SNPs) do not change the predicted reading frame they were considered 'noise' without functional effects. However, detailed analyses of synonymous exonic and intronic mutations revealed a strong effect on splicing that leads to frameshifts, subsequent loss of protein function and human disease (15). Recent array analysis suggests that a large number of SNPs associate with changes in splicing (16). An increased awareness concerning the role of alternative splicing in the etiology of human diseases has led to a strong increase in the number of diseases reported to be associated with changes in alterative splicing (reviewed in (17–21)). It is estimated that up to 50% of the mutations that cause human disease alter the efficiency and pattern of splicing (16). Analysis of splicing mutations causing cystic fibrosis revealed that the splicing pattern caused by a SNP can differ between individuals. This most likely reflects that splice site selection is regulated by multiple factors that work in combination and that a mutated allele responds only to a certain combination. These findings suggest that alternative splicing is a genetic modifier (18, 22) and imply that a large fraction of mutations affect exon usage. A list of single nucleotide mutations in splicing regulatory elements is given in **Table 10.1**. The widespread influence of SNPs on alternative splicing is the reason for their further bioinformatics and experimental analysis.

1.4. Bioinformatics Resources for Alternative Splicing

The completion of numerous genomic sequencing projects has provided a wealth of information revealing the abundant usage of alternative splicing in metazoan organisms. A number of groups have created resources by collecting splice variants and alternative transcript structures. These resources can be classified into two categories: computer and manually generated databases on alternative splicing events and computational tools to decipher the

Table 10.1
Examples of single nucleotide mutations change splicing pattern. The table lists examples of mutations in the intronic or exonic sequences that cause aberrant splicing

Gene	Mutation	Effect	References
ADA	CTGTCCAC**G**CC→CTGTCCAC**A**CC	Exon 7 skipping	(31)
ATM	ATT**C**GAGTG→ATT**T**GAGTG	Exon 9 skipping	(32)
	ACT**C**AACAT→ACT**T**AACAT	Exon 9 skippig	
	AG**g**taa→Ag**t**taa	Intron 12 retention	
	ttta**g**GT→ttta**a**GT	Exon 18 skipping	
	AAG**G**TTTTA→AAG**A**TTTTA	Exon 26 skipping	
	CTC**G**AAACA→CTC**A**AAACA	Exon 44 skipping	
ATP7A	GAT**G**GAATC→GAT**C/A**GAATC	Exon 4 skipping	(33)
BRCA1	ATC**T**TAGAG→ATC**G**TAGAG	Exon 18 skipping	(34)
FAH	ATGAA**C**GAC→ATCAA**T**GAC	Exon 8 skipping	(35, 36)
	AAGC**A**GGAC→AAGC**G**GGAC	Exon 9 skipping	
HEXB	GCGC**C**GGGC→GCGC**T**GGGC	Exon 11 skipping	(37)
HPRT1	CAT**G**GACTA→CAT**A**GACTA	Exon 8 skipping	(38)
	GAAC**G**TCTT→GAAC**A**TCTT		
	ATT**G**TGGAA→ATT**A**TGGAA		
IKBKAP	gtaag**t**gc→gtaag**c**gc	Exon 20 skipping	(39)
MAPT (tau)	ATTAA**T**AAG→ATTAA**G**AAG	Exon 10 inclusion	(40, 41)
	GGCA**G**Tgtg→GGCA**A**Tgtg		
	GATCT**T**AGC→GATCT**C**AGC		
	ATAA**T**ATCA→ATAA**C**ATCA		
MLH1	GAG**A**AGAGA→GAG**T**AGAGA	Exon 12 skipping	(42)
NF1	CTTAA**G**AAC→CTTAA**A**AAC	Exon 7 skipping	(43)
SMN2	GGTTT**T**AGAC→GGTTT**C**AGAC	Exon 7 skipping	(44)

Large letters indicate exonic sequences; small letters indicate intronic sequences. The sequence on the left side of the arrow is wild type, and the right side is mutant. The bold letter indicates the single nucleotide mutation.

splicing signals. Most of the databases are based on the continuously increasing amount of expressed sequence tag (EST) data. These EST data provide the major information source for computational detection of alternative splicing patterns.

In order to predict alternative splicing events with accuracy, several algorithms were specially devised. The first computational approach is based on EST and mRNA comparison (23). However, EST-mRNA comparison has limited power because the intronic information is not included (24). To address the problems caused by EST-mRNA comparison, algorithms based on genome-EST pair-wise alignment are used in computational prediction programs. Several popular programs, such as BLAT, are designed using alignment of cDNA and genomic segments. These programs perform both genomic mapping and alignment (25). However, genome-EST comparison algorithms do not provide direct exon–intron gene structures and may not be reliable. Hence, an algorithm based on EST-genome multiple alignment comparison has been devised in order to overcome this limitation. The algorithm minimizes the false splice site prediction due to incorrect EST-gene alignment that is not supported by the majority of EST data (26). **Table 10.2** lists database resources for alternative splicing.

Splicing regulatory RNA sequences and their *trans*-acting factors have been individually studied in great detail, both by experimental and bioinformatics approaches. The result of these studies let to the development of programs that predict splice sites, splicing regulatory sequences, their binding partners, as well as possible RNA secondary structure. **Table 10.3** lists these computational tools. It should be emphasized that due to the complexity of splice site selection the programs are currently fairly inaccurate. However, the usage of several programs allows the generation of hypotheses that can be experimentally tested.

1.5. Reporter Gene Analysis of Splicing Events

The most common technique to analyze exon usage and especially the effect of a mutation on splice site selection is reporter minigene analysis. In this method, an exon of interest, as well as its flanking exons are cloned into an expression vector and analyzed after transfecting this construct into eukaryotic cells (reviewed in (27, 28)). Currently about 200 such constructs have been reported in the literature. The comparison of two minigenes that were mutated to reflect naturally occurring alleles allows one to determine how a SNP influences alternative splicing.

The system can be expanded for the analysis of *trans*-acting factors by cotransfecting an increasing amount of factor-expressing cDNA constructs with the reporter minigene. To facilitate cloning of splicing reporter constructs, a recombination-based system that allows rapid generation of minigenes from PCR products containing the alternative exon has been developed (29). The system allows generating reporter minigenes within 1 week (*see* **Fig. 10.3**). Reporter minigenes then permit experimental tests concerning the influence of a mutation on splicing.

2. Materials

The computational analysis requires only internet access, as all programs are freely available.

2.1. Cloning of the Minigenes

1. Bacterial strain DB3.1 (Invitrogen, *E. coli* F$^-$ gyrA462 endA1 (sr1-recA) mcrB mrr hsdS20(r_B-, m_B) supE44 ara14

Table 10.2
Overview of existing databases on alternative splicing events

Database	Description	Reference sequence	Species covered	Additional analysis tool	Integration with other resource	URL	References
ASAP	Database of human tissue-specific regulation of alternative splicing information through a genome-wide analysis of expressed sequence tags (ESTs)	ESTs	Human	Microarray analysis tutorial download	Unigene, genbank database	**http://bioinfo.mbi.ucla.edu/ASAP/**	(45)
ASD (Alternative Splicing Database)	Database of manually annotated and computational generated data on alternative splicing events and the resultant isoform splice patterns of genes from model species. Database Alt-Splice and AEdb are included	Genomic sequence	Human and mouse	Intron analysis, scoring ATG-context sequence, MZEF-SPC exon finder, regulatory sequences	Ensemble genome annotation project	**http://www.ebi.ac.uk/asd/**	(46, 47)
ASDB (Alternative Splicing Database)	Database of alternatively spliced gene, their products and expression patterns	Genomic sequence	Human		Genbank database, Swiss-Prot protein knowledge-base	**http://hazelton.lbl.gov/~teplitski/alt/**	(48, 49)

Table 10.2 (continued)

Database	Description	Reference sequence	Species covered	Additional analysis tool	Integration with other resource	URL	References
ASTRA (Alternative Splicing and Transcription Archives)	Database of elementary patterns of alternative splicing and transcriptional initiation	Genomic sequence	Human, mouse, drosophila, C. elegans, arabidopsis, and Japanese rice		Genbank database	**http://alterna.cbrc.jp/index.php?sp=mm#search**	(50, 51)
ECgene	Database that combined the genome-based EST clustering and transcript assembly procedures	Genomic sequence	Human	Gene structure and function analysis, tissue-specific transcripts expression level analysis	Genbank, HUGO gene nomenclature committee, Swiss-Prot	**http://genome.ewha.ac.kr/ECgene/**	(52, 53)
H-DBAS (Human-transcriptome Database for Alternative Splicing)	Database of genome-wide representative alternative splicing variants generated from H-Inv full-length cDNAs and all transcripts datasets	Genomic sequence	Human	Motif sequence search, ESE prediction, protein subcellular localization and transmembrane domain prediction	HUGO gene nomenclature committee, ensemble genome annotation project, Genbank database	**http://www.h-invitational.jp/h-dbas/**	(54–57)

Table 10.2 (continued)

Database	Description	Reference sequence	Species covered	Additional analysis tool	Integration with other resource	URL	References
Hollywood	Database built upon genomic annotation of splicing patterns of known genes derived from spliced alignment of cDNAs and ESTs, and provide various analysis tools	Genomic sequence	Human and mouse	Tool for searching alternative conserved exons (ACEs), splice site score, ESEs and ESSs	Ensemble and genbank database	**http://hollywood.mit.edu/Login.php**	(58–60)
MAASE (the Manually Annotated Alternatively Spliced Events Database)	Manually annotated alternatively spliced events database, designed for supporting splicing microarray applications	Genomic sequence	Human, mouse and Drosophila	Oligonucleotides design for microarray	Putative alternative splicing database, Swiss-Prot protein knowledgebase	**http://maase.genomics.purdue.edu/**	(61)
PALS db (Putative Alternative Splicing Database)	A collection of all available putative alternative splicing information hidden in biological sequence databases	mRNA	Human, mouse and worm	Putative alternative splice site finder	Unigene, HUGO gene nomenclature committee and Cancer Genome Anatomy project (CGAP)	**http://ymbc.ym.edu.tw/palsdb/**	(62)

Table 10.2 (continued)

Database	Description	Reference sequence	Species covered	Additional analysis tool	Integration with other resource	URL	References
ProSplicer	Database of putative alternative splicing information which are produced from variant proteins and expression patterns of genes	Genomic sequence	Human		Unigene, Ensemble and Swiss-Prot	**http://prosplicer.mbc.nctu.edu.tw/**	(63)
SpliceInfo	Database provides information on tissue-specific alternative splicing events	Genomic sequence	Human	Motif discover tool for ESE, ESS and intronic splicing motifs	ProSplicer, ensemble, InterPro and gene onotology	**http://spliceinfo.mbc.nctu.edu.tw/**	(63)

Table 10.2 (continued)

Database	Description	Reference sequence	Species covered	Additional analysis tool	Integration with other resource	URL	References
SpliceNest	A web based graphical tool to visualize splicing, based on a mapping on the ESR consensus sequences from GeneNest database to the complete genome	Genomic sequence	Human, mouse, drosophila and arabidopsis		Unigene, and HUGO golden path assembly	**http://splicenest.molgen.mpg.de/**	(64) (65),
TassDB	Database stores information about alternative splicing events at GYNGYN donors and NAGNAG acceptors. TassDB allows searching genes containing tandem splice sites with specific features location in the UTR or in the CDS	Genomic sequence	Human, mouse, rat, dog, chicken, drosophila, and *C. elegans*			**http://helios.informatik.uni-freiburg.de/TassDB/**	(66)

Table 10.3
Computational tools to predict and analyze alternative splicing events

Tools	Description	URL	References
Alifold	Tool predicts consensus secondary structures for set of aligned RNA and DNA sequences	**http://rna.tbi.univie.ac.at/cgi-bin/RNAalifold.cgi**	(67)
ASD (Alternative Splicing Database)	Provide various tools for intron analysis, donor/acceptor score analysis, Polypyrimidine Tract (PPT) position analysis, branch point position and score analysis. ASD also can be used to identify potential exons in human genes, to identify and score alternative Open Reading Frames (ORF), to identify binding sites for splicing factor in RNA sequences and to check the known regulatory motifs in the sequences	**http://www.ebi.ac.uk/asd-srv/wb.cgi**	(46, 47)
AST (Analyzer Splice Tool)	Calculate splice site score, number of H-bond between U1and the 5′ splice site, and the ΔG of U1/5′ splice site pairing	**http://ast.bioinfo.tau.ac.il/SpliceSiteFrame.htm**	(68, 69)
BDGP (Berkeley Drosophila Genome Project)	Splice site prediction using neural network recognizers Provide single nucleotide polymorphism (SNP) maps	**http://www.fruitfly.org/seq_tools/splice.html**	(70)
CMfinder	RNA motif prediction tool. Perform well on unaligned sequences with long extraneous flanking regions	**http://wingless.cs.washington.edu/htbin-post/unrestricted/CMfinderWeb/CMfinderInput.pl**	N/A
ESEfinder	Analysis of sequence to find ESE motifs	**http://rulai.cshl.edu/cgi-bin/tools/ESE3/esefinder.cgi**	(71–76)
GENE BEE	RNA secondary structure prediction using energy model	**http://www.genebee.msu.su/services/rna2_reduced.html**	(77)
H-DBAS (Human-transcriptome Database for Alternative Splicing)	Coding-sequence (CDS) prediction, and ESE search	**http://www.h-invitational.jp/h-dbas/adv_search.jsp**	(55, 56)

Table 10.3 (continued)

Tools	Description	URL	References
Hollywood	5′ and 3′ splice site score calculation, ESE, ESS motif search in both human and mouse	**http://hollywood.mit.edu/Dexon.php**	(58, 78, 79),
HSF (Human Splicing Finder)	Analyses of mutation, branch point sequence, splice site, and multiple transcripts	**http://www.umd.be/HSF/**	(72, 80–82)
ImageJ	Image analysis tool	**http://rsb.info.nih.gov/ij/**	N/A
NIPU web server	Analyses of splicing regulatory motifs and single-stranded regions	**http://biwww2.informatik.uni-freiburg.de/Software/NIPU/**	(83)
Primer 3	Optimization of PCR primers	**http://frodo.wi.mit.edu/**	N/A
RegRNA	A regulatory RNA motifs and elements finder, include motifs in 5′ and 3′ UTR, motifs involved in mRNA splicing, motifs involved in transcriptional regulation, riboswitches, splice site prediction, RNA structural features, and miRNA target sites	**http://regrna.mbc.nctu.edu.tw/php/browse.php**	(84)
RNA-Bioinformatics	Collection of minigenes	**http://www.stamms-lab.net/minigenes.htm**	N/A
RNAMST (RNA Motif Search Tool)	An efficient and flexible RNA motif search tool for RNA structural homologs	**http://bioinfo.csie.ncu.edu.tw/~rnamst/search.php**	(85)
Splice site score calculation	Calculation of 5′ splice site and 3′ splice site score. The statistical data were calculated using the sequence compilation for GENIE program	**http://rulai.cshl.edu/new_alt_exon_db2/HTML/score.html**	(86)
SplicePredictor	Identify potential splice in plant pre-mRNA using Bayesian statistical models	**http://deepc2.psi.iastate.edu/cgi-bin/sp.cgi**	(87)
StrataSplice	A human splice site predictor software that combines local GC content with a first-order dependence weight array model to predict splice sites	**http://www.sanger.ac.uk/Software/analysis/stratasplice/**	(87)

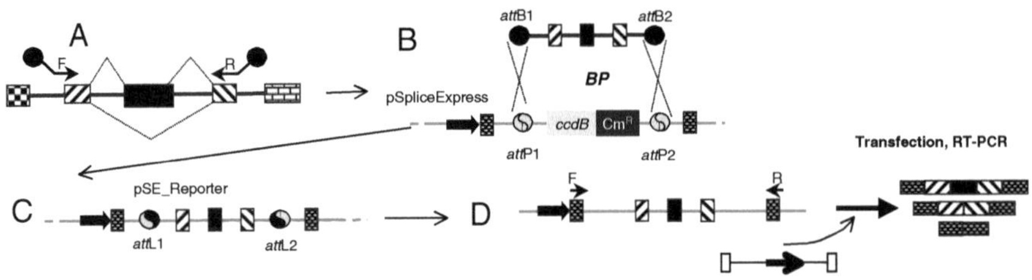

Fig. 10.3. Construction and analysis of splicing reporter genes using pSpliceExpress. A PCR product encompassing the gene region of interest is directly converted in to a splicing reporter gene by cloning it into pSpliceExpress. **a** Amplification of the region of interest. Two primers F and R are used to amplify a part of the genomic DNA that harbors the alternative exon (*black*, *splicing patterns* are indicated). The primers have recombination sites that are indicated by *circles*. **b** Construction of the splicing reporter using pSpliceExpress. The PCR fragment is recombined in vitro with pSpliceExpress vector. The vector contains Cm and ccdB selection markers that are used to isolate recombined clones. **c** Structure of the final construct using pSpliceExpress. The inserted DNA is flanked by two constitutive rat insulin exons, indicated by a *doted pattern*. The transcript is driven by a CMV promoter (*Arrow*) and the subcloned genomic fragment is flanked by attL sites, generated by the recombination of attB and attP sites. **d** The analysis of the reporter occurs in cotransfection assays using expression constructs for splicing factors, siRNAs and other regulatory factors. The analysis is done by RT-PCR using primers in the constitutive insulin exons that are indicated by *small pointed arrows*.

galK2 lacY1 proA2 rpsL20(Sm^R) xyl5 Δleu mtl1) to clone inserts that contain the bacterial ccdB marker.

2. Bacterial strain TOP10: (Invitrogen, *E. coli* F^- mcrA Δ(mrr-hsdRMS-mcrBC) Δ80lacZΔM15 ΔlacX74 recA1 araD139 Δ(ara-leu)7697 galU galK rpsL (Str^R) endA1 nupG) for all other constructs.
3. Primers

attB1F	5′-GGGG-ACAAGTTTCTACAAAAAAGCAGGCT – (template specific sequence)-3′
attB2R	5′-GGGG-ACCACTTTGTACAAGAAAGCTGGGT – (template specific sequence)-3′
attB1nestedF	5′-AAAAAGCAGGCT-template-specific sequences-3′
attB2nestedR	5′-AGAAAGCTGGGT-template-specific sequences-3′
attB1adapterF	5′-GGGGACAAGTTTGTACAAAAAAGCAGGCT -3′
attB2adapterR	5′-GGGGACCACTTTGTACAAGAAAGCTGGGT -3′

The primers are used as 10 pmol/μL working solutions in dH_2O. Store at –20°C.

4. *Pfx* DNA polymerase (5 U/μL) and buffer (Invitrogen). Store at –20°C.
5. BP clonase enzyme mix (Invitrogen). Store at –20°C.
6. *DpnI* restriction enzyme (20 U/μL) and buffer (New England Biolabs). Store at –20°C.
7. Proteinase K (Invitrogen, 20 mg/mL). Store at –20°C.

2.2. Transfection

1. Dulbeco's Modified Eagle's Medium (DMEM) (Gibco) supplemented with 10% fetal calf serum (Gibco).
2. 1 M $CaCl_2$.
3. 2× Hepes-buffered saline (2× HBS): 50 mM Hepes, 280 mM NaCl, 1.5 mM Na_2HPO_4, pH 6.95. Store at 4°C.

2.3. In Vivo Splicing Assay

1. RNAeasy kit (Qiagen).
2. Superscript III reverse transcriptase (200 U/μL) and buffer (Invitrogen). Store at –20°C.
3. *DpnI* restriction enzyme (20 U/μL) and buffer (New England Biolabs). Store at –20°C.
4. *Taq* DNA polymerase (5 U/μL) and buffer (New England Biolabs). Store at –20°C.
5. dNTP mix (Invitrogen, 100 mM)

3. Methods

As an example, we show the analysis of a mutation in an exon. In the first step, the effect of the mutation on alternative splicing is investigated bioinformatically. In the second step of analysis, these predictions are validated experimentally using transfection assays. The analysis strategy is shown in **Fig. 10.4**.

3.1. Bioinformatics Analysis

1. The sequence of interest is entered into several prediction programs. We routinely employ the "splicing rainbow" that predicts binding to regulatory factors and the "ESResearch" tool that predicts exonic regulatory elements using algorithms developed by three different laboratories.

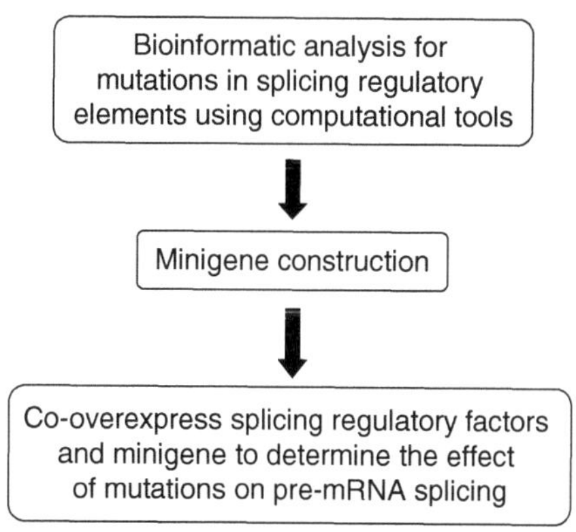

Fig. 10.4. Flowchart of the analysis.

2. In addition, we use the "NIPU" server that predicts whether a nucleotide is in an enhancer or silencer by neighborhood interference (NI) and whether this nucleotide is in a single stranded or double stranded region (PU: probability unpaired).
3. Finally, we determine the splice site strength of the exon hosting the mutation using "splice site score calculation." The internet links to the programs are listed in **Table 10.3** and screenshots of the programs are shown in **Fig. 10.5**. Depending on the outcome of these predictions, we use additional programs, such as "ESE finder," which determines binding to a subset of splicing regulatory proteins. Since different algorithms frequently give conflicting results, we combine the output of numerous programs.
4. If several programs indicate that a nucleotide exchange influences a splicing regulatory sequence, we test these predictions experimentally using reporter gene analysis.

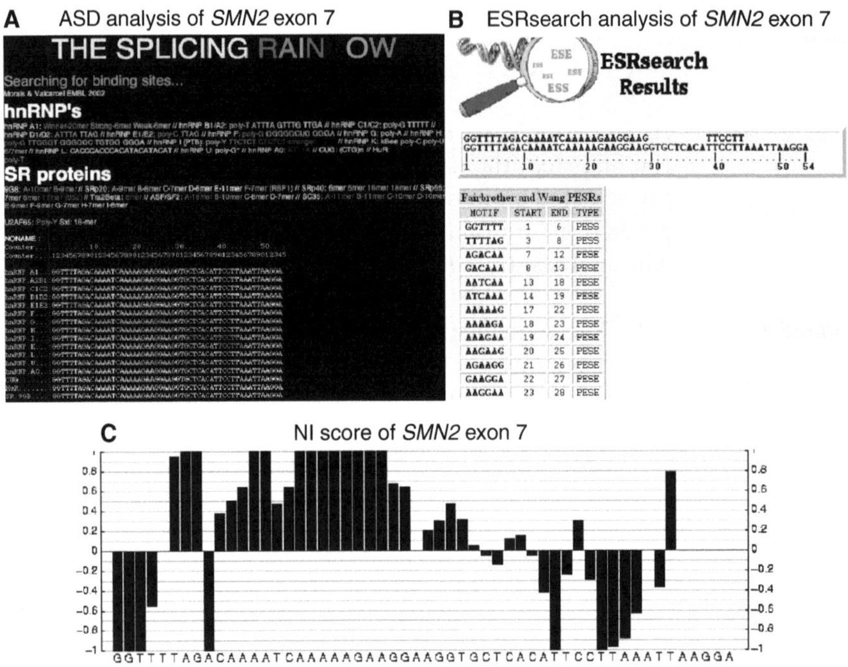

Fig. 10.5. Example of exonic elements prediction using computational tools. The sequence of SMN2 exon 7 is used as a model sequence. The internet addresses of the programs are listed in **Table 10.3**. Analyses were done with (**a**) "splicing rainbow," (**b**) "ESRsearch," (**c**) NIPU.

3.2. Experimental Testing of the Bioinformatics Prediction

Most splicing reporter genes (minigenes) are constructed by cloning the alternative exon flanked by its constitutive exons into a eukaryotic expression vector. The resulting construct is transfected into cells and the splicing products are analyzed by

RT-PCR; frequently, an exon-trap vector is used. These vectors already contain two constitutive exons that flank a multiple cloning site. An exon of interest is inserted into this site and the construct is analyzed via transfection assays. Exon-trap vectors can be used when the exon of interest is flanked by large introns. The construction and analysis of minigenes has been previously reviewed (27, 28, 30). A list of currently employed minigenes is annotated on the web (*see* **Table 10.3** for address).

The cloning of reporter constructs is time-consuming and a major impediment of the technique. We therefore developed a cloning system that relies on site-specific recombination and allows generation of reporter minigenes within 1 week (29). The system is based on pSpliceExpress, a vector that contains two strong, constitutively used insulin exons. The insulin exons ensure that pre-mRNA splicing occurs in these constructs. The system is fast, allowing to generate reporter minigenes within 1 week. Numerous comparisons between conventional cloned minigenes and reporter genes with pSpliceExpress have shown that both systems behave similar (29). An overview of the technique is shown in **Fig. 10.3**.

3.2.1. Generation of Vectors with pSpliceExpress

1. Set up a standard PCR reaction using a proofreading DNA polymerase such as Pfx DNA polymerase and genomic DNA or a cloned piece of genomic DNA as template. For amplification primers, AttB1F and AttB2R are used (*see* **Section 2.1** for a list of primers) (*see* **Note 1**).
2. Add 5–10 units of *DpnI* to the PCR reaction and incubate at 37°C for 2 h to remove contaminating DNA originating from the genomic clone (*see* **Note 2**).
3. Set up a reaction to clone the PCR fragment into pSpliceExpress by mixing:
   a. 20–30 fmoles of the attB containing PCR product
   b. 25 fmoles of pSpliceExpress vector
   c. 1 μL of 5-fold BP clonase reaction buffer mixture
   d. TE buffer, pH 8–5 μL
   The reaction is incubated at 25°C for 1 h (preferably overnight for fragments larger than 3 kb).
4. Add 0.5 μL of Proteinase K (2 mg/mL) solution to the reaction in order to inactivate the enzyme. Incubate at 37°C for 10 min.
5. Use the recombination mixture to transform Top10 bacteria. Any *recA*, *endA* *E. coli* strain including OmniMAX^{TM} 2-T1R, TOP10, DH5α^{TM}, DH10B^{TM}or equivalent can be used for transformation; however, no strains with the F ́episome should be used.

6. Isolate colonies, inoculated in LB amp-medium, and extract DNA by standard minipreparation. The recombination site is flanked by *KpnI* sites. Digesting the minipreparation DNA with *KpnI* or its isoschizomer Asp718I is used to identify clones with inserts. All constructs subject to further analysis should be verified by sequencing.

3.2.2. Transfection and Analysis of Minigenes (see Note 3)

1. Use 1–2 μg of the minigene plasmid to transfect eukaryotic cells (*see* **Note 4**). Cells are seeded in 6-well plates and transfection is performed 24 h after plating (*see* **Notes 5** and **6**).
2. After incubation for 14–17 h at 3% CO_2, isolate total RNA from the cells using RNA columns (RNAeasy kit).
3. Set up a reverse transcription reaction for RT-PCR using 400 ng of RNA. The reverse primer used for RT is specific for the vector in which the minigene was cloned. This prevents amplification of endogenous RNA.
4. Add 5–10 units of *Dpn I* to the PCR reaction and incubate at 37°C for 2 h to avoid the problem of the amplification of minigene DNA (*see* **Note 2**). A control reaction with H_2O instead of RNA served as a contamination control.
5. 1/8 of the reverse transcription reactions is used for PCR with minigene-specific primers. The primers are selected to amplify alternatively spliced minigene products. A control reaction with no template (RNA instead of cDNA) is included in the PCR. The PCR programs should be optimized for each minigene in trial experiments. We alter the annealing temperature, elongation time, and cycle number (*see* **Notes 7** and **8**).
6. PCR reactions are resolved on a 0.3–0.4 cm thick 1–2% agarose TBE gel and the image are analyzed using "ImageJ" analysis software (*see* **Table 10.3** for internet address). A typical analysis is shown in **Fig. 10.6**.

4. Notes

1. Since the amplification primers contain significant amounts of non-target sequences, in some genes we encountered undesired PCR products. This problem was especially apparent when we used genomic DNA and can often be avoided by using BAC clones. If the problem persists, we perform a two-step PCR procedure. First, the reaction is performed with a primer that is template specific and contains a part of the *attB* sequence at the 5′ end. The first PCR is then used as a template for the second PCR with

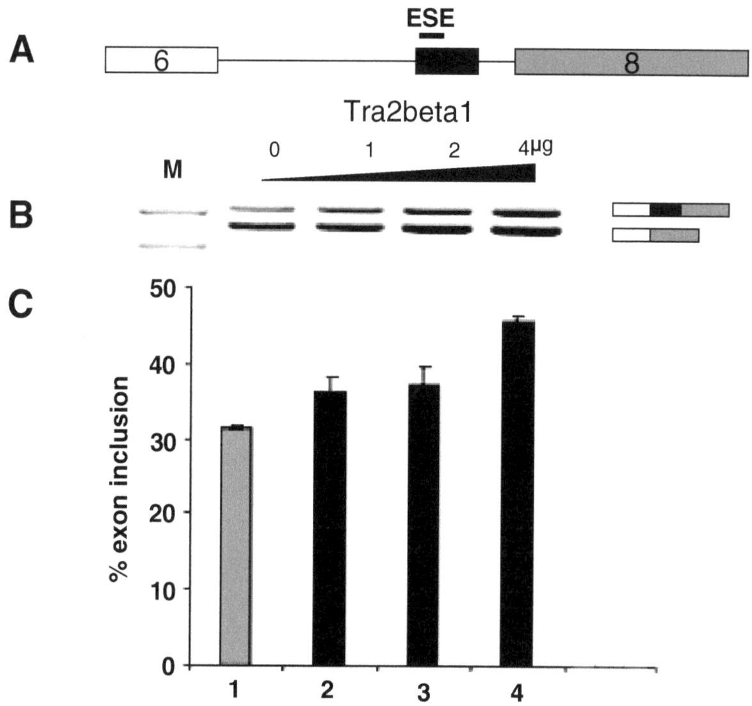

Fig. 10.6. Example for a minigene analysis. (**a**) Structure of the SMN2 minigene. (**b**) Cotransfection analysis of the SMN2 minigene with an increasing amount of the splicing factor tra2-beta1. (**c**) Quantification of the results.

adapter primers having a complete *att*B sequence. Template-specific primers for the first PCR reaction are designed with twelve bases of the attB1 or attB2 site on the 5′ end of each primer (attB1nestedF and attB1nestedR, see primers). For the second PCR reaction, adapter primers are designed to generate the complete *att*B sequences (attB1adapterF and attB2adapter). The identity between adapter primers and template-specific primers has been underlined in **Table 10.1**. This alternative method allows smaller primers to be synthesized. Only the first set of primers (template-specific primers) is specific for a new minigene. The second set of primers (adapter primers) is used repeatedly for different minigene cloning projects.

2. The *DpnI* treatment degrades the contaminating plasmid DNA as *DpnI* recognizes methylated GATC sites. The treatment reduces background in the subsequent BP recombination reaction associated with template contamination. Purification of the PCR-amplified DNA is not required if a strong single band is obtained. In those cases where there is a high background, PCR purification of the products is performed by agarose gel electrophoresis followed by crystal violet staining and gel isolation of the relevant PCR product.

3. More detailed experimental details have been published (30).
4. In order to determine the effect of a single mutation, two versions of the minigene are generated and compared in transfection assays. Often, splicing patterns are cell-type dependent and we therefore test variant minigenes in different cell lines. The influence of predicted *trans*-acting factors can be assessed by cotransfecting an expression construct together with the reporter minigene. The expression construct can either encode a splicing factor or an shRNA targeted against a splicing factor. Usually, a concentration-dependent effect is analyzed. The expression construct is transfected in increasing amounts, in the range of 0–3 μg. To avoid "squelching" effects, the "empty" parental expression plasmid containing the same promoter is added in decreasing amounts to ensure a constant amount of transfected DNA.
5. To obtain the best result, cells should be in optimal physiological conditions. HEK293 cells should be 60–80% confluent at the day of transfection.
6. The pH of transfection reagent 2× HBS is crucial. It should be 6.95 and tested with a pH meter. After filtering the transfection reagents under sterile conditions, these reagents should be tested by transfecting empty EGFP vectors into HEK293 cells. 24 h later, the transfection rate will be determined by observing the green cells ratio under fluorescent microscope.
7. In order to prevent amplification of endogenous genes, a vector-specific primer should be applied in RT reaction, and a gene-specific primer and a vector-specific primer should be used in PCR reaction.
8. We adjust the annealing temperature, which can be calculated using free online "Primer 3 program" (*see* **Table 10.3**), and the elongation time is if there are any difficulties of PCR amplification.

Acknowledgment

This work was supported by the EURASNET (European Alternative Splicing Network of Excellence), NIH (National Institutes of Health; P20 RR020171 from the National Center for Research Resources), BMBF (Federal Ministry of Education and Research, Germany), and the DFG (Deutsche Forschungsgemeinschaft; SFB 473).

References

1. Pan, Q., Shai, O., Lee, L. J., Frey, B. J., Blencowe, B. J. (2008) Deep surveying of alternative splicing complexity in the human transcriptome by high-throughput sequencing. *Nat Genet* 40, 1413–1415.
2. Wang, E. T., Sandberg, R., Luo, S., Khrebtukova, I., Zhang, L., Mayr, C., Kingsmore, S. F., Schroth, G. P., Burge, C. B. (2008) Alternative isoform regulation in human tissue transcriptomes. *Nature* 456, 470–476.
3. Stamm, S., Ben-Ari, S., Rafalska, I., Tang, Y., Zhang, Z., Toiber, D., Thanaraj, T. A., Soreq, H. (2005) Function of alternative splicing. *Gene* 344C, 1–20.
4. Ast, G. (2004) How did alternative splicing evolve? *Nat Rev Genet* 5, 773–782.
5. Zavolan, M., Kondo, S., Schonbach, C., Adachi, J., Hume, D. A., Hayashizaki, Y., Gaasterland, T. (2003) Impact of alternative initiation, splicing, and termination on the diversity of the mRNA transcripts encoded by the mouse transcriptome. *Genome Res* 13, 1290–1300.
6. Jurica, M. S., Moore, M. J. (2003) Pre-mRNA splicing: awash in a sea of proteins. *Mol Cell* 12, 5–14.
7. Nilsen, T. W. (2003) The spliceosome: the most complex macromolecular machine in the cell?. *Bioessays* 25, 1147–1149.
8. Hertel, K. J. (2008) Combinatorial control of exon recognition. *J Biol Chem* 283, 1211–1215.
9. Black, D. L. (2003) Mechanisms of alternative pre-messenger RNA splicing. *Annu Rev Biochem* 72, 291–336.
10. Smith, C. W., Valcarcel, J. (2000) Alternative pre-mRNA splicing: the logic of combinatorial control. *Trends Biochem Sci* 25, 381–388.
11. Maniatis, T., Reed, R. (2002) An extensive network of coupling among gene expression machines. *Nature* 416, 499–506.
12. Cooper, D. N., Stenson, P. D., Chuzhanova, N. A. (2006) The Human Gene Mutation Database (HGMD) and its exploitation in the study of mutational mechanisms. *Curr Protoc Bioinformatics* **Chapter 1**, Unit 1.13, 1.13.1 to 1.13.20.
13. Cooper, T. A., Mattox, W. (1997) The regulation of splice-site selection, and its role in human disease. *Am J Hum Genet* 61, 259–266.
14. Hull, J., Campino, S., Rowlands, K., Chan, M. S., Copley, R. R., Taylor, M. S., Rockett, K., Elvidge, G., Keating, B., Knight, J., Kwiatkowski, D. (2007) Identification of common genetic variation that modulates alternative splicing. *PLoS Genet* 3, e99.
15. Pagani, F., Stuani, C., Tzetis, M., Kanavakis, E., Efthymiadou, A., Doudounakis, S., Casals, T., Baralle, F. E. (2003) New type of disease causing mutations: the example of the composite exonic regulatory elements of splicing in CFTR exon 12. *Hum Mol Genet* 12, 1111–1120.
16. Graveley, B. R. (2008) The haplo-spliceo-transcriptome: common variations in alternative splicing in the human population. *Trends Genet* 24, 5–7.
17. Orengo, J. P., Cooper, T. A. (2007) Alternative splicing in disease. *Adv Exp Med Biol* 623, 212–223.
18. Nissim-Rafinia, M., Kerem, B. (2002) Splicing regulation as a potential genetic modifier. *Trends Genet* 18, 123–127.
19. Buratti, E., Baralle, M., Baralle, F. E. (2006) Defective splicing, disease and therapy: searching for master checkpoints in exon definition. *Nucleic Acids Res* 34, 3494–3510.
20. Faustino, N. A., Cooper, T. A. (2003) Pre-mRNA splicing and human disease. *Genes Dev* 17, 419–437.
21. Tazi, J., Bakkour, N., Stamm, S. (2009) Alternative splicing and disease. *Biochim Biophys Acta* 1792, 14–26.
22. Rave-Harel, N., Kerem, E., Nissim-Rafinia, M., Madjar, I., Goshen, R., Augarten, A., Rahat, A., Hurwitz, A., Darvasi, A., Kerem, B. (1997) The molecular basis of partial penetrance of splicing mutations in cystic fibrosis. *Am J Hum Genet* 60, 87–94.
23. Brett, D., Hanke, J., Lehmann, G., Haase, S., Delbrück, S., Krueger, S., Reich, J., Bork, P. (2000) EST comparison indicates 38% of human mRNAs contain possible alternative splice forms. *FEBS Lett* 474, 83–86.
24. Heber, S., Alekseyev, M., Sze, S. H., Tang, H., Pevzner, P. A. (2002) Splicing graphs and EST assembly problem. *Bioinformatics* 18 Suppl 1, S181–S188.
25. Kent, W. J. (2002) BLAT—the BLAST-like alignment tool. *Genome Res* 12, 656–664.
26. Grasso, C., Modrek, B., Xing, Y., Lee, C. (2004) Genome-wide detection of alternative splicing in expressed sequences using partial order multiple sequence alignment graphs. *Pac Symp Biocomput* 29–41.
27. Tang, Y., Novoyatleva, T., Benderska, N., Kishore, S., Thanaraj, T. A., Stamm, S. (2005) in Analysis of Alternative Splicing in Vivo using Minigenes. Chapter 46 (Westhof, E., Bindereif, A., Schön, A., Hartmann, K. eds.), *Handbook of RNA Biochemistry*. Wiley-VCH, Weinheim, pp. 755–782,.

28. Cooper, T. A. (2005) Use of minigene systems to dissect alternative splicing elements. *Methods* 37, 331–340.
29. Kishore, S., Khanna, A., Stamm, S. (2008) Rapid generation of splicing reporters with pSpliceExpress. *Gene* 427, 104–110.
30. Stoss, O., Stoilov, P., Hartmann, A. M., Nayler, O., Stamm, S. (1999) The in vivo minigene approach to analyze tissue-specific splicing. *Brain Res Protoc* 4, 383–394.
31. Ozsahin, H., Arredondo-Vega, F. X., Santisteban, I., Fuhrer, H., Tuchschmid, P., Jochum, W., Aguzzi, A., Lederman, H. M., Fleischman, A., Winkelstein, J. A., Seger, R. A., Hershfield, M. S. (1997) Adenosine deaminase deficiency in adults. *Blood* 89, 2849–2855.
32. Teraoka, S. N., Telatar, M., Becker-Catania, S., Liang, T., Onengut, S., Tolun, A., Chessa, L., Sanal, O., Bernatowska, E., Gatti, R. A.,et al. (1999) Splicing defects in the ataxia-telangiectasia gene, ATM: underlying mutations and consequences. *Am J Hum Genet* 64, 1617–1631.
33. Das, S., Levinson, B., Whitney, S., Vulpe, C., Packman, S., Gitschier, J. (1994) Diverse mutations in patients with Menkes disease often lead to exon skipping. *Am J Hum Genet* 55, 883–889.
34. Yang, Y., Swaminathan, S., Martin, B. K., Sharan, S. K. (2003) Aberrant splicing induced by missense mutations in BRCA1: clues from a humanized mouse model. *Hum Mol Genet* 12, 2121–2131.
35. Ploos van AmstelJ. K., , BergmanA. J., van BeurdenE. A., RoijersJ. F., PeelenT., van den BergI. E., Poll-TheB. T., KvittingenE. A., BergerR (1996) Hereditary tyrosinemia type 1: novel missense, nonsense and splice consensus mutations in the human fumarylacetoacetate hydrolase gene; variability of the genotype-phenotype relationship. *Hum Genet* 97, 51–59.
36. Dreumont, N., Poudrier, J. A., Bergeron, A., Levy, H. L., Baklouti, F., Tanguay, R. M. (2001) A missense mutation (Q279R) in the fumarylacetoacetate hydrolase gene, responsible for hereditary tyrosinemia, acts as a splicing mutation. *BMC Genet* 2, 9.
37. Wakamatsu, N., Kobayashi, H., Miyatake, T., Tsuji, S. (1992) A novel exon mutation in the human beta-hexosaminidase beta subunit gene affects 3′ splice site selection. *J Biol Chem* 267, 2406–2413.
38. Torres, R. J., Mateos, F. A., Molano, J., Gathoff, B. S., O'Neill, J. P., Gundel, R. M., Trombley, L., Puig, J. G. (2000) Molecular basis of hypoxanthine-guanine phosphoribosyltransferase deficiency in thirteen Spanish families. *Hum Mutat* 15, 383.
39. Carmel, I., Tal, S., Vig, I., Ast, G. (2004) Comparative analysis detects dependencies among the 5′ splice-site positions. *RNA* 10, 828–840.
40. Lee, V. M., Goedert, M., Trojanowski, J. Q. (2001) Neurodegenerative tauopathies. *Annu Rev Neurosci* 24, 1121–1159.
41. Wang, J., Gao, Q.-S., Wang, Y., Lafyatis, R., Stamm, S., Andreadis, A. (2004) Tau exon 10, whose missplicing causes frontotemporal dementia, is regulated by an intricate interplay of cis elements and trans factors. *J Neurochem* 88, 1078–1090.
42. Stella, A., Wagner, A., Shito, K., Lipkin, S. M., Watson, P., Guanti, G., Lynch, H. T., Fodde, R., Liu, B. (2001) A nonsense mutation in MLH1 causes exon skipping in three unrelated HNPCC families. *Cancer Res* 61, 7020–7024.
43. Fahsold, R., Hoffmeyer, S., Mischung, C., Gille, C., Ehlers, C., Kucukceylan, N., Abdel-Nour, M., Gewies, A., Peters, H., Kaufmann, D., et al. (2000) Minor lesion mutational spectrum of the entire NF1 gene does not explain its high mutability but points to a functional domain upstream of the GAP-related domain. *Am J Hum Genet* 66, 790–818.
44. Lefebvre, S., Burglen, L., Reboullet, S., Clermont, O., Burlet, P., Viollet, L., Benichou, B., Cruaud, C., Millasseau, P., Zeviani, M., et al. (1995) Identification and characterization of a spinal muscular atrophy-determining gene . *Cell* 80, 155–165.
45. Kim, N., Alekseyenko, A. V., Roy, M., Lee, C. (2007) The ASAP II database: analysis and comparative genomics of alternative splicing in 15 animal species. *Nucleic Acids Res* 35, D93–D98.
46. Stamm, S., Riethoven, J. J., Le Texier, V., Gopalakrishnan, C., Kumanduri, V., Tang, Y., Barbosa-Morais, N. L., Thanaraj, T. A. (2006) ASD: a bioinformatics resource on alternative splicing. *Nucleic Acids Res* 34, D46–D55.
47. Thanaraj, T. A., Stamm, S., Clark, F., Riethoven, J. J., Le Texier, V., Muilu, J. (2004) ASD: the Alternative Splicing Database. *Nucleic Acids Res* 32, Database issue, D64–D69.
48. Dralyuk, I., Brudno, M., Gelfand, M. S., Zorn, M., Dubchak, I. (2000) ASDB: database of alternatively spliced genes. *Nucleic Acids Res* 28, 296–307.
49. Gelfand, M. S., Dubchak, I., Dralyuk, I., Zorn, M. (1999) ASDB: database of

alternatively spliced genes. *Nucleic Acids Res* 27, 301–312.
50. Nagasaki, H., Arita, M., Nishizawa, T., Suwa, M., Gotoh, O. (2005) Species-specific variation of alternative splicing and transcriptional initiation in six eukaryotes. *Gene* 364, 53–62.
51. Nagasaki, H., Arita, M., Nishizawa, T., Suwa, M., Gotoh, O. (2006) Automated classification of alternative splicing and transcriptional initiation and construction of visual database of classified patterns. *Bioinformatics* 22, 1211–1216.
52. Kim, N., Shin, S., Lee, S. (2005) ECgene: genome-based EST clustering and gene modeling for alternative splicing. *Genome Res* 15, 566–576.
53. Kim, P., Kim, N., Lee, Y., Kim, B., Shin, Y., Lee, S. (2005) ECgene: genome annotation for alternative splicing. *Nucleic Acids Res* 33, D75–D79.
54. Takeda, J., Suzuki, Y., Nakao, M., Barrero, R. A., Koyanagi, K. O., Jin, L., Motono, C., Hata, H., Isogai, T., Nagai, K., et al. (2006) Large-scale identification and characterization of alternative splicing variants of human gene transcripts using 56,419 completely sequenced and manually annotated full-length cDNAs. *Nucleic Acids Res* 34, 3917–3928.
55. Takeda, J., Suzuki, Y., Nakao, M., Kuroda, T., Sugano, S., Gojobori, T., Imanishi, T. (2007) H-DBAS: alternative splicing database of completely sequenced and manually annotated full-length cDNAs based on H-Invitational. *Nucleic Acids Res* 35, D104–D109.
56. Takeda, J., Suzuki, Y., Sakate, R., Sato, Y., Seki, M., Irie, T., Takeuchi, N., Ueda, T., Nakao, M., Sugano, S., et al. (2008) Low conservation and species-specific evolution of alternative splicing in humans and mice: comparative genomics analysis using well-annotated full-length cDNAs. *Nucleic Acids Res* 36, 6386–6395.
57. Imanishi, T., Itoh, T., Suzuki, Y., O'Donovan, C., Fukuchi, S., Koyanagi, K. O., Barrero, R. A., Tamura, T., Yamaguchi-Kabata, Y., Tanino, M. (2004) Integrative annotation of 21,037 human genes validated by full-length cDNA clones. *PLoS Biol* 2, e162.
58. Fairbrother, W. G., Yeh, R. F., Sharp, P. A., Burge, C. B. (2002) Predictive identification of exonic splicing enhancers in human genes. *Science* 297, 1007–1013.
59. Fairbrother, W. G., Yeo, G. W., Yeh, R., Goldstein, P., Mawson, M., Sharp, P. A., Burge, C. B. (2004) RESCUE-ESE identifies candidate exonic splicing enhancers in vertebrate exons. *Nucleic Acids Res* 32, W187–W190.
60. Fairbrother, W. G., Holste, D., Burge, C. B., Sharp, P. A. (2004) Single nucleotide polymorphism-based validation of exonic splicing enhancers. *PLoS Biol* 2, E268.
61. Zheng, C. L., Nair, T. M., Gribskov, M., Kwon, Y. S., Li, H. R., Fu, X. D. (2004) A database designed to computationally aid an experimental approach to alternative splicing. *Pac Symp Biocomput* 78–88.
62. Huang, Y. H., Chen, Y. T., Lai, J. J., Yang, S. T., Yang, U. C. (2002) PALS db: putative alternative splicing database. *Nucleic Acids Res* 30, 186–190.
63. Huang, H. D., Horng, J. T., Lin, F. M., Chang, Y. C., Huang, C. C. (2005) SpliceInfo: an information repository for mRNA alternative splicing in human genome. *Nucleic Acids Res* 33, D80–D85.
64. Gupta, S., Zink, D., Korn, B., Vingron, M., Haas, S. A. (2004) Strengths and weaknesses of EST-based prediction of tissue-specific alternative splicing. *BMC Genomics* 5, 72.
65. Gupta, S., Vingron, M., Haas, S. A. (2005) T-STAG: resource and web-interface for tissue-specific transcripts and genes. *Nucleic Acids Res* 33, W654–W658.
66. Hiller, M., Nikolajewa, S., Huse, K., Szafranski, K., Rosenstiel, P., Schuster, S., Backofen, R., Platzer, M. (2007) TassDB: a database of alternative tandem splice sites. *Nucleic Acids Res* 35, D188–D192.
67. Hofacker, I. L. (2003) Vienna RNA secondary structure server. *Nucleic Acids Res* 31, 3429–3431.
68. Norton, P. A. (1994) Polypyrimidine tract sequences direct selection of alternative branch sites and influence protein binding. *Nucleic Acids Res* 22, 3854–3860.
69. Coolidge, C. J., Seely, R. J., Patton, J. G. (1997) Functional analysis of the polypyrimidine tract in pre-mRNA splicing. *Nucleic Acids Res* 25, 888–896.
70. Reese, M. G., Eeckman, F. H., Kulp, D., Haussler, D. (1997) Improved splice site detection in Genie. *J Comput Biol* 4, 311–323.
71. Blencowe, B. J. (2000) Exonic splicing enhancers: mechanism of action, diversity and role in human genetic diseases. *Trends Biochem Sci* 25, 106–110.
72. Cartegni, L., Chew, S. L., Krainer, A. R. (2002) Listening to silence and understanding nonsense: exonic mutations that affect splicing. *Nat Rev Genet* 3, 285–298.
73. Caceres, J. F., Kornblihtt, A. R. (2002) Alternative splicing: multiple control mechanisms

and involvement in human disease. *Trends Genet* 18, 186–193.
74. Hastings, M. L., Krainer, A. R. (2001) Pre-mRNA splicing in the new millennium. *Curr Opin Cell Biol* 13, 302–309.
75. Graveley, B. R. (2000) Sorting out the complexity of SR protein function. *RNA* 6, 1197–1211.
76. Graveley, B. R. (2001) Alternative splicing: increasing diversity in the proteomic world. *Trends Genet* 17, 99–108.
77. Brodsky, L. I., Drachev, A. L., Leontovich, A. M., Feranchuk, S. I. (1993) A novel method of multiple alignment of biopolymer sequences. *Biosystems* 30, 65–79.
78. Wang, Z., Rolish, M. E., Yeo, G., Tung, V., Mawson, M., Burge, C. B. (2004) Systematic identification and analysis of exonic splicing silencers. *Cell* 119, 831–845.
79. Lim, S. R., Hertel, K. J. (2001) Modulation of SMN pre-mRNA splicing by inhibition of alternative 3′ splice site pairing. *J Biol Chem* 2, 2.
80. Garcia-Blanco, M. A., Baraniak, A. P., Lasda, E. L. (2004) Alternative splicing in disease and therapy. *Nat Biotechnol* 22, 535–546.
81. Wang, G. S., Cooper, T. A. (2007) Splicing in disease: disruption of the splicing code and the decoding machinery. *Nat Rev Genet* 8, 749–761.
82. Wang, Z., Burge, C. B. (2008) Splicing regulation: from a parts list of regulatory elements to an integrated splicing code. *RNA* 14, 802–813.
83. Stadler, M. B., Shomron, N., Yeo, G. W., Schneider, A., Xiao, X., Burge, C. B. (2006) Inference of splicing regulatory activities by sequence neighborhood analysis. *PLoS Genet* 2, e191.
84. Huang, H. Y., Chien, C. H., Jen, K. H., Huang, H. D. (2006) RegRNA: an integrated web server for identifying regulatory RNA motifs and elements. *Nucleic Acids Res* 34, W429–W434.
85. Chang, T. H., Huang, H. D., Chuang, T. N., Shien, D. M., Horng, J. T. (2006) RNAMST: efficient and flexible approach for identifying RNA structural homologs. *Nucleic Acids Res* 34, W423–W428.
86. Stamm, S., Zhu, J., Nakai, K., Stoilov, P., Stoss, O., Zhang, M. Q. (2000) An alternative-exon database and its statistical analysis. *DNA Cell Biol* 19, 739–756.
87. Brendel, V., Xing, L., Zhu, W. (2004) Gene structure prediction from consensus spliced alignment of multiple ESTs matching the same genomic locus. *Bioinformatics* 20, 1157–1169.

Chapter 11

S1 Nuclease Analysis of Alternatively Spliced mRNA

Martin Lützelberger and Jørgen Kjems

Abstract

The characterization of alternatively spliced RNA is a frequently performed task in the molecular biology laboratory. Several methods have been established to characterize specific transcripts, of which microarrays, northern analysis, RT-PCR and nuclease protection assays are the most frequently performed methods in the laboratory. Here, we describe the analysis of alternatively spliced RNA by using 5′-end labelled DNA oligonucleotide probes and S1 nuclease. The method is sensitive, allowing detection of as little as a few hundred femtograms of a specific RNA, and useful for the quantitation of alternatively spliced mRNA isoforms. Because of its insensitivity towards RNA secondary structures and partially degraded RNA, it may perform better in the quantitation of RNA than northern analysis or RT-PCR, especially when long transcripts are studied.

Key words: Alternative splicing, pre-mRNA, S1 nuclease, nuclease protection assay, liquid hybridization, RNA quantitation.

1. Introduction

Alternative splicing is a highly regulated process in the higher eukaryotic cell. By recent estimates, the primary transcripts of about 30–70% of human genes are alternatively spliced (1). In complex genes alternative splicing can generate dozens or even hundreds of different mRNA isoforms from a single transcript (2). Thus, the characterization of alternatively spliced RNA transcribed from a single gene can be a complicated and tedious task. A number of practical approaches have been developed to characterize specific transcripts, of which microarrays (3), northern analysis (4, 5), RT-PCR (6–8) and nuclease protection (9) are probably the most commonly used methods in the laboratory.

H. Nielsen (ed.), *RNA*, Methods in Molecular Biology 703,
DOI 10.1007/978-1-59745-248-9_11, © Springer Science+Business Media, LLC 2011

Northern analysis generally provides information about the size and number of different transcripts expressed by a gene, but this method is often compromised by limited resolution of agarose gels and the inability to load large amounts of RNA onto the gel. The inefficient transfer of the RNA to the membrane and non-specific cross-hybridization between probe and transcripts is another obstacle to be overcome. Furthermore, complete hybridization to the probe is often difficult to achieve, since some of the membrane-bound RNA may not be fully accessible to the probe. Thus, northern analysis is at best a semi-quantitative approach for the analysis of alternatively spliced RNA and often not suitable for transcripts with a low abundance.

To a great extent, RT-PCR has replaced northern analysis because it allows characterization of transcripts in greater detail and with high sensitivity by using gene-specific primers. However, it requires reverse transcription of the RNA which is difficult to optimize when it has to proceed over long distances, or has to pass intense RNA secondary structures. Furthermore, the quantitation of RNA by RT-PCR requires careful normalization of the RNA samples and appropriate controls, in order to prevent forged results caused by the exponential nature of PCR amplification. Thus, hybridization techniques appear to be a better choice for mRNA quantitation, since they detect RNA directly and do not involve reverse transcription or amplification.

The procedure given here is a variation of the protocol described by Quarless and Heinrich (10), which allows to detect as little as a few hundred femtograms of a specific mRNA, corresponding to $2 \times 10^{-5}\%$ of the poly(A)$^+$ fraction in 100 μg of total RNA. The method is outlined in **Fig. 11.1** and consists of the following steps: Purified RNA is hybridized in solution with a labelled oligonucleotide probe to form hybrid molecules. Any RNA that does not participate in the formation of a hybrid molecule is digested away by the treatment with single-strand-specific S1 nuclease. The hybrid molecules are then ethanol-precipitated and analysed by electrophoresis. The size and intensity of the bands detected by autoradiography is a direct measure for their steady-state level in vivo.

The key advantage of this method is that it does not require a reverse transcription step as with RT-PCR. It is often difficult to perform reverse transcription on long transcripts or RNA molecules that contains extensive secondary structures. S1 nuclease analysis is also more tolerant against partially degraded RNA. A single break in a transcript has the consequence that the molecule is not being detected by northern analysis because it migrates as bands of different size. If the break occurs upstream of the primer, it cannot be detected by RT-PCR either, since the reverse transcriptase will not reach the 5′-end of the RNA molecule.

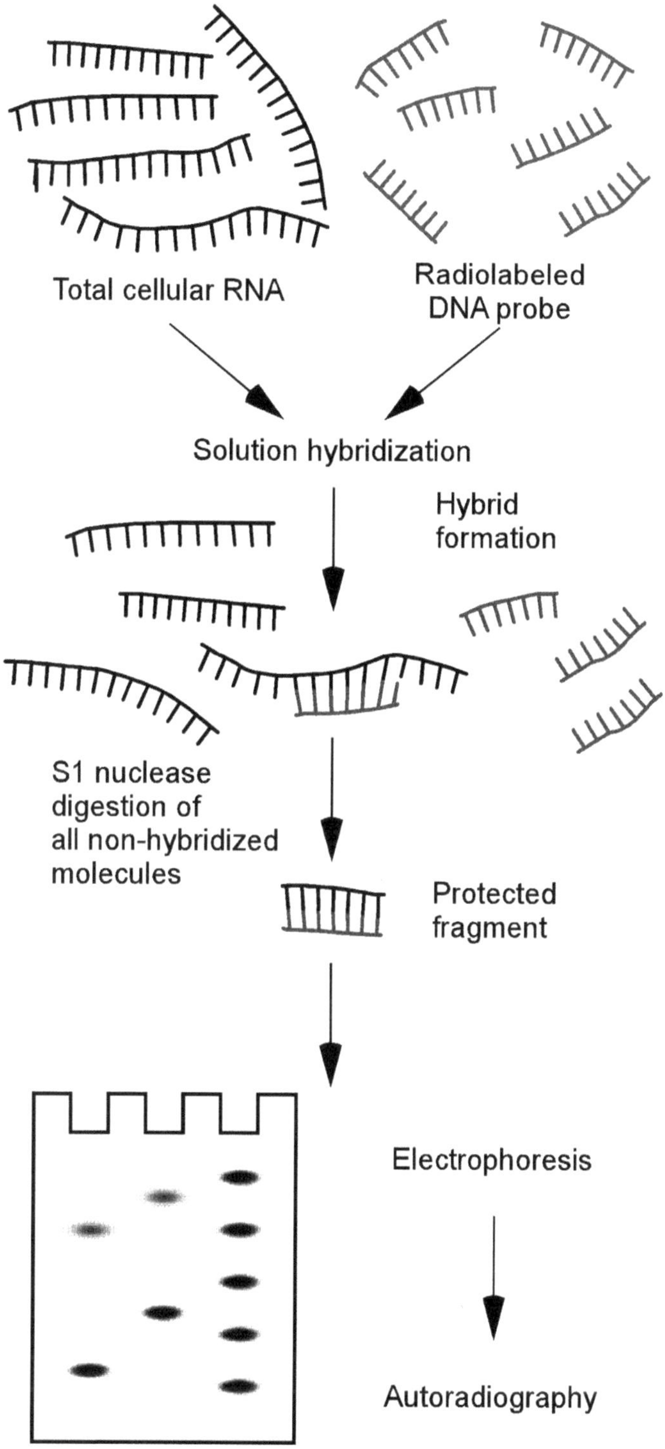

Fig. 11.1. Schematic representation of the S1 nuclease assay quantifying a specific RNA. The purified RNA is hybridized in solution with a labelled probe to form hybrid molecules. Probe molecules or RNA molecules that do not hybridize are removed by S1 nuclease digestion. The hybrid molecules are ethanol-precipitated and separated on a denaturing polyacrylamide-gel followed by autoradiography. The size and abundance of the protected molecules is a direct measure of the steady-state level for a specific RNA.

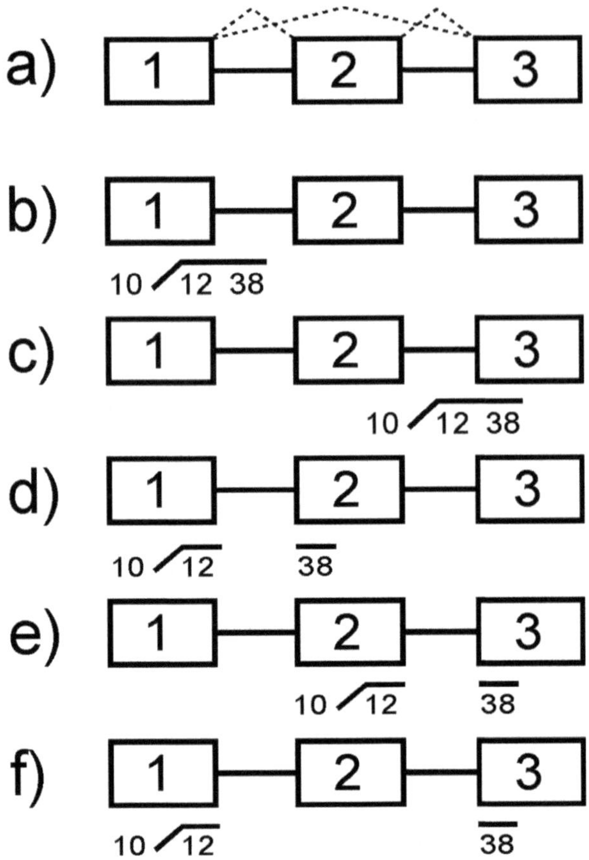

Fig. 11.2. Design strategy of oligonucleotides for the detection and quantification of alternatively spliced mRNA. (**a**) Schematic representation of a three-exon transcript in which the middle exon is alternatively spliced, i.e. included or skipped (**b–f**).

In order to detect alternative splicing events with S1 nuclease analysis, the oligonucleotides used as a probe must be designed as outlined in **Fig. 11.2**. They should either span exon–exon or intron–exon junctions so that the unspliced and (alternatively) spliced mRNA can be identified from the size of the S1 reaction product. A typical result of such an S1 nuclease assay is shown in **Fig. 11.3**, analysing splicing of HIV-1 exons 1, 1A and exon 2. In this experiment, total RNA was prepared from HIV-1 infected SupT1 cells after 0–3 days of infection and analysed with two different oligonucleotides. During the course of infection, increasing amounts of spliced and unspliced HIV-1 RNA become visible, which demonstrates that this method is both suitable for the detection and quantitation of alternatively spliced RNA species.

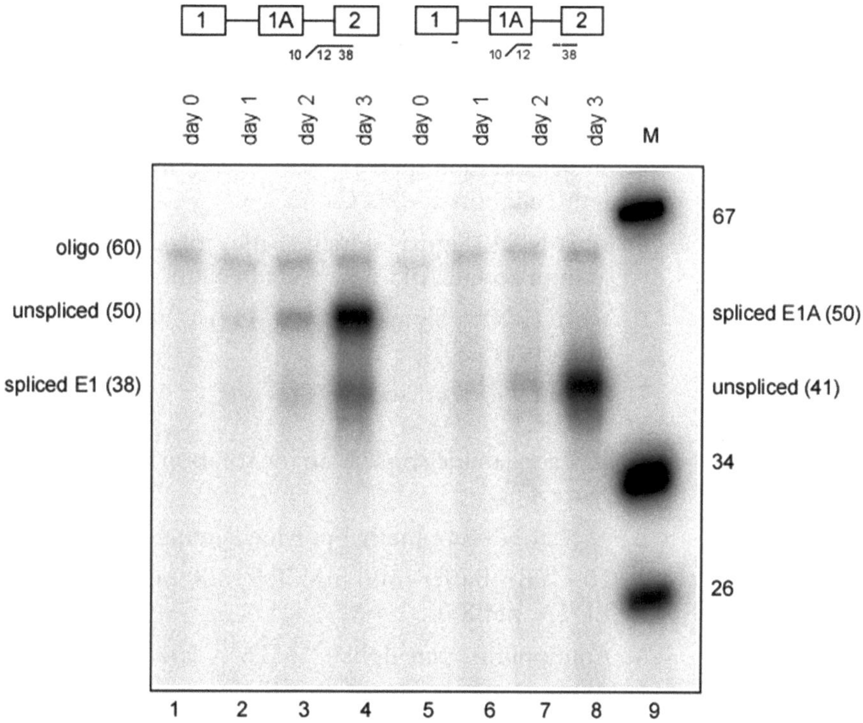

Fig. 11.3. S1 nuclease analysis of total RNA prepared from HIV-1 infected SupT1 cells 0–3 days post infection. The oligonucleotides used for this experiment are drawn on the *top* of the autoradiograph. Exons and introns are represented by *open boxes* and *horizontal lines*, respectively. The two oligonucleotides were designed to hybridize to the 38 nt at the 5′-end of HIV-1 exon 2 and 12 nt of the intron upstream or the 3′-end of HIV-1 exon 1A. Due to sequence repetition between the last nucleotides in exon and intron sequences some of the protected bands are extended with 3 nt (marked with *dotted lines*). M, size marker (^{32}P labelled pUC18 *Hpa*II fragments).

2. Materials

2.1. Labelling of the Oligonucleotide

1. T4 Polynucleotide Kinase (NEB, 10 U/μL).
2. [γ-^{32}P] ATP (10 mCi/mL, 7,000 Ci/mmol).
3. 10× T4 PNK buffer (700 mM Tris-Cl, pH 7.6, 100 mM $MgCl_2$, 50 mM DTT).
4. 2 pmol DNA oligonucleotide, 40–80 nt in length (*see* **Note 1**).
5. Sephadex® G-50 micro-spin column (Amersham-Pharmacia).

2.2. S1 Nuclease Assay

1. 80% FA-PIPES buffer (80% deionized formamide, 40 mM PIPES, pH 6.4, 400 mM NaCl, 1 mM EDTA, pH 8.0). Store in 1-mL aliquots at –80°C (*see* **Note 2**).

2. 4× S1 nuclease buffer (1.12 M NaCl, 0.2 M sodium acetate, pH 4.5, 1.8 mM $ZnSO_4$, *see* **Note 3**).
3. S1 stop buffer (4 M ammonium acetate, 20 mM EDTA, pH 8.0, 40 μg/mL tRNA).
4. Sheared salmon sperm DNA, 10 mg/mL, in water. Store in 1-mL aliquots at –20°C.
5. Yeast RNA (Sigma-Aldrich, 1 mg/mL). Prepare in 100 mM sodium acetate, pH 5.2. Store in aliquots at –20°C.
6. Yeast tRNA (Sigma-Aldrich, 11 μg/μL). Store in 1-mL aliquots at –20°C.
7. 70 and 99.5% ethanol.

2.3. Denaturing Polyacrylamide Gel Electrophoresis

1. 40% acrylamide/bisacrylamide solution (19:1), 8 M urea (*see* **Note 4**).
2. N,N,N,N'-Tetramethyl-ethylendiamine (TEMED).
3. 10× TBE buffer (500 mM Tris, 500 mM boric acid, 1 mM EDTA, pH 8.0).
4. Ammonium persulphate (APS), 10%, freshly prepared. Aliquots may be stored at –20°C.
5. 2× formamide loading buffer (50% formamide, 2× TBE buffer, 0.2% bromphenole blue).

3. Methods

3.1. Design of the Oligonucleotide Probe

We recommend using oligos not shorter than 60 nt (*see* **Note 5**) to allow hybridization at 30°C under high stringency conditions in presence of 80% formamide and 400 mM NaCl. The oligonucleotides should be designed so, that they span the splice-sites to be analysed asymmetrically, as outlined in **Fig. 11.2**. This ensures that the difference between the annealing temperature of the full and partially annealed oligo is limited to a few degrees Celsius (*see* **Note 6**). This is particularly important, if the ratio between spliced and unspliced RNA isoforms needs to be calculated.

The 3′-ends of the oligos shown in **Fig. 11.2** consist of a 12/38 nt-long part which spans an intron/exon (**Fig. 11.2b, c**) or exon/exon (**Fig. 11.2d, e, f**) junction. To the 3′-end, a 10-nt-long overhang is added, which cannot hybridize with any part of the target RNA. Since the oligonucleotide probe is added in molar excess over the RNA to be studied (*see* **Note 7**), the non-hybridized material is not always completely removed by S1 digestion. Without an overhang, the free oligonucleotide would comigrate with the protected band and increase its signal.

The oligonucleotide can be positioned at any exon/exon or exon/intron junction of the transcript to be studied.

3.2. Finding the Optimal Hybridization Temperature

The hybridization temperature is a direct function of the type of probe (RNA or DNA), probe length, G+C content, NaCl, and formamide concentration (11). The typical range for the temperature defined in this protocol is 30–65°C. For the oligonucleotides described above, 30°C are a good starting point; however, the optimal hybridization temperature must be empirically determined. Incomplete hybridization would otherwise cause the removal of the 5′-end label of the oligonucleotide and decrease the sensitivity of the S1-reaction dramatically.

Although S1 nuclease is a relatively thermostable enzyme, we do not recommend incubation above 37°C. Instead, hybridization stringency should be controlled by adjusting the formamide concentration.

3.3. Finding the Optimal Enzyme and Probe Concentration

S1 nuclease has been described as being an aggressive enzyme which is often difficult to control (12). It has its optimal activity at a pH of 4.5 and at high ionic strength in presence of 200 mM NaCl. The enzyme is resistant to denaturants such as formamide, sodium dodecyl sulphate and urea. Most vendors sell S1 nuclease in concentrations ranging from 20 to 500 U/μL. Excessive amounts of the enzyme will result in the rapid degradation of hybrid molecules, which is particularly a problem for AU-rich regions. Thus, if bands lower than the expected sizes occur on the gel, the optimal enzyme concentration should be titrated.

For precise quantitation of the RNA it is necessary to add the oligonucleotide probe in substantial molar excess. In the protocol described below, we use 50–100 fmol, which is usually sufficient to give reliable quantitation results. However, we strongly recommend to test if the probe is present in excess, by assaying a constant mass of probe against various quantities of a given RNA sample. If the probe is provided in excess over the RNA target, the signal should change proportionally.

3.4. Labelling of Oligonucleotides Using T4 PNK

1. Place in a reaction tube the following components:
2. 2.0 μL DNA oligo, 1 pmol/μL
3. 3.0 μL [γ-^{32}P] ATP (7,000 Ci/mmol)
4. 2.5 μL 10× T4 PNK buffer
5. 16.5 μL H_2O
6. 1.0 μL T4 PNK
7. Incubate at 37°C for 30–60 min. Terminate the reaction by incubating the reaction for 10 min in a heating block at 80°C.

8. Remove the free nucleotides by using a Sephadex G-50 micro-spin column. Open the bottom of the column and place it into an Eppendorf tube. Remove the buffer by centrifugation for 2 min with 720×*g*. Ensure that no air bubbles are trapped in the resin. Place the column into a fresh tube and load the whole T4 PNK reaction on top of the resin. Centrifuge for 2 min with 720×*g* and discard the column. The volume may slightly increase, but the oligonucleotide will be ready to use. Precipitation of the oligonucleotide is not necessary.

3.5. S1 Nuclease Assay

1. Wear gloves! Purified RNA samples are always susceptible to RNase degradation. Use nuclease-free DEPC-treated water for the preparation of all solutions. Sterile-filtrate solutions which cannot be autoclaved.
2. Place in a reaction tube the following components:
3. 1 μL 5′-end labelled oligonucleotide (*see* **Note 7**).
4. 15 μg total RNA (*see* **Note 8).**
5. 10 μL 3 M sodium acetate, pH 5.2.
6. to 100 μL H_2O.
7. 250 μL absolute ethanol.
8. Precipitate, wash with 70% ethanol and dry the RNA (*see* **Note 9**).
9. Redissolve the pellet in 80% FA-PIPES buffer.
10. Denature the RNA in a heating block for 10 min at 85°C (*see* **Note 10**).
11. Place the reaction immediately into a waterbath equilibrated to the desired hybridization temperature (*see* **Note 11**).
12. Hybridize over night (*see* **Note 12**).
13. On the next morning, prepare 300 μL of the following solution for each hybridization reaction:
14. 75 μL 4× S1 nuclease buffer.
15. 3 μL ssDNA, 10 mg/mL.
16. 100 U S1 nuclease (*see* **Note 13**).
17. 222 μL H_2O.
18. Incubate at 30°C for 60 min in a waterbath.
19. Terminate the reaction with 80 μL S1 stop buffer. Add 1 mL absolute ethanol for precipitation.
20. Wash the pellet with 70% ethanol and dry it for 5 min in a Speed Vac evaporator.
21. Resuspend the pellet in 5 μL 2× formamide loading buffer.

22. Heat each sample to 95°C for 3 min and then cool on ice for 2 min.
23. Electrophorese on a denaturing polyacrylamide/ bisacrylamide gel with 8 M urea in 1× TBE buffer at 18 W (*see* **Note 14**) until the dye has reached the bottom or is close to the end of the gel.
24. When electrophoresis is complete, disassemble the gel, dry down the gel on Whatman 3MM paper and perform autoradiography at –70°C with and intensifying screen, or alternatively expose on a phosphoimager screen.

4. Notes

1. 2 pmol of labelled oligonucleotide are usually sufficient for 50–100 S1 nuclease reactions.
2. The rapid oxidation of formamide requires deionization before use for optimal efficacy in hybridization. Deionization can be achieved by adding 1 g of AG 501-X8 mixed-bed resin (BioRad) to each gram of formamide to be deionized (13). Deionization requires about 1 h, although the mixture can be left overnight. The resin can be removed by filtration or by centrifugation.
3. This is a generic S1 nuclease buffer. If the enzyme was supplied in $ZnCl_2$, then the zinc sulphate should be substituted in the generic digestion buffer.
4. Acrylamide is a neurotoxin and possibly a carcinogen and teratogen. Avoid skin contact and aerosol formation when casting gels. Wear gloves and safety glasses! Work in the fume-hood.
5. Although most vendors are able to synthesize 60 nt-long DNA oligonucleotides of good quality, we recommend ordering HPLC-purified oligos for best results.
6. If DNA oligonucleotides are used to study pre-mRNA splicing, design a probe with 10 non-matching nucleotides at the 3′-end, and 12/38 nt bridging the splice-site to be studied. To avoid hybridization effects, the Tm difference between spliced variants may not be more than 1°C. The use of 80% FA PIPES usually compensates for these differences.
7. To allow quantitation, it is critical that the oligonucleotide probe is in five- to tenfold molar excess over the mRNA to be studied. We routinely use 50–100 fmol, which is usually sufficient to give reliable quantitation results with most

transcripts. With the given reaction conditions and average labelling efficiency, 50–100 fmol correspond to approx. 1×10^6 cpm labelled probe.

8. Total RNA prepared from cell cultures or tissues with Invitrogen's TRIZOL® reagent, or RNA extracted several times with phenol/chloroform is usually of sufficient quality for the applications described here. However, we recommend checking the concentration, purity and integrity of the RNA before processing multiple samples, for example, by measuring the $A_{260/280}$ ratio and/or running a small aliquot of the RNA on a denaturing agarose gel. Contaminating DNA may be removed by treating the samples with RNase-free DNase I, but great care must be taken to inactivate the enzyme afterwards, otherwise the probe will be degraded. 10 μg of total cellular RNA are usually sufficient to detect many mRNAs. In case of low abundant mRNAs it may be necessary to in crease the amount of RNA to as much as 100 μg. Then, the concentration of the RNA must be kept close to the limit of solubility (5 mg/mL or 100 μg/20 μL). The volume of the hybridization buffer may then be increased to 30 μL.
9. Do not overdry the pellet; otherwise the RNA becomes virtually impossible to dissolve.
10. It is very important to denature the RNA completely because secondary structures of the RNA will have an adverse effect on the hybridization efficiency.
11. We generally recommend doing all hybridization steps in a waterbath because of a better heat transfer to the samples and a more accurate temperature control in comparison to a heating block, which are often prone to produce a temperature gradient. Do not let the samples cool down to room temperature before placing them into the waterbath.
12. The hybridization time may be reduced to only 3 h, given that carrier RNA to a final mass of 100 μg is provided.
13. The optimal amount of S1 nuclease must be titrated. Between 100 and 1,000 units are usually sufficient for 15–100 μg RNA. However, do not add as much S1 nuclease to the reaction that the final glycerol concentration reaches 5%. This will inhibit the enzyme. Instead of raising the enzyme concentration, try first to extend the incubation time.
14. The reaction products of the S1 nuclease reaction are 50 and 38 nt in length, if oligonucleotides are used as discussed in **Section 2.1**. A 16% gel should be appropriate to characterize the protected fragments. It is highly recom-

mended to load 5′-end labelled oligonucleotides with the same length as the expected reaction products onto the gel as a size standard.

References

1. Venter, J. C., Adams, M. D., Myers, E. W., Li, P. W., Mural, R. J., Sutton, G. G., Smith, H. O., Yandell, M., Evans, C. A., Holt, R. A., Gocayne, J. D., Amanatides, P., Ballew, R. M., Huson, D. H., Wortman, J. R., Zhang, Q., Kodira, C. D., Zheng, X. H., Chen, L., Skupski, M., Subramanian, G., Thomas, P. D., Zhang, J., Gabor Miklos, G. L., Nelson, C., Broder, S., Clark, A. G., Nadeau, J., McKusick, V. A., Zinder, N., Levine, A. J., Roberts, R. J., Simon, M., Slayman, C., Hunkapiller, M., Bolanos, R., Delcher, A., Dew, I., Fasulo, D., Flanigan, M., Florea, L., Halpern, A., Hannenhalli, S., Kravitz, S., Levy, S., Mobarry, C., Reinert, K., Remington, K., Abu-Threideh, J., Beasley, E., Biddick, K., Bonazzi, V., Brandon, R., Cargill, M., Chandramouliswaran, I., Charlab, R., Chaturvedi, K., Deng, Z., Di Francesco, V., Dunn, P., Eilbeck, K., Evangelista, C., Gabrielian, A. E., Gan, W., Ge, W., Gong, F., Gu, Z., Guan, P., Heiman, T. J., Higgins, M. E., Ji, R. R., Ke, Z., Ketchum, K. A., Lai, Z., Lei, Y., Li, Z., Li, J., Liang, Y., Lin, X., Lu, F., Merkulov, G. V., Milshina, N., Moore, H. M., Naik, A. K., Narayan, V. A., Neelam, B., Nusskern, D., Rusch, D. B., Salzberg, S., Shao, W., Shue, B., Sun, J., Wang, Z., Wang, A., Wang, X., Wang, J., Wei, M., Wides, R., Xiao, C., Yan, C., et al. (2001) The sequence of the human genome. *Science* 291, 1304–1351.
2. Schmucker, D., Clemens, J. C., Shu, H., Worby, C. A., Xiao, J., Muda, M., Dixon, J. E., Zipursky, S. L. (2000) Drosophila Dscam is an axon guidance receptor exhibiting extraordinary molecular diversity. *Cell* 101, 671–684.
3. Srinivasan, K., Shiue, L., Hayes, J. D., Centers, R., Fitzwater, S., Loewen, R., Edmondson, L. R., Bryant, J., Smith, M., Rommelfanger, C., Welch, V., Clark, T. A., Sugnet, C. W., Howe, K. J., Mandel-Gutfreund, Y., Ares, M., Jr. (2005) Detection and measurement of alternative splicing using splicing-sensitive microarrays. *Methods* 37, 345–359.
4. Alwine, J. C., Kemp, D. J., Parker, B. A., Reiser, J., Renart, J., Stark, G. R., Wahl, G. M. (1979) Detection of specific RNAs or specific fragments of DNA by fractionation in gels and transfer to diazobenzyloxymethyl paper. *Methods Enzymol* 68, 220–242.
5. Alwine, J. C., Kemp, D. J., Stark, G. R. (1977) Method for detection of specific RNAs in agarose gels by transfer to diazobenzyloxymethyl-paper and hybridization with DNA probes. *Proc Natl Acad Sci USA* 74, 5350–5354.
6. Mullis, K., Faloona, F., Scharf, S., Saiki, R., Horn, G., Erlich, H. (1986) Specific enzymatic amplification of DNA in vitro: the polymerase chain reaction. *Cold Spring Harb Symp Quant Biol* 51(Pt 1), 263–273.
7. Mullis, K. B., Faloona, F. A. (1987) Specific synthesis of DNA in vitro via a polymerase-catalyzed chain reaction. *Methods Enzymol* 155, 335–350.
8. Saiki, R. K., Scharf, S., Faloona, F., Mullis, K. B., Horn, G. T., Erlich, H. A., Arnheim, N. (1985) Enzymatic amplification of beta-globin genomic sequences and restriction site analysis for diagnosis of sickle cell anemia. *Science* 230, 1350–1354.
9. Gilman, M. (ed.) (2001) Ribonuclease protection assay, in (Ausubel, F. M., Brent, R., Kingston, R. E., Moore, D. D., Seidman, J. A., eds.), *Current Protocols in Molecular Biology*. Wiley, New York, NY, 4.7.1–4.7.8.
10. Quarless, S. A., Heinrich, G. (1986) The use of complementary RNA and S1 nuclease for the detection and quantitation of low abundance mRNA transcripts. *BioTechniques* 4, 434–438.
11. Casey, J., Davidson, N. (1977) Rates of formation and thermal stabilities of RNA:DNA and DNA:DNA duplexes at high concentrations of formamide. *Nucleic Acids Res* 4, 1539–1552.
12. Ando, T. (1966) A nuclease specific for heat-denatured DNA in isolated from a product of *Aspergillus oryzae*. *Biochim Biophys Acta* 114, 158–168.
13. Samanta, H. K., Engel, D. (1987) Deionization of formamide with Biorad AG501-X(D). *J Biochem Biophys Methods* 14, 261–266.

Chapter 12

Promises and Challenges in Developing RNAi as a Research Tool and Therapy

Mouldy Sioud

Abstract

Small interfering RNA (siRNAs), the main effector of RNA interference (RNAi), are now routinely used to assess gene function, both in vitro and in vivo, and many innovative screens have been reported on the use of RNAi to identify potential drug targets. Despite several technical advances, however, there are still many challenges in determining the ideal design of siRNA sequence, the activation of the immune system, off-target effects, and competition with endogenous microRNAs for cellular miRNA-processing machinery. Therefore, the translation of RNAi technology into the clinic depends on resolving these challenges. This chapter summarizes recent progress in siRNA design, sensing by the immune system, and discusses some of the promising approaches that are currently being explored in separating siRNA unwanted effects from gene silencing.

Key words: RNA interference, siRNA, innate immunity, Toll-like receptors, chemical modifications.

1. Introduction

In 1998, Fire and colleagues discovered that exogenous delivery of double-stranded RNA (dsRNA) into *Caenorhabditis elegans* (C) induced sequence-specific RNA degradation mechanism targeted against any cellular mRNA that shared sequence homology with the introduced dsRNA molecules (1). There was little or no effect with either sense or antisense single-stranded (ss) RNA. This process, referred to as RNA interference (RNAi), was later found to occur in many other eukaryotes including fruit fly, mouse, and human. Obviously, a similar mechanism termed

H. Nielsen (ed.), *RNA*, Methods in Molecular Biology 703,
DOI 10.1007/978-1-59745-248-9_12, © Springer Science+Business Media, LLC 2011

post-transcriptional gene silencing (PTGS) had been described years ago in plants, and now it is believed to function as a surveillance system for blocking the function of harmful RNAs such as viral RNAs (2). Remarkably, RNAi is systemic in both plants and nematodes, spreading from cell to cell. In *C. elegans*, RNAi is also heritable: silencing can be transferred to the progeny of the worm originally injected with the trigger dsRNA. Viral infection, inverted repeat transgenes, or aberrant transcription products all lead to the production of dsRNA that can be converted to siRNAs.

It should be noted that the work of Hamilton and Baulcombe has provided the first decisive findings showing that a distinct population of 21- to 24-nucleotide (nt)-long RNAs with antisense sequence of silenced genes invariably accumulated in co-suppressed plants (3). Their work also suggests that 21–24 nt RNA duplexes could provide sequence specificity to the machinery that degrades homologous mRNAs in RNAi/PTGS (3). Unlike invertebrates, vertebrates react to long dsRNA by activating the interferon pathway (4). However, it has been shown that chemically made small RNAs, known as siRNAs, with features of Dicer cleavage products were sufficient to mediate RNAi in human cells without activating the IFN response (5). Subsequent to this landmark finding by Tuschl and colleagues in 2001, siRNA technology is now used as a standard tool in most laboratories for gene function analysis and drug-target discovery and validation. Furthermore, siRNAs have emerged as a powerful tool for therapeutic gene silencing (6, 7). In principle, the mRNA encoding any protein that is associated with a disease can be cleaved selectively by siRNAs. However, recent data make it clear that siRNA faces some major hurdles before it can used as a drug.

2. RNAi Pathway

In general the RNAi pathway is initiated by the enzyme Dicer, which cleaves long dsRNA into double-stranded siRNA. Dicer is a dsRNA-specific RNAase III family ribonuclease, which generates siRNA duplexes containing of 20–25 nt in length. Dicer leaves 2-nt 3′-overhangs and 5′-phosphate groups in each strand. These siRNA duplexes are then incorporated into a multiprotein complex, the RNA-induced silencing complex (RISC). Notably, synthetic siRNAs enter directly into the RNAi pathway (*see* Fig. **12.1**). Subsequent to strand separation, the antisense (guide) strand guides the RISC to recognize and cleave target mRNA sequences. The catalytic activity of RISC is mediated by Argonaute 2 (Ago2) protein, the only Ago family member that is

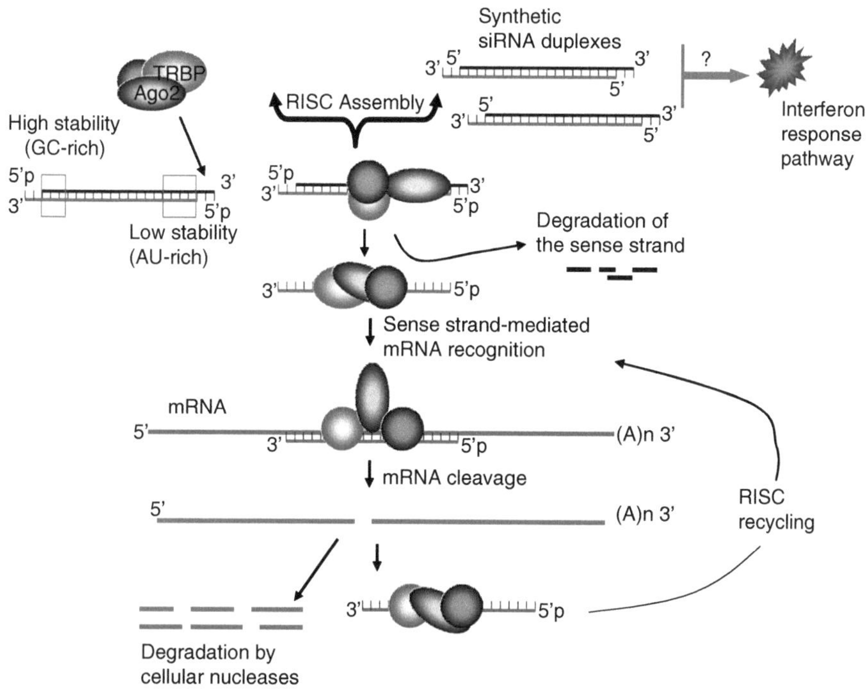

Fig. 12.1. Schematic representation of gene silencing by siRNAs. In contrast to long double-stranded RNAs, siRNAs are directly loaded into a multi-protein complex termed RNA-induced silencing complex (RISC), where the sense (passenger) strand with high 5′-stability) is cleaved by the nuclease AGO2. This will lead to strand separation. Subsequently, the RISC containing the antisense (guide) strand seeks out and binds to complementary mRNA sequences. Bound mRNA molecules are then cleaved by AGO2 and cleaved mRNA fragments are rapidly degraded by cellular nucleases. Following dissociation, the active RISC is able to recycle and cleave additional mRNA molecules.

cleavage competent (8, 9). The protein members of the argonaute family are highly basic proteins that contain two common domains, PAZ and PIWI domains. While the PA2 domain is responsible for RNA binding, the PIWI domain mediates the interaction with Dicer and contains the nuclease activity that cleaves of target mRNAs. Ago2 is also responsible for cleavage of the passenger siRNA strand, thus facilitating the formation of functional RISC complexes (10, 11). Analysis of the crystal structures of a siRNA guide strand associated with Ago2 PIWI domain revealed that nucleotides 2–8 form a seed sequence that directs target mRNA recognition by RISC (12). While mammals and *C. elegans* each have a single Dicer that makes both miRNAs and siRNAs, *Drosophila* has two Dicers (13): Dicer-1 makes miRNAs, whereas Dicer-2 is specialized for siRNA production. Although the discovery of RNAi has provided a new tool to study gene function and drug target validation, recent studies have shown that siRNA duplexes can activate innate immunity and silence a variety of genes in addition to the intended target gene (14–17). Therefore, the development of strategies that block siRNA unwanted effects is crucial to their therapeutic use. Also, there remain other important obstacles for effective therapeutic use of siRNA, including stability and delivery.

3. Design Rules

Gene silencing by siRNA varies markedly in mammalian cells, where the gene-silencing effectiveness depends very much on the target sequence positions (sites) selected from the target gene (18). Since individual siRNAs Vary extremely in their effeciency to induce down-regulation of the target, several siRNAs have to be designed for each target gene and evaluated for their efficacy and lack of side effects (see below). The earliest guidelines for siRNA design were proposed by Elbashir and colleagues (5). They suggested that synthesizing siRNA duplexes of 19 base-paired nucleotides with 2-nt 3′-overhangs at either end (referred to as 21-nt siRNA duplexes) mediates efficient cleavage of the target mRNA. Preferably, the 19-nt target sequence should be flanked in the mRNA with two adenosines at the 5′-end. Regions at the mRNA to select the target site are preferably in the coding region, 100 bp from the AUG start codon. Despite these early rules, the siRNA efficacy is highly dependent on target sequence and siRNA efficacy cannot be predicted from the primary sequence. Therefore, a successful siRNA research project may require the design of various distinct siRNAs at a high cost. More recently, some studies showed that the 5′-end of the antisense strand might be incorporated into the RNA-induced silencing complex and strand incorporation may depend on weaker base pairing, indicating an A-U terminus may lead to more strand incorporation than a G-C terminus (19, 20). Therefore, it was concluded that the relative thermodynamic stability of the 5′-ends of the two siRNA strands in the duplex determines the identity of the selected strand, either the antisense (guide) or the sense (passenger) strand. Other factors reported to be involved in gene-silencing efficacy are GC content, point-specific nucleotides, specific motif sequences, and secondary structures of mRNAs (21). Based on the published data, several siRNA design rules/guidelines using efficacy-related factors have been reported. Although there are common preferred and unpreferred nucleotides at both position 1 and 19 in some guidelines, there is the problem of inconsistencies for nucleotide frequency of each position suggesting that some rules from guidelines are sequence-dependent.

Despite these design limitations, the experimental data indicated four apparent features of the siRNA sequence to possibly serve to discriminate active siRNAs from those that are non-active (21). First, the 5′-end of the antisense strand (AS) of active siR-

NAs may always be A or U, with the counterpart of non-active siRNAs may always be C or G. Second, the 5′-end of the sense strand (SS) of active siRNAs are preferably G or C, with the counterpart of non-active siRNAs being A or U. Third, in the case of active siRNAs, the 5′-terminal AS are A/U-rich whereas the corresponding region of non-active siRNAs are G/C rich. Fourth, position 10 of the target site should be U.

In addition to the siRNA sequence, mRNA secondary structure is also considered to be an important factor in predicting siRNA efficacy. However, there are conflicting results concerning the effects of secondary structures on siRNA functionality. On one hand, some studies suggested that the secondary structure of the mRNA plays a role in determining the efficacy of gene silencing. On the other hand, other studies did not find any correlation between functionality of the siRNA and second structure of the target mRNA. Therefore, this issue still requires further investigation.

4. Detection of Exogenous RNAs by Innate Immune Receptors

Notably, the immune system has evolved cellular and molecular strategies to discriminate between foreign and self nucleic acids. Among the cytoplasmic sensors of long dsRNA is the dsRNA-dependent protein kinase (PKR) that phosphorylates serine and threonine residues of target proteins (22). Most human cells constitutively express a low level of PKR that remain inactive. However, upon binding to dsRNA, PKR forms a homodimer resulting in its autophosphorylation and activation. Activated PKR phosphorylates various substrates including the protein synthesis initiation factor elF-2α and blocks the translation of viral and cellular proteins, an essential step in antiviral resistance. It should be noted that PKR's binding to dsRNAs is sequence-independent and the presence of interferon upregulates its expression.

Although PKR is implicated in antiviral immunity, it is mainly IFN effector and not absolutely required for IFN production. More recently two additional intracellular helicases, retinoic-acid-inducible gene I (RIG-I) and/or melanoma differentiation-associated gene 5 (MDA-5) were identified (23). RIG-I encodes a caspase recruitment domain (CARD) at the N terminus, in addition to an RNA helicase domain. The RNA helicase domain requires ATPase activity and is responsible for viral dsRNA recognition and induction of conformational changes leading to the interaction of the RIG-I CARD domain with another CARD-containing adaptor protein, known as IPS-1, MAVS, Cardif, or VISA (24). IPS-1 is an outer mitochondrial membrane binding protein, which activates IRF3 and IFR-7 through TBK1/IKKi, resulting in the production IFN-β production. Mitochondrial retention of IPS-1 is essential for IRF3, IRF7, and NF-κB activation by RIG-1 (24).

Although RIG-I seems to be an essential sensor of viral RNAs, microbial nucleic acids are also recognized by toll-like receptors (TLRs), especially in immune cells (25, 26). Whereas most TLRs are expressed in the plasma membrane for detecting bacterial components, TLR3, TLR7, TLR8, and TLR9, are expressed in intracellular compartments (endosomes, lysosomes) (25). The immune function of this cellular localization is to sense viral RNAs. TLR3 is also expressed on the cell surface and it is believed to recognize extracellular viral dsRNAs (27). TLR7 and TLR8 recognize viral ssRNA and small synthetic antiviral compounds referred to as imidazoquinolines. TLR9 recognizes unmethylated CpG-DNA motifs that exist in both viral and bacterial DNA, but are suppressed or methylated in the vertebrate genomes (28, 29). It should be noted that intracellular NOD-like receptors detect bacteria, whereas viruses are mainly detected by TLRs and RIG-like receptors. The virus-detecting TLRs operate mainly in plasmacytoid dendritic cells by responding to viral nucleic acids that enter the cell via endocytosis. In these cells, the major immune response is the production of type 1 interferon (30).

Although siRNAs were initially thought to bypass the IFN response because they are too short to be recognized by dsRNA sensors (5), we and others have shown that they could activate innate immunity in mammalian cells (14–17). Early studies indicated sequence-independent activation of PKR and TLR3 signaling pathways by siRNAs (17, 31). However, recently it was demonstrated that PKR and TLR3 do not represent the major pathways by which chemically synthesized siRNAs activate immunity in immune cells (32–35). Indeed, a group of siRNA sequences stimulated monocytes or dendritic cells to produce proinflammatory cytokines and type I interferons. This response is mainly mediated through TLR7 in mice and TLR7/8 in humans. Under our experimental conditions, ss siRNAs were more effective than ds siRNAs in activating TLR7/8 in human monocytes and peripheral blood mononuclear cells (PBMCs) (32, 35). The extent of siRNA unwanted effects has recently been confirmed by expression profiling using microarrays, which identified over 400 siRNA-affected genes in PBMCs. Genes encoding for proinflammatory cytokines, interferons, and Mx proteins are among the genes that are significantly induced (36). Mx proteins are IFN-induced GTPases that form complexes with dynamin disrupting trafficking or activity of viral polymerases, thereby interfering with viral replication.

5. What Is the Nature of IFN-Inducing Motif Present in One Sequence But Absent in Another?

TLR7 and TLR8 recognize certain siRNA sequences, provided they are delivered to the endosomes via cationic liposomes. Initial experiments indicate that some types of secondary structures and/or specific nucleotides are responsible for the activation of NF-κB signaling pathway by siRNAs in human monocytes (14). Judge and colleagues found that the 5′-UGUGU motif was indispensable for the immune activation by a siRNA in human blood cells (34). However, Hornung and colleagues identified a 9-nt motif RNA motif (5′-GUCCUUCAA) that is recognized by TLR7 in the context of siRNA duplexes and the activity does not depend on GU content (33). Collectively, our data indicated that interferon induction by siRNAs can not be easily suppressed by selecting siRNA sequences without the GU dinucleotides. Indeed, several siRNA sequences without GU induced TNF-α production in human PBMC and monocytes (32). Although the precise nature of the RNA motifs responsible of innate immune activation is not known, we showed that the ability of ssRNA or dsRNAs to activate TNF-α production is largely depend on the uridine content (35). Indeed, their replacement with adenosines abrogated immune activation by either ss or ds siRNAs. A recent study by Goodchild and colleagues (37), where they analyzed the effects of 250 siRNA sequences, confirms the importance of uridines for siRNA stimulation. Also, their data supported our proposed model in which ds siRNAs are dissociated in the endosomes, leading to the activation of TLR7/8 by ss siRNAs (32).

6. The Molecular Basis of RNA Sensing by RIG-I

As indicated above, recent studies on the immune response to chemically made siRNAs have highlighted the involvement of the endosomes. Indeed, cytoplasmic delivery of synthetic siRNA by electroporation into human blood cells did not induce either inflammatory cytokines or interferons, whereas the same sequences when delivered by lipid did (32). Thus, synthetic siRNAs are not detected by cytoplasmic sensors for viral RNAs in immune cells. However, the data do not explain why ds siRNAs were not sensed by RIG-1, a major viral RNA sensor. It has been demonstrated that chemically made siRNA duplexes harboring 2-nt 3′-overhangs cannot engage innate immunity activation, whereas the same siRNA sequence with blunt ends did (38). The authors showed that RIG-1 can bind to siRNAs with or without

2-nt 3′-overhangs, but only siRNAs with blunt ends could activate RIG-1. These findings imply that endogenous shRNAs or microRNA harboring the Dicer signature, 2-nt 3′-overhang, are not an ideal stimulator of RIG-I. Thus, the structures of the 5′-ends between shRNAs (substrate for Dicer) and non-self dsRNAs such as viral RNAs are critical for self and non-self discrimination.

It should be noted that the predominant form of naturally occurring dsRNAs in mammalian cells is derived from endogenously expressed miRNAs that constitute a large class of noncoding small RNAs involved in gene regulation in a variety of organisms ranging from plants to mammalians (39). Presently, more than 1,000 potential human miRNAs have been identified and numerous have been experimentally validated. Usually miRNAs are transcribed from endogenous genes by RNA polymerase II as long RNA precursor called a primary miRNA (pri-miRNA), containing one or more distinct miRNAs. In the nucleus the RNA precursors are processed by Drosha to 60–80 nt RNA hairpin intermediate, bearing 2-nt 3′-overhang, called a pre-miRNA. Interestingly, the Drosha cleavage site was shown to be 11 base pairs from the stem single-stranded RNA junction (40). Processed pre-miRNAs are then transported from the nucleus to the cytoplasm by exportin-5, where its 2-nt 3′-overhang is recognized by Dicer, which generates the mature miRNA that can evade the detection by innate immune receptors such as RIG-I (*see* Fig. **12.2**).

During our studies, we have also found that synthetic ss siRNAs (21 nt) do not activate innate immunity when delivered to the cytoplasm via electroporation (32). These used RNAs do not contain 2-nt 3′-overhang because they are single-stranded. To further examine the contribution of RIG-I in sensing exogenous RNAs, we have transfected monocytes with either T7 RNA polymerase (RNAP) transcribed siRNAs or chemically made siRNAs. The inhibition of endosome maturation by chloroquine abrogated the immuostimulatory activity of chemically made siRNAs, but not the T7 RNAP-made siRNAs (41). In addition, the immunostimulatory effect of the T7 RNAP-made siRNAs was not inhibited with 2-aminopurine, a specific inhibitor of PKR. Therefore, which cytoplasmic factors are able to sense in vitro transcribed RNA? Additional studies from other investigators showed that RIG-I senses ssRNA-bearing 5′-triphosphate, a specific signature of viral and in vitro transcribed RNAs (42). Interestingly, artificial capping or base modifications of the 5′-triphosphate bearing RNA abolished immune response. In general, self-RNA undergo several modifications to eliminate or mask the 5′-triphosphate group. However, the reported data do not explain why certain endogenous RNA with 5′-triphosphates (e.g., 7SL RNA, an abundant cytoplasmic RNA) escape RIG-I recognition. Natural 2′ribose-modifications might protect

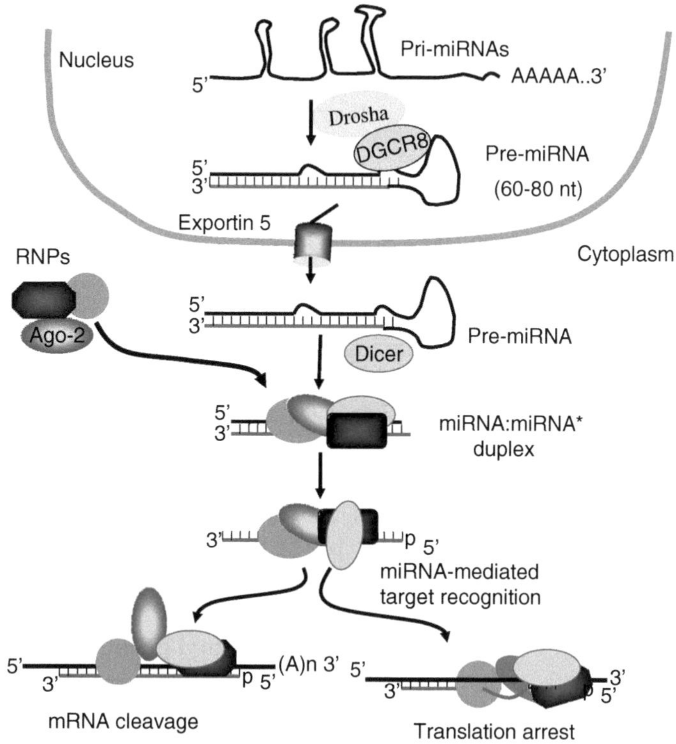

Fig. 12.2. Gene regulation by miRNAs. MiRNAs are derived from genome transcribed primary transcripts (pri-miRNAs) that are predicted to form multiple stems and hairpin structures. Pri-miRNAs are processed by the ribonuclease Drosha to 70–80 nucleotide pre-miRNAs that are transported to the cytoplasm by exportin 5. Subsequently, they are processed by Dicer into mature 22–24 nucleotide miRNAs, which are then incorporated into a RNP complex that can direct either RNA cleavage (perfect complementarity with mRNA) or translation arrest (mismatches with target mRNA).

endogenous RNA-bearing 5′-triphosphate from being detected by RIG-I and other RNA sensors. MDA5, the most closely related protein of RIG-I, is also an IFN-inducible protein. However, the exact mechanisms of RNA sensing by MDA5 has yet to be defined but seems to sense dsRNA structures from certain viruses.

7. Separation of siRNA Unwanted Effects from Gene Silencing

Much of the recent interest in the mechanisms involved in RNA sensing or tolerance by the immune system was generated by the observation that siRNA can activate innate immunity (14, 17). Considering the high frequency of uridines in messenger RNAs it is more likely that a high proportion of self and non-self chemically made siRNA sequences will activate innate immunity. Therefore, it would be desirable to develop strategies that evade immune activation. At least three distinct ways to avoid immune activation by siRNAs can be used. The first would be to use delivery agents that avoid the delivery and/or retention of siRNA within the endosomes. The second way relies on the use of modified nucleotides. In this respect, Morrissey and colleagues showed that the incorporation of various 2′-modified nucleotides in siRNA sequence abrogated their immunostimulatory potency (43). However, the chemical modifications that block immune activation must be chosen carefully so as not to inhibit siRNA silencing potency. Thus, finding the appropriate chemical modifications for interfering with siRNA immune activation will be important for exploring their therapeutic applications.

Fortunately, we have shown that replacement of only uridines with their 2′-fluoro, 2′-deoxy, or 2′-*O*-methyl modified counterparts can abrogate immune recognition of ss siRNA or ds siRNAs by TLRs without reducing their silencing potency (35, *see* Fig. **12.3**). In accordance with our data, Judge and colleagues demonstrated that the incorporation of 2′-*O*-methyl-uridine or 2′-*O*-mehtyl guanosine residues into siRNAs abrogates their immunostimulatory potency (44). Collectively, the published data offer the possibility of choosing the appropriate modifications that evade immune activation without reducing siRNA-silencing. I recommend the use of 2′-deoxy uridine or thymidine

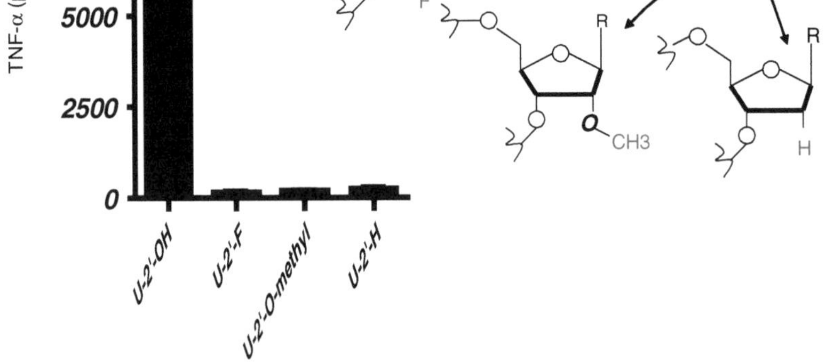

Fig. 12.3. A representative example of TNF-α production in human monocytes in response to unmodified or 2′-uridine modified RNAs. Cells were transfected with either unmodified or 2′-uridine modified single-stranded siRNA. Subsequent to 18 h transfection time, secreted TNF-α in culture supernatants was measured by ELISA.

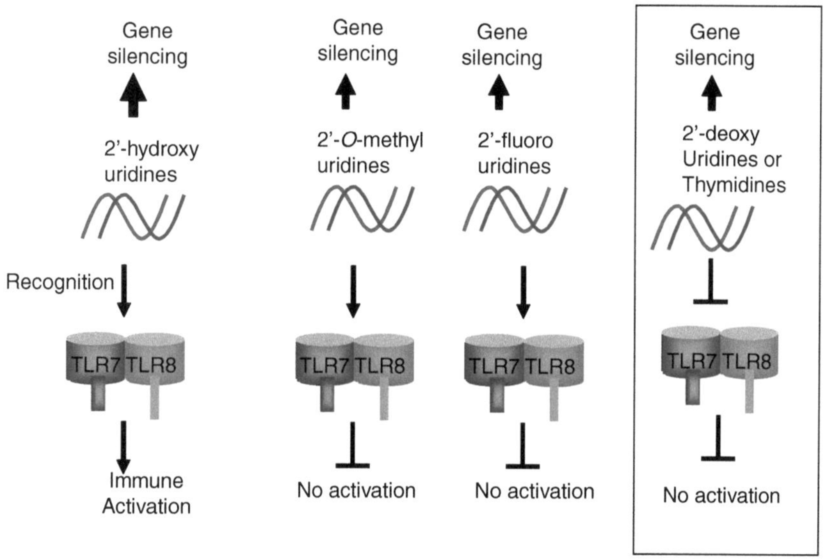

Fig. 12.4. An overview of the effects of chemical modifications on gene silencing and activation of endosomal TLR7/8. For details see the text.

modified siRNAs because they showed no significant immunostimulatory effects and binding to TLR7/8 (*see* Fig. **12.4**).

The finding that 2′-modified RNAs can evade immune activation suggests that naturally modified RNAs are not sensed by TLR7/8. Support of this view has been provided by Karikó and colleagues, who demonstrated that modifications that are frequently found in mammalian RNA (such as pseudouridine, 2′-*O*-methyl) can interfere with the capacity of RNA to activate TLR-7 in dendritic cells (45). Thus, unmodified RNA corresponding to mammalian sequences would be expected to activate TLR7 more effectively than native RNAs provided they are delivered to the endosomes (32).

The finding that unmodified, but not 2′-modified RNA, are potent triggers of innate immunity also raised questions about the differences in their structures that might be relevant to binding to TLR7/8. Which step is affected by 2′-modifications, and why can't 2′-modified RNAs trigger immune activation?. One way to address the first question is to assess whether 2′-modified RNA antagonize with immunostimulatory RNAs to trigger TLR7/8 signaling. Studies of transfected human monocytes show that 2′-*O*-methyl modified RNAs abrogates the activation of TLR7 by immunostimulatory RNAs (46). Of considerable interest, is that 2′-*O*-methyl modified RNAs suppressed immune activation at very low concentrations (47). In addition, we have shown that they can effectively inhibit immune activation by a variety of immunostimulatory sequences including bacterial and mitochondrial RNAs. Also, chemically modified RNA can antagonize with immunostimulatory RNA to activate indoleamine 2,3-dioxygenase, an immunosuppressive protein, in human monocytes (48). In accordance with our data, Robbins and colleagues have reported that 2′-modified immunostimulatory RNAs can function as TLR7/8 antagonist by inhibiting TLR7 signaling by immunostimulatory RNAs and loxoribine in both murine and human cells (49). Suppressive 2′-modified RNAs should represent a new class of agents that may be useful in the treatment of autoimmunity triggered by TLR7/8 signaling. Some studies have described a number of different classes of non-stimulatory DNA sequences including mutated CpG sequences and repeats of the TTAGGG motifs present in mammalian telomeres (50). Interestingly, the linkage of a G-rich DNA oligonucleotide (10-mer) to a 2′-modified immunomodulatory RNA (10-mer) led to the design of a sequence capable of blocking TLR7, TLR8, and TLR9 signaling (Sioud M, unpublished data). The development of single tri-functional oligonucleotide may be required for the treatment of autoimmune disease where TLRs are involved.

8. Sequence-Dependent Off-Target Effects

When first reported, siRNAs seemed highly specific, as a single mutation in the target site could be demonstrated to completely abolish silencing (5). However, it was soon demonstrated that mutations in the target site of several siRNAs did not always inhibit knockdown (18). Furthermore, siRNAs with only partial complementarily to mRNAs can also cause a reduction in the RNA levels of a large number of transcripts (15). Another potential source of siRNA toxicity is therefore the destruction of cellular mRNAs that share partial homology to the siRNA sequences. Because the cellular pathways activated by miRNAs and siRNAs are overlapping, it is more likely that each siRNA sequence will exhibit a miRNA activity. The most commonly used strategy to ensure siRNA target specificity is the basic local alignment search tool BLAST. However, short sequence stretches may not be detected by BLAST program. In addition, the identification of such sequences does not necessarily indicate the occurrence of off-target effects. Similarly, the absence of short homologies will not rule out off-target effects. The best way to deal with this problem is to analyze global gene expression, specifically when siRNAs are going to be used in functional genomics or to develop therapeutics. During our studies with siRNAs, we have found that 2′-uridine modifications of siRNAs not only evade the activation of TLR7/8 but they can block most of TLR-independent effects including off-target effects (36). Although the evading mechanisms remain to be investigated, it is probable that the interaction of siRNA sequences with unintended cellular mRNAs is affected by chemical modifications. Modifications of RNA might be particularly disruptive for siRNA binding to mismatched sequences. In accordance with these observations, Jackson and colleagues found that the incorporation of 2′-*O*-methyl group at the second position of the siRNA guide strand can reduce most off-target gene silencing effects without affecting siRNA silencing of the intended target gene (51). Collectively, these studies offer a simple strategy for reducing off-target effects. Therefore, I suggest the incorporation of few 2′-modified bases in siRNA sequences, particulary thymidines.

References

1. Fire, A., Xu, S., Montgomery, M. K., Kostas, S. A., Driver, S.E, Mello, C. C. (1998) Potent and specific genetic interference by double-stranded RNA in *Caenorhabditis elegans. Nature* 391, 806–811.
2. Jorgensen, R. (1990) Altered gene expression in plants due to trans interactions between homologous genes. *Trends Biotechnol* 8, 340–344.
3. Hamilton, A.J, Baucombe, D. C. (1999) A species of small antisense RNA in posttranscriptional gene silencing in plants. *Science* 286, 950–952.
4. Sen, G. C. (2001) Viruses and interferons. *Annu Rev Microbiol* 55, 255–281.
5. Elbashir, S. M., Harborth, J., Lendeckel, W., Yalcin, A., Weber, K., Tuschl, T. (2001) Duplexes of 21-nucleotide RNAs mediate RNA interference in cultured mammalian cells. *Nature* 411, 494–498.
6. Sioud, M. (2004) Therapeutic siRNAs. *Trends Pharmacol Sci* 25, 22–28.
7. Hannon, G. J., Rossi, J. J. (2004) Unlocking the potential of the human genome with RNA interference. *Nature* 431, 371–378.
8. Liu, J., Carmell, M. A., Rivas, F. V., Marsden, C. G., Thomson, J. M., Song, J. J., Hammond, S. M., Joshua-Tor, L., Hannon, G. J. (2004) Argonaute2 is the catalytic engine of mammalian RNAi. *Science* 305, 1437–1441.
9. Song, J. J., Smith, S. K., Hannon, G. J., Joshua-Tor, L. (2004) Crystal structure of Argonaute and its implications for RISC slicer activity. *Science* 305, 1434–1437.
10. Matranga, C., Tomari, Y., Shin, C., Bartel, D. P., Zamore, P. D. (2005) Passenger-strand cleavage facilitates assembly of siRNA into Ago2-containing RNAi enzyme complexes. *Cell* 123, 607–620.
11. Rand, T. A., Petersen, S., Du, F., Wang, X. (2005) Argonaute2 cleaves the anti-guide strand of siRNA during RISC activation. *Cell* 123, 621–629.
12. Ma, J. B., Yuan, Y. R., Meister, G., Pei, Y., Tuschl, T., Partel, D. J. (2005) Structural basis for 5′-end-specific recognition of guide RNA by the A. fulgidus Piwi protein. *Nature* 434, 666–670.
13. Lee, Y. S., et al. (2004) Distinct roles for Drosophila Dicer-1 and Dicer-2 in the siRNA/miRNA silencing pathway. *Cell* 117, 69–81.
14. Sioud, M., Sørensen, D. R. (2003) Cationic liposome-mediated delivery of siRNAs in adult mice. *Biochem Biophys Res Commun* 312, 1220–1225.
15. Jackson, A. L., Bartz, S. R., Schelter, J., Kobayashi, S. V., Burchard, J., Mao, M., Li, B., Cavet, G., Linsley, P. S. (2003) Expression profiling reveals off-target gene regulation by RNAi. *Nat Biotechnol* 21, 635–637.
16. Semizarov, D., Frost, L., Sarthy, A., Kroeger, P., Halbert, D. N., Fesik, S. W. (2003) Specificity of short interfering RNA determined through gene expression signatures. *Proc Natl Acad Sci USA* 100, 6347–6352.
17. Sledz, C. A., Holko, M., de Veer, M. J., Silverman, R. H., Williams, B. R. (2003) Activation of the interferon system by short-interfering RNAs. *Nat Cell Biol* 5, 834–839.
18. Holen, T., Amrzguioui, M., Wiiger, M. T., Babaie, E., Prydz, H. (2002) Positional effects of short interfering RNAs targeting the human coagulation tissue factor. *Nucleic Acids Res* 30, 1757–1766.
19. Schwarz, D. S, et al. (2003) Asymmetry in the assembly of the RNAi enzyme complex. *Cell* 115, 199–208.

20. Khvorova, A., Reynolds, A., Jayasena, S. D. (2003) Functional siRNAs and miRNAs exhibit strand bias. *Cell* 115, 209–216.
21. Ui-Tei, K., Naito, Y., Saigo, K. (2007) Guidelines for the selection of effectives short-interferin RNA sequences for functional genomics. *Methods Mol Biol* 361, 201–216.
22. Li, S., Peters, G. A., Ding, K., Zhang, X., Qin. J., Sen, G. C. (2006) Molecular basis for PKR activation by PACT or dsRNA. *Proc Natl Acad Sci USA* 103, 1005–1010.
23. Kato, H., Takeuchi, O., Sato, S., Yoneyama, M., Yamamoto, M., Matsui, K., Uematsu, S., Jung, A., Kawai, T., Ishii, K. J., Yamaguchi, O., Otsu, K., Tsujimura, T., Koh, C. S., Reis e Sousa, C., Matsuura, Y., Fujita, T., Akira, S. (2006) Differential roles of MDA5 and RIG-I helicases in the recognition of RNA viruses. *Nature* 441, 101–105.
24. Meylan, E., Tschopp, T., Karin, M. (2006) Intracellular pattern recognition receptors in the host response. *Nature* 442, 39–44.
25. Takeda, K., Akira, S. (2005) Toll-like receptors in innate immunity. *Int Immunol* 17, 1–14.
26. Sioud, M. (2005) Innate sensing of self and non-self RNAs by Toll-like receptors. *Trends Mol Med* 12, 167–176.
27. Alexopoulou, L., Holt, A. C., Medzhitov, R., Flavell, R. A. (2001) Recognition of double-stranded RNA and activation of NF-kappaB by Toll-like receptor 3. *Nature* 413, 732–738.
28. Heil, F., Hemmi, H., Hochrein, H., Ampenberger, A., Kirschning, C., Akira, S., Lipford, G., Wagner, H., Bauer, S. (2004) Species-specific recognition of single-stranded RNA via toll-like receptor 7 and 8. *Science* 303, 1526–1529.
29. Krieg, A. M. (2002) CpG motifs in bacterial DNA and their immune effects. *Annu Rev Immunol* 20, 709–760.
30. Cao, W., Liu, Y. J. (2007) Innate immune functions of plasmacytoid dendritic cells. *Curr Opin Immunol* 19, 24–30.
31. Kariko, K., Bhuyan, P., Capodici, J., Weissman, D. (2004) Small interfering RNAs mediate sequence-independent gene suppression and induce immune activation by signaling through toll-like receptor 3. *J. Immunol* 172, 6545–6549.
32. Sioud, M. (2005) Induction of inflammatory cytokines and interferon responses by double-stranded and single-stranded siRNAs is sequence-dependent and requires endosomal localization. *J Mol Biol* 348, 1079–1090.
33. Hornung, V., Guenthner-Biller, M., Bourquin, C., Ablasser, A., Schlee, M., Uematsu, S., Noronha, a., Manoharan, M., Akira, S., de Fougerolles, A., Endres, S., Hartmann, G. (2005) Sequence-specific potent induction of IFN-alpha by short interfering RNA in plasmacytoid dendritic cells through TLR7. *Nat Med* 11, 263–270.
34. Judge, A. D., Sood, V., Shaw, J. R., Fang, D., McClintock, K., MacLachlan, I. (2005) Sequence-dependent stimulation of the mammalian innate immune response by synthetic siRNA. *Nat Biotechnol* 23, 457–462.
35. Sioud, M. (2006) Single-stranded small interfering RNA are more immunostimulatory than their double-stranded counterparts: a central role for 2′-hydroxyl uridines in immune responses. *Eur J Immunol* 36, 1222–1230.
36. Cekaite, L., Furset, G., Hovig, E., Sioud, M. (2007) Gene expression analysis in blood cells in response to unmodified and 2′-modified siRNAs reveals TLR-dependent and independent effects. *J Mol Biol* 365, 90–108.
37. Goodchild, A., Nopper, N., King, A., Doan, T., Tanudji, M., Arndt, G. M., Poidinger, M., Rivory, L. P., Passioura, T. (2009) Sequence determinants of innate immune activation by short interfering RNAs. *BMC Immunol* 10, 40–48.
38. Marques, J. T., Devosse, T., Wang, D., Zamanian-Daryoush, M., Serbinowski, P., Hartmann, R., Fujita, T., Behlke, M. A., Williams, B. R. (2006) A structural basis for discriminating between self and nonself double-stranded RNAs in mammalian cells. *Nat Biotechnol* 24, 559–565.
39. Bartel, D. P. (2004) MicroRNAs: genomics, biogenesis, mechanism, and function. *Cell* 116, 281–294.
40. Han, J., Lee, Y., Yeom, K. H., Nam, J. W., Heo, I., Rhee, J. K., Sohn, S. Y., Cho, Y., Zhang, B. T., Kim, V. N. (2006) Molecular basis for the recognition of primary microRNAs by the Drosha-DGCR8 complex. *Cell* 125, 887–901.
41. Furset, G., Sioud, M. (2007) Design of bifunctional siRNAs: combining immunostimulation and gene-silencing in one single siRNA molecule. *Biochem Biophys Res Commun* 352, 642–649.
42. Hornung, V., Ellegast, J., Kim, S., Brzozka, K., Jung, A., Kato, H., Poeck, H., Akira, S., Conzelmann, K. K., Schlee, M., Endres. S., Hartmann, G. (2006) 5′-Triphosphate RNA is the ligand for RIG-I. *Science* 314, 994–997.
43. Morrissey, D. V., Lockridge, J. A., Shaw, L., Blanchard, K., Jensen, K., Breen, W.,

Hartsough, K., Machemer, L., Radka, S., Jadhav, V., Vaish, N., Zinnen, S., Vargeese, C., Bowman, K., Shaffer, C. S., Jeffs, L. B., Judge, A., MacLachlan, I., Polisky, B. (2005) Potent and persistent in vivo anti-HBV activity of chemically modified siRNAs. *Nat Biotechnol* 23, 1002–1007.

44. Judge, A. D., Bola, G., Lee, A. C., MacLachlan, I. (2006) Design of non-inflammatory synthetic siRNA mediating potent gene silencing in vivo. *Mol Ther* 13, 494–505.

45. Kariko, K., Buckstein, M., Ni, H., Weissman, D. (2005) Suppression of RNA recognition by Toll-like receptors: the impact of nucleoside modification and the evolutionary origin of RNA. *Immunity* 23, 165–175.

46. Sioud, M. (2007) RNA interference and innate immunity. *Adv Drug Deliv Rev* 59, 153–163.

47. Sioud, M., Furset, G., Cekaite, L. (2007) Suppression of immunostimulatory siRNA-driven innate immune activation by 2′-modified RNAs. *Biochem Biophys Res Commun* 361, 122–126.

48. Furset, G., Floisand, Y., Sioud, M. (2007) Impaired expression of indoleamine 2, 3-dioxygenase in monocyte-derived dendritic cells in response to Toll-like receptor-7/8 ligands. *Immunology* 123, 263–271.

49. Robbins, M., Judge, A., Liang, L., McClintock, K., Yaworski, E., MacLachlan, I. (2007) 2′-*O*-methyl-modified RNAs act as TLR7 antagonists. *Mol Ther* 15, 1663–1669.

50. Zhu, F. G., Reich, C. F., Pisetsky, D. S. (2002) Inhibition of murine dendritic cell activation by synthetic phosphorothioate oligodeoxynucleotides. *J Leukoc Biol* 72, 1154–1163.

51. Jackson, A. L., Burchard, J., Leake, D., Reynolds, A., Schelter, J., Guo, J., Johnson, J. M., Lim, L., Karpilow, J., Nichols, K., Marshall, W., Khvorova, A., Linsley, P. S. (2006) Position-specific chemical modification of siRNAs reduces “off-target” transcript silencing. *RNA* 12, 1197–1205.

Chapter 13

Inhibition of Gene Function in Mammalian Cells Using Short-Hairpin RNA (shRNA)

Jørn Remi Henriksen, Jochen Buechner, Cecilie Løkke, Trond Flægstad, and Christer Einvik

Abstract

RNAi is now the preferred method for silencing gene expression in a variety of systems. In this chapter we describe the procedure for applying short-hairpin RNA (shRNA) to study gene function. Detailed descriptions of target site selection, shRNA construction, shRNA transfection and target knockdown validation are included.

Key words: RNAi, short-hairpin RNA, gene silencing, target knockdown validation, MYCN, neuroblastoma.

1. Introduction

Gene silencing by antisense technology is now being used as a powerful molecular tool to study gene functions in living organisms. The antisense agents bind to target messenger RNA (mRNA), thus inactivating the target gene expression. The inhibitory effects on protein production from the corresponding gene may result in phenotypic changes. Thereby, the function of the gene can be understood. To date, there are a number of antisense molecules that can affect efficient post-transcriptional gene silencing. They include antisense oligonucleotides (ON), antisense 'third-generation' nucleic acid analogues (Peptide Nucleic Acid (PNA), Locked Nucleic Acid (LNA) or morpholinos), ribozymes, small-interfering RNAs (siRNAs), short-hairpin RNAs (shRNAs) and microRNAs (miRNAs). These antisense molecules cause specific gene inhibitory effects through different mechanisms. In this chapter, we demonstrate the characteristics of gene

H. Nielsen (ed.), *RNA*, Methods in Molecular Biology 703,
DOI 10.1007/978-1-59745-248-9_13, © Springer Science+Business Media, LLC 2011

silencing using RNAi-based short-hairpin RNA (shRNA) technology.

RNAi is a highly conserved gene silencing mechanism that plays an important role in regulation of gene expression. In addition, the RNAi system is important in protecting the host cell from viral infections and invasion by mobile genetic elements (1).

The RNAi pathway takes place in the cytoplasm and can be subdivided into two phases; an initiation and an effector phase. In the initiation phase, double-stranded RNAs are cleaved by the RNase III-like nuclease Dicer to produce 21–23 nt duplex RNAs, called small interfering RNAs (siRNAs). During the effector phase, the siRNA molecule is incorporated into the RNA-induced silencing complex (RISC), where an ATP-dependent RNA helicase activity unwinds the duplex. The siRNA strand which is antisense to the target RNA (guide strand) is incorporated into RISC, while the complementary passenger strand is destroyed. The guide strand permits highly sequence-specific recognition of the complementary mRNA, which is then cleaved by Argonaute 2, a component of RISC. This results in inhibition of protein synthesis from the mRNA (2).

Unlike miRNA and other long dsRNAs, shRNAs transcribed by RNA pilymerase III from exogenously introduced DNA do not require initial processing in the nucleus by the RNase III enzyme, Drosha. These tight hairpin turn RNA structures are transported directly to the cytoplasm via exportin-5 where they are cleaved by Dicer into siRNA molecules, which then follow the RNAi pathway for gene silencing (3).

RNAi is now a well-established method for high-throughput analysis as well as for functional studies in vitro, including mammalian cells (4). Two different methods are commonly used to deliver siRNA molecules for gene silencing in mammalian cell lines: (a) synthetic siRNAs (5) and (b) RNA polymerase III transcribed shRNAs from plasmids or viral vectors (6, 7).

Plasmid vector based shRNA expression is a low-cost and easy-to-perform method for studying gene function in mammalian cells. In addition, this strategy offers the possibility of inducible siRNA expression in the case where gene silencing is expected to have deleterious effects on the target cell (8).

2. Materials

2.1. Oligonucleotides and Plasmids Used for Cloning

1. Oligonucleotides:

Name	Sequence (5′-3′)	Description
ON 3	GTTTTCCCAGTCACGACGTTGTA	M13 forward sequencing primer
ON10	CGGGATCCAAAAAAGGTCTGGG TCCTTGCAGACCACGCCCGACCAAGC TTCGCCGGGCATGATCTGCAAGAACCCAG ACCGGTGTTTCGTCCTTTCCACAA	Reverse antiMYCN-27 shRNA cloning primer. Contains a BamHI-site
ON11	CGGGATCCAAAAAAGCTAGTGC TCCTCGGCCTAGAAGGAGTAGCAAGC TTCCCACTCCCTCCAGGCCGAGGAGCA CCAGCGGTGTTTCGTCCTTTCCACAA	Reverse antiMYCN-1291 shRNA cloning primer. Contains a BamHI-site
ON19	CGGGATCCAAAAAAGAATCACT CAGAGTGTCCCCTCCGGAAGTGAAGCT TGACCTCCGGAGAGGACACCCTGAGCG ATTCGGTGTTTCGTCCTTTCCACAA	Reverse antiMYCN-760 shRNA cloning primer. Contains a BamHI-site
ON20	CGGGATCCAAAAAAGTTCTTG AGACACACAGCGATGGTAAATGGAAGCT TGCATTCACCATCACTGTGCGTCCCAA GAACGGTGTTTCGTCCTTTCCACAA	Reverse antiMYCN-887 shRNA cloning primer. Contains a BamHI-site
ON22	ATAAGAATGCGGCCGCAAGGTC GGGCAGGAAGAGGGCC	U6 forward primer. Contains a NotI-site
ON51	CGGGATCCAAAAAAGAGCGTTCG GAGCTGATGGCCATAAATACGAAGCTT GGTACTTATGACCACCAACTCCGAA CGCTCGGTGTTTCGTCCTTTCCACAA	Reverse cloning primer for SCR shRNA. Contains a BamHI-site
ON 56	ATTTGGGTCGCGGTTCTTG	UBC forward QPCR primer
ON 57	TGCCTTGACATTCTCGATGGT	UBC reverse QPCR primer
ON 87	CACCCTGAGCGATTCAGATGA	MYCN forward QPCR primer
ON 89	CCGGGACCCAGGGCT	MYCN reverse QPCR primer
ON 58	GCAGCTACTCCTCCAGCTCT	NFL forward QPCR primer
ON 59	ACTTGAGGTCGTTGCTGATG	NFL reverse QPCR primer
ON 60	TCCAGCCCAGAGACACTGATT	NPY forward QPCR primer
ON 61	AGGGTCTTCAAGCCGAGTTCT	NPY reverse QPCR primer
ON 96	AAGTTCTACGGTGACGAGGAG	CRT forward QPCR primer
ON 97	GTCGATGTTCTGCTCATGTTTC	CRT reverse QPCR primer
ON 100	AGATCCCGGAGTTGGAAAAC	c-MYC forward QPCR primer
ON 101	AGCTTTTGCTCCTCTGCTTG	c-MYC reverse QPCR primer
ON 170	TCACCCACACTGTGCCCATCTACGA	β-actin forward QPCR primer
ON 171	CAGCGGAACCGCTCATTGCCAATGG	β-actin reverse QPCR primer
ON 176	TGACACTGGCAAAACAATGCA	HPRT1 forward QPCR primer

Name	Sequence (5′-3′)	Description
ON 177	GGTCCTTTTCACCAGCAAGCT	HPRT1 reverse QPCR primer
ON 304	CGAGAGCGAGCGGATGA	CHGB forward QPCR primer
ON 305	GGCGTGTCTTCACTTCTTCAGA	CHGB reverse QPCR primer

2. Plasmids:

   pSHAG-Ffl (9) encodes an U6-driven anti-luciferase (anti-luc) shRNA homologous to nucleotides 1,340–1,368 of the coding sequence of the firefly luciferase gene (NCBI accession number U47296).

2.2. Cell Culture and Transfection

1. Neuroblastoma cell line SK-N-BE(2) – MYCN amplified.
2. 6-well multiwell culture plates (Falcon).
3. RPMI-1640 supplemented with 10% fetal bovine serum (FBS).
4. Phosphate-buffered saline (PBS).
5. Trypsin solution: 0.25% trypsin and 0.05% ethylenediamine tetraacetic acid (EDTA) in PBS, pH 7.5.
6. Cells were maintained in a humidified 37°C incubator with 5% CO_2, supplied with fresh complete medium every 3 days and subcultured before confluence was reached.
7. Lipofectamin2000 (Invitrogen).

2.3. Cellular Protein Isolation and Western Blot Analysis

1. Tropix lysis buffer, Protease inhibitor (Roche), DTT.
2. XCell SureLock™ Mini-Cell (Invitrogen).
3. NuPAGE Novex 4–12% Bis-Tris Gel – 1.0 mm × 10 well (Invitrogen).
4. NuPAGE LDS Sample Buffer (4×) (Invitrogen).
5. Molecular markers. MagicMark™XP western standard (Invitrogen) as a protein size marker. Kaleidoscope Prestained Standard (Bio-RAD) for visualization of size during electrophoresis and protein transfer efficiency from gel to membrane during blotting.
6. Methanol.
7. Running buffer; NuPAGE MOPS SDS Running Buffer (20×) (Invitrogen).
8. XCell II™ Blot Module (Invitrogen).
9. Immobilon-FL PVDF transfer membrane (Millipore).
10. Whatman 3MM chromatography paper.

11. Transfer buffer: NuPAGE Transfer Buffer (20×) (Invitrogen.)
12. Blocking buffer: Odyssey® Blocking buffer (LI-COR Biosciences).
13. Primary Ab:
    anti-N-myc mouse mAb (Calbiochem)
    anti-actin rabbit pAb (Sigma)
14. Secondary Ab:
    Alexa Fluor® 680 conjugated goat anti-mouse IgG (Invitrogen)
    IRDye800CW conjugated goat anti-rabbit IgG (Rockland)
15. PBST: PBS containing 0.1% Tween-20.
16. Odyssey Infrared Imaging System (LI-COR).

2.4. Total RNA Isolation and cDNA Synthesis

1. RNeasy Plus Mini Kit (Qiagen)
2. QIAshredder™ (Qiagen)
3. Superscript™ III reverse transcriptase (200 U/μL) (Invitrogen)
4. Oligo-dT$_{20}$ primer (50 μM)
5. dNTPs (2.5 mM each)
6. RNase inhibitor (20 U/μL)
7. Thermocycler

2.5. Real-Time RT-PCR Analysis

1. Power SYBR Green PCR Master Mix (Applied Biosystems)
2. MicroAmp Optical 96-well Reaction Plate (Applied Biosystems)
3. MicroAmp Optical Adhesive Film (Applied Biosystems)
4. 7,300 Real Time PCR System (Applied Biosystems)
5. 7,300 System Sequence Detection Software v1.4 (Applied Biosystems)
6. qBase (**http://medgen.ugent.be/qbase/**)

3. Methods

This section gives a detailed description of the design and cloning of anti-MYCN shRNA constructs. A detailed procedure for cell transfection and evaluation of shRNA treated neuroblastoma cells, using Western blot and real-time RT-PCR analysis, is also described.

3.1. Selection of Anti-MYCN shRNA Target Sites

There are many factors affecting shRNA efficiency. Among the most important are shRNA and target mRNA structures. Several software programmes for siRNA target prediction have been developed, but no single standard exists for predicting the best siRNA target sequence (10). Functional studies are required to evaluate the efficiencies of any shRNA constructed.

We chose four different target sites in the MYCN cDNA (GeneBank accession NM_005378) sequence (*see* **Fig. 13.1a**). Two sites (antiMYCN-27 and antiMYCN-1291) were picked at random. The other two target sites (antiMYCN-760 and antiMYCN-887) and a scrambled shRNA sequence were selected using Genescript siRNA Target Finder and Genescript siRNA Sequence Scrambler (**http://www.genscript.com/tools.html**), respectively (11).

All antiMYCN shRNA sequence candidates were BLASTed (NCBI database) to ensure that only the MYCN mRNA was targeted.

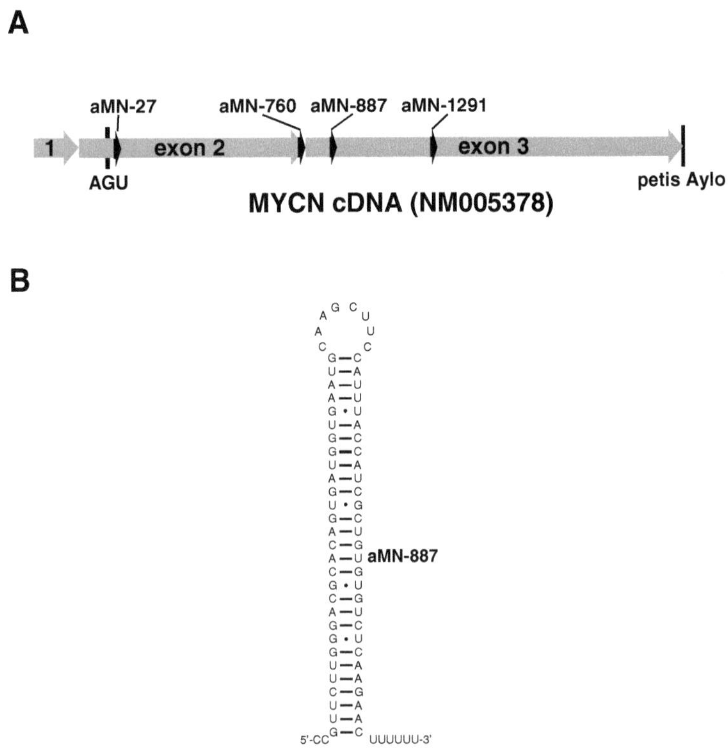

Fig. 13.1. **a** Schematic representation of MYCN cDNA. Exons are shown as *grey arrows*. Localization of antiMYCN shRNA target sites are indicated by *black arrowheads*. **b** Sequence and secondary structure representation of antiMYCN-887 (aMN-887) shRNA.

3.2. Designing Reverse Primers for Anti-MYCN shRNA Cloning

The PCR-based cloning strategy used to construct the U6 expressed shRNAs from the pSHAG plasmid, requires a reverse primer containing the complete shRNA sequence.

The following steps describe how ON-20, the reverse primer for amplifying antiMYCN-887, was designed:

1. Pick a 29-nt target sequence which ends with a C from the MYCN cDNA sequence:

   5′–**CATTCACCATCACTGTGCGTCCCAAGAAC**–3′

2. Reverse complement the target sequence to create the 'antisense' strand:

   5′–**GTTCTTGGGACGCACAGTGATGGTGAATG**–3′

3. Add a HindIII-containing 'loop' sequence to the 3′-end:

   5′–GTTCTTGGGACGCACAGTGATGGTGAATG **CAAGCTTC**–3′

4. Add the reverse complement of the 'antisense' sequence to the 3′-end of the 'loop' sequence:

   5′–GTTCTTGGGACGCACAGTGATGGTGAATGC AAGCTTC**CATTCACCATCACTGTGCGTCCCAAG AAC**–3′

   This sequence represents the shRNA molecule (*see* **Fig. 13.1b**).

5. Change 4 nt in the 'sense' strand to create G-U basepairs in the shRNA stem sequence (*see* **Note 1**):

   5′-GTTCTTGGGACGCACAGTGATGGTGAATGC AAGCUUCCATT**T**ACCATC**G**CTGTG**T**GTC**T**CAAG AAC – 3′

6. Add 6 thymidines to create the RNA polymerase III transcription termination sequence:

   5′ – GTTCTTGGGACGCACAGTGATGG /………/ CCATCGCTGTGTGTCTCAAGAAC**TTTTTT** – 3′

7. Reverse complement the sequence:

   5′ –**AAAAAAGTTCTTGAGACACACAGCGATGG**/ ……../**CCATCACTGTGCGTCCCAAGAAC** – 3′

8. Add a 21 nt downstream U6 promoter binding sequence **(GGTGTTTCGTCCTTTCCACAA) to** the 3′-end:

   5′ –AAAAAAGTTCTTGAGAC/……../CCATCACTG TGCGTCCCAAGAAC**GGTGTTTCGTCCTTTCCAC AA** – 3′

9. Add a BamHI restriction enzyme site to the 5′ end for cloning purposes to finish the reverse cloning primer ON-20 (*see* **Section 2.1**) A CG-dinucleotide is added in front of the BamHI-site to enable efficient restriction enzyme digestion at the end of the resulting PCR product :

   5′ –**CGGGATCC**AAAAAAGTTCTTGAGAC/……../ CCATCACTGTGCGTCCCAAGAACGGTGTTTCGTCC TTTCCACAA – 3′

3.3. Construction of Anti-MYCN shRNA Expressing Plasmids

Plasmids containing different anti-MYCN shRNA sequences, expressed from a U6 promoter, are made using a PCR-based strategy. In this strategy, 272 bp of the U6 promoter from pSHAG-Ffl is amplified using a NotI-containing U6 forward primer (ON22) in combination with different BamHI-containing reverse primers including the complete anti-MYCN shRNA sequences (*see* **Note 2**).

PCR products are digested with NotI/BamHI and purified from agarose gels, before ligation into NotI/BamHI digested pSHAG-Ffl plasmids. Reverse primers ON10, ON11, ON19, ON20 and ON51 are used to construct plasmids pantiMYCN-27, pantiMYCN-1291, pantiMYCN-760, pantiMYCN-887 and pScr-shRNA, respectively. Numbers in plasmid names indicate the first position of the shRNA target recognition site in the MYCN cDNA sequence. All plasmid constructs are verified by DNA sequencing using ON-3.

3.4. Transient Transfection of Anti-MYCN shRNAs into a MYCN-Amplified Neuroblastoma Cell Line

1. Day 1: Seed 1.3×10^5 SK-N-BE(2) cells into each well of a 6-well tissue culture plate.
2. Day 2: Cells are transfected with 3 μg plasmids pantiMYCN-27, pantiMYCN-1291, pantiMYCN-760, pantiMYCN-887 or pScr-shRNA using Lipofectamin2000 (4 μL) in a total of 2 mL media in each well according to the manufacturers protocol.
3. Day 5: Isolate total cellular RNA and protein extracts.

   Transfection efficiencies typically vary between 50 and 80%.

3.5. Total Cellular Protein Isolation

1. Wash cells with 2 mL PBS, make sure to remove all supernatant.
2. Add trypsin solution.
3. When cells detach from the culture dish (few minutes at room temperature), add 1.0 mL RPMI1640 with 10% serum, resuspend and transfer to 1.5 mL Eppendorf tubes.
4. Wash cells once in 0.5 mL PBS.
5. Resuspend cells in 40 μL Tropix lysis buffer containing protease inhibitor and 1 mM DTT (*see* **Note 3**).
6. Leave on ice for 5 min. Centrifuge at maximum speed for 5 min. Collect the supernatants containing total cellular proteins in fresh tubes.
7. Measure total protein concentrations.

3.6. Western Immunoblot Analysis

To investigate the effect of anti-MYCN shRNA knockdown at the protein level, we use Western immunoblot analysis. The XCell

SureLock^{TM} Mini-Cell and XCell II^{TM} Blot Module (Invitrogen) is used to resolve the proteins by Bis-Tris polyacrylamide gel electrophoresis and to transfer the resolved proteins from the gel to a membrane support, respectively. The procedures were followed according to the specifications of the producer and briefly include the following steps:

Separation of proteins by Bis-Tris polyacrylamide gel electrophoresis:

1. Assemble the electrophoresis chamber with a 4–12% Bis-Tris Gel and running buffer.
2. Load 25 μg total protein in sample buffer and incubate for 10 min at 70°C to each sample well (total 20 μL). Include MagicMark (1 μL) and Kaleidoscope (7 μL) markers in separate wells.
3. Run the gel at constant 200 V for 1 h.
   Transfer of resolved protein from the gel to a membrane support:
4. Prepare the membrane by briefly soaking it for 10 s in 100% methanol, 5 s in water and store it in transfer buffer until used.
5. Assemble the gel/blot sandwich from the cathode core in the following order: 2× blotting pads, 1 Whatman 3MM filter paper, gel, membrane, 1× Whatman 3MM filter paper and 3× blotting pads (*see* **Note 4**).
6. Assemble the Mini-Cell Blot Module with the gel/blot sandwich.
7. Electroblot at 30 V for 1 h. The kaleidoscope marker colours should be transferred to the membrane.
   Processing of the blot for detection of specific proteins with an antibody:
8. Wash membrane in PBS for 5 min.
9. Block membrane in blocking buffer for 1 h.
10. Add primary antibodies anti-N-myc (1:400) and anti-actin (1:1000) diluted in blocking buffer containing 0.01% SDS and 0.01% Tween-20. Incubate over night at 4°C (*see* **Note 5**).
11. Wash membrane 4 × 5 min at room temperature in PBST.
12. Add secondary antibodies diluted 1:5,000 in blocking buffer containing 0.01% SDS and 0.01% Tween-20. Incubate 1 h at room temperature. Cover in aluminium to protect from light.

13. Wash membrane 4 × 5 min at room temperature in PBST. Protect membrane from light.
14. Scan membrane on an infrared imaging system to develop the final Western immunoblot.

3.7. Total RNA Isolation

We use the RNeasy Plus Mini Kit to isolate total RNA samples. This kit includes gDNA Eliminator Mini Spin Columns for efficient DNA removal and do not require additional DNase treatment. Procedures are according to that recommended by the manufacturer and include the following steps:

1. Cells from one transfected well of a 6-well culture dish are disrupted by addition of 350 μL of Buffer RLT Plus.
2. Homogenize cell lysates using QIAshredder™ (*see* **Note 6**).
3. Closely follow steps 4–12 in the Qiagen protocol for purification of total RNA from animal cells. Samples are usually eluted in 35 μL RNase-free water both in Steps 11 and 12.

3.8. cDNA Synthesis

Superscript™ III reverse transcriptase is used to reverse transcribe total RNA to cDNA.

1. For each RNA sample prepare the following in a 0.5-μL microcentrifuge tube:

Component	Volume
Oligo-dT_{20} primer (50 μM) (*see* **Note 7**)	1 μL
dNTP (2.5 mM each)	1 μL
$MgCl_2$ (25 mM)	1 μL
1.4–2.0 μg total RNA (*see* **Note 8**)	× μL
RNase-free water	(13 – ×) μL
Total	13 μL

2. Incubate at 65°C for 5 min, then on ice for 1 min.
3. Add the following to RNA-containing solution from Step 1.

Component	Volume
(RNA-containing solution from Step 1)	(13 μL)
5× First Strand buffer	4 μL
DTT (0.1 M)	1 μL
RNase inhibitor	1 μL
SuperScript™ III (*see* **Note 9**)	1 μL
TOTAL:	20 μL

4. Incubate in a thermocycler:

Temperature	Time (min)
50°C	60
70°C	15
4°C	until PCR setup is ready

3.9. Real-Time PCR Analysis

1. Prepare stocks of master reaction mixes. Each reaction includes the following:

Component	Volume
Nuclease-free water	6.5 μL
2× SYBR Green Master Mix	12.5 μL
cDNA template (30 × diluted)	5.0 μL
Primer mix (5 μM each primer) (*see* **Note 10**)	1.0 μL
TOTAL:	25 μL

2. Add 25 μL of the reaction mix to each well of an optical 96-well reaction plate.
3. Seal the plate with adhesive film. Make sure all edges are properly sealed.
4. Spin the reaction plate at 420×*g* for 1 min.
5. Start the PCR software on the real-time PCR system and assign each well with the correct sample/control
6. Insert the reaction plate and execute the PCR with the following program:
   Pre-incubation:

Temperature	Time (min)
50°C	2 min
95°C	10 min

Amplification (40 cycles):

Temperature	Time (min)
95°C	15 s
60°C	1 min

Melting curve:

Temperature	Time (min)
95°C	15 s
60°C	1 min
95°C	15 s
60°C	15 s

7. Calculate relative gene expressions (*see* **Note 11**).

3.10. Results

Western immunoblot analysis was used to evaluate the effect of different anti-MYCN shRNA constructs. As can be seen from **Fig. 13.2**, the MYCN protein levels are reduced to different degrees with the tested shRNAs. aMN-887 is the most efficient anti-MYCN shRNA. β-actin protein levels remain unaffected by the different transfections. These results show that the knock-down effect by shRNAs at the RNA level is reflected at the MYCN protein level.

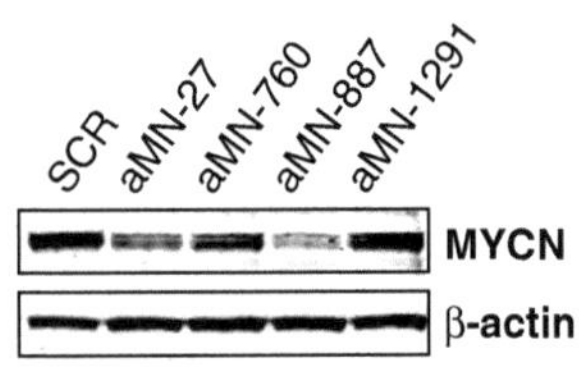

Fig. 13.2. Western immunoblotting analysis of MYCN in antiMYCN shRNA transfected SK-N-BE(2) cells. Knockdown effects from 4 different antiMYCN shRNAs (aMN-27, aMN-760, aMN-887 and aMN-1291) were compared. aMN-887 is the most efficient antiMYCN shRNA.

To measure the direct effect of different anti-MYCN shRNAs on MYCN mRNA, we use real-time RT-PCR analysis. **Figure 13.3a** shows relative expression levels of MYCN mRNA. The reductions in MYCN mRNA levels are consistent with the decrease of MYCN protein observed on Western immunoblots. aMN-887 is the most efficient shRNA showing a 70% reduction of MYCN mRNA in transient transfection experiments. To support the observed MYCN knockdown effects, we quantified c-MYC mRNA levels in the transfected cells using quantitative real-time RT-PCR. Previous studies have shown that there is a reverse correlation of MYCN and c-MYC expression in neuroblastoma cells (12, 13). **Figure 13.3b** shows a close reverse correlation between MYCN mRNA and c-MYC mRNA expression, supporting the knockdown efficiencies of the tested anti-MYCN shRNAs. Based on the measured CT values in these experiments, MYCN mRNA levels exceed c-MYC mRNA levels by a factor of 10^4 in the MYCN-amplified neuroblastoma cell line SK-N-BE(2).

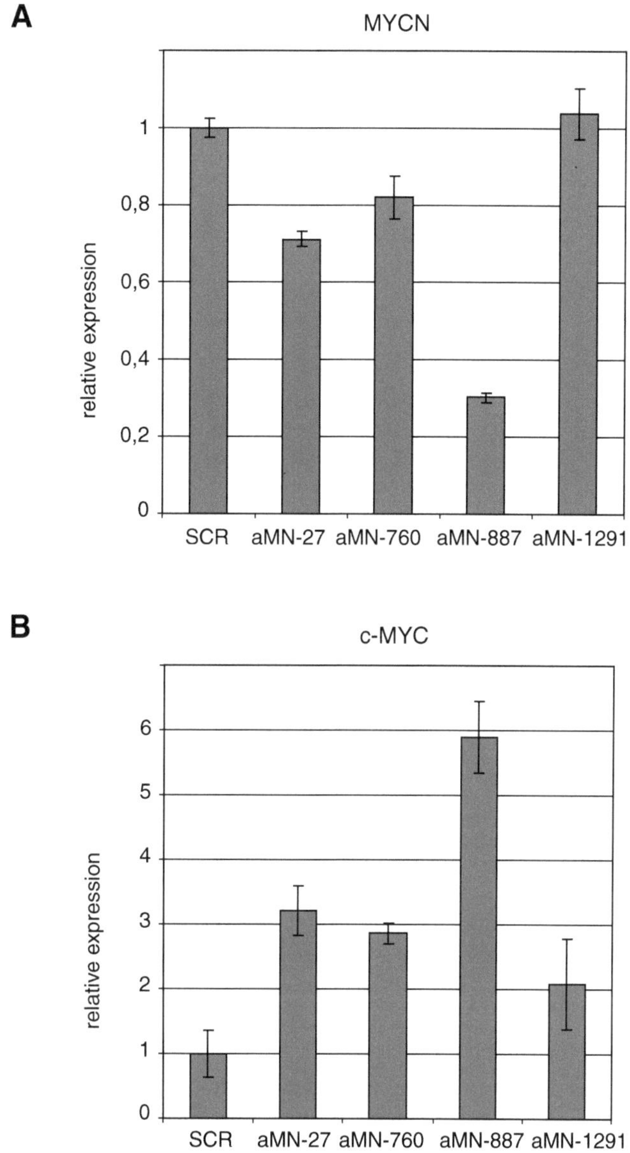

Fig. 13.3. Real-time RT-PCR analysis of MYCN (**a**) and c-myc (**b**) mRNA levels in shRNA transfected SK-N-BE(2) cells. Knockdown effects from four different antiMYCN shRNAs (aMN-27, aMN-760, aMN-887 and aMN-1291) were compared. aMN-887 is the most efficient antiMYCN shRNA, showing approximately 70% reduction in MYCN mRNA compared to the scrambled control (SCR). A close reverse correlation is observed between MYCN mRNA and c-MYC mRNA expression levels. .

To further verify MYCN downregulation upon anti-MYCN shRNA treatment, we used real-time RT-PCR to confirm neuronal differentiation observed by immunostaining confocal microscopy (data not shown). We chose a panel of 4 neuronal differentiation markers, Neuropeptid Y –NPY (14),

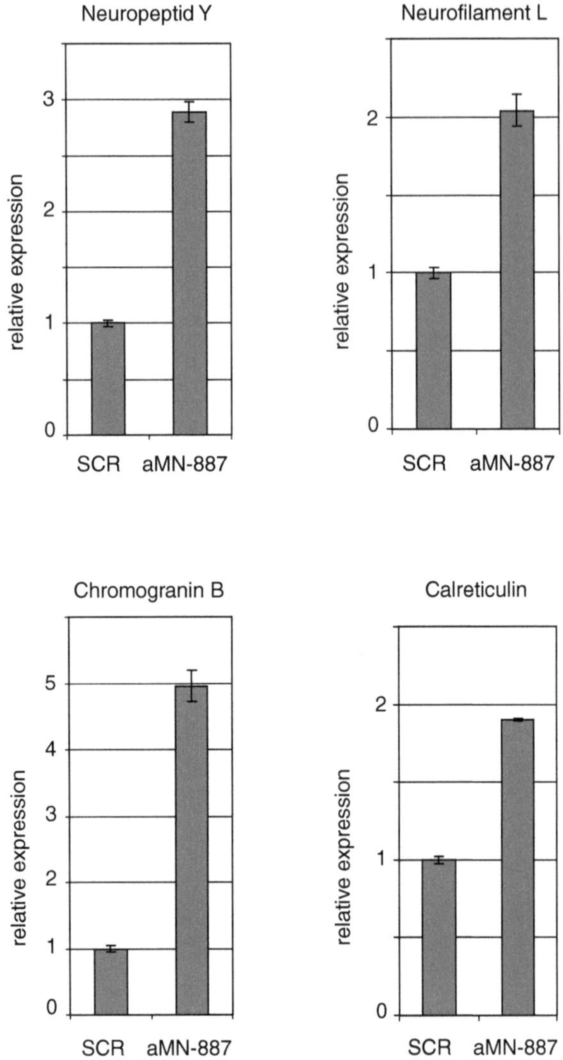

Fig. 13.4. Real-time RT-PCR analysis of known neuronal differentiation markers, neuropeptid Y, neurofilament L, chromogranin B and calreticulin, in shRNA transfected SK-N-BE(2) cells. All markers show increased expression when transfected with an antiMYCN shRNA (aMN-887) compared to a scrambled control shRNA (SCR).

Calreticulin – CRT (15, 16), Chromogranin B – CHGB (17), Neurofilament L – NFL (18) and investigated mRNA levels in aMN-887 treated neuroblastoma cells. All markers were significantly increased in anti-MYCN shRNA treated cells (*see* **Fig. 13.4**), confirming the observed neuronal differentiation.

Hypoxanthine phosphoribosyltransferase 1 (HPRT1), β-actin and Ubiquitin C (UBC) were used as housekeeping genes in all real-time PCR experiments (*see* **Note 12**).

4. Notes

1. The G-U basepairs added are not central for the efficiency of the shRNA, but they are thought to stabilize the shRNA molecule (19). Furthermore, the wobble basepairs aid sequencing of the shRNA construct and is also believed to reduce unwanted immunostimulation.
2. When using very long primers (the reverse primer including the entire shRNA construct) in PCR, it is recommended to add DMSO to the PCR reaction mix. We use 2 μL DMSO in a 50-μL PCR reaction mix.
3. Lysis of cells can be done directly in the well, but this requires larger volume of lysis buffer, and thus results in lower concentration of protein in the final solution. This might cause problems since there is only room for ca. 20 μL sample in each well of the western gel.
4. Blotting pads, Whatman paper and membrane should be soaked thoroughly in transfer buffer before assembly. Avoid air bubbles.
5. We perform primary antibody staining in 50-mL centrifuge tubes to reduce the amounts of blocking buffer and antibody (total 5 mL). Membranes are rolled in oven mesh sheets to ensure complete exposure to the antibody solution.
6. Passing cell lysates five times through a RNase-free 20-gauge needle works as well.
7. Random hexamer primers work as well.
8. If RNA concentration is below the amount required to get 1.4 μg in 10 μL, we use Microcon Ultracel YM-100 filters (Millipore) to concentrate the RNA.
9. A No-RT control reaction, lacking reverse transcriptase, is always used as a control for DNA removal. Using primer sets located within an exon sequence (ex: UBC) will give PCR amplification whenever DNA is present in this reaction.
10. If the same cDNA template is being analysed for multiple genes, it is more efficient to add primers separately. Mix the other components of the master reaction mix in the desired amount (number of parallels multiplied with number of genes being analysed) Apply 1 μL of primer mix to the well intended for its respective genes. Then add 24 μL of the reaction mix to the wells.
11. We use qBase v1.3.5 to calculate relative expression levels with 2–3 housekeeping genes (20).

12. There is not one single housekeeping gene suitable for all experiments. It is recommended to test the consistency of expression of several housekeeping genes in your own experiments. We have used the Human Endogenous Control Gene Panel (tataa Biocenter) to find the housekeeping genes most suitable for our experimental setup. We always use 2–4 different housekeeping genes in each real-time PCR experiment.

References

1. Obbard, D. J., Gordon, K. H., Buck, A. H., Jiggins, F. M. (2008) Review. The evolution of RNAi as a defence against viruses and transposable elements. *Philos Trans R Soc Lond B Biol Sci* 364, 99–115.
2. Shrivastava, N., Srivastava, A. (2008) RNA interference: an emerging generation of biologicals. *Biotechnol J* 3, 339–353.
3. Yi, R., Qin, Y., Macara, I. G., Cullen, B. R. (2003) Exportin-5 mediates the nuclear export of pre-microRNAs and short hairpin RNAs. *Genes Dev* 17, 3011–3016.
4. Scherr, M., Eder, M. (2007) Gene silencing by small regulatory RNAs in mammalian cells. *Cell Cycle* 6, 444–449.
5. Watts, J. K., Deleavey, G. F., Damha, M. J. (2008) Chemically modified siRNA: tools and applications. *Drug Discov Today* 13, 842–855.
6. Wang, Q. Z., Lv, Y. H., Diao, Y., Xu, R. (2008) The design of vectors for RNAi delivery system. *Curr Pharm Des* 14, 1327–1340.
7. Walchli, S., Sioud, M. (2008) Vector-based delivery of siRNAs: in vitro and in vivo challenges. *Front Biosci* 13, 3488–3493.
8. Henriksen, J. R., Lokke, C., Hammero, M., Geerts, D., Versteeg, R., Flaegstad, T., Einvik, C. (2007) Comparison of RNAi efficiency mediated by tetracycline-responsive H1 and U6 promoter variants in mammalian cell lines. *Nucleic Acids Res* 35, e67.
9. Paddison, P J., Caudy, A. A., Bernstein, E., Hannon, G. J., Conklin, D. S. (2002) Short hairpin RNAs (shRNAs) induce sequence-specific silencing in mammalian cells. *Genes Dev* 16, 948–958.
10. Li, W., Cha, L. (2007) Predicting siRNA efficiency. *Cell Mol Life Sci* 64, 1785–1792.
11. Wang, L., Mu, F. Y. (2004) A Web-based design center for vector-based siRNA and siRNA cassette. *Bioinformatics* 20, 1818–1820.
12. Breit, S., Schwab, M. (1989) Suppression of MYC by high expression of NMYC in human neuroblastoma cells. *J Neurosci Res* 24, 21–28.
13. Westermann, F., Muth, D., Benner, A., Bauer, T., Henrich, K. O., Oberthur, A., Brors, B., Beissbarth, T., Vandesompele, J., Pattyn, F., et al. (2008) Distinct transcriptional MYCN/c-MYC activities are associated with spontaneous regression or malignant progression in neuroblastomas.. *Genome Biol* 9, R150.
14. Jalava, A., Heikkila, J., Lintunen, M., Akerman, K., Pahlman, S. (1992) Staurosporine induces a neuronal phenotype in SH-SY5Y human neuroblastoma cells that resembles that induced by the phorbol ester 12-O-tetradecanoyl phorbol-13 acetate (TPA). *FEBS Lett* 300, 114–118.
15. Johnson, R. J., Liu, N., Shanmugaratnam, J., Fine, R. E. (1998) Increased calreticulin stability in differentiated NG-108-15 cells correlates with resistance to apoptosis induced by antisense treatment. *Brain Res Mol Brain Res* 53, 104–111.
16. Hsu, W. M., Hsieh, F. J., Jeng, Y. M., Kuo, M. L., Chen, C. N., Lai, D. M., Hsieh, L. J., Wang, B. T., Tsao, P. N., Lee, H., et al. (2005) Calreticulin expression in neuroblastoma—a novel independent prognostic factor.. *Ann Oncol* 16, 314–321.
17. Jogi, A., Vallon-Christersson, J., Holmquist, L., Axelson, H., Borg, A., Pahlman, S. (2004) Human neuroblastoma cells exposed to hypoxia: induction of genes associated with growth, survival, and aggressive behavior. *Exp Cell Res* 295, 469–487.
18. Breen, K. C., Anderton, B. H. (1991) Temporal expression of neurofilament polypeptides in differentiating neuroblastoma cells. *Neuroreport* 2, 21–24.
19. Paddison, P. J., Caudy, A. A., Sachidanandam, R., Hannon, G. J. (2004) Short hairpin activated gene silencing in mammalian cells. *Methods Mol Biol* 265, 85–100.
20. Hellemans, J., Mortier, G., De, P. A., Speleman, F., Vandesompele, J. (2007) qBase relative quantification framework and software for management and automated analysis of real-time quantitative PCR data. *Genome Biol* 8, R19.

Chapter 14

Validation of RNAi by Real Time PCR

Knud Josefsen and Ying C. Lee

Abstract

Real time PCR is the analytic tool of choice for quantification of gene expression, while RNAi is concerned with downregulation of gene expression. Together, they constitute a powerful approach in any loss of function studies of selective genes. We illustrate here the use of real time PCR to verify luciferase mRNA silencing.

Key words: RNA isolation, reverse transcription, real time PCR, LightCycler, luciferase, shRNA, RNAi.

1. Introduction

Real time PCR and RNAi (1) together constitute a powerful approach in any loss of function studies of selective genes. Real time PCR is essentially a cycle by cycle kinetic analysis of a signal generated by a fluorescent reporter used in the amplification reaction. It detects the DNA amplification during the PCR reaction with a broad dynamic range, and analyses each sample at its optimal point of amplification. The increase in fluorescence is directly proportional with the amount of PCR product accumulated in the reaction. The threshold cycle (C_T) is the key parameter in real time PCR and is defined as the cycle number when measurable amplification product has accumulated. Thus, the more specific cDNA a sample contains, the faster the amplification reaches the detection level and the lower C_T is obtained.

The most cost-effective and simplest detection chemistry for real time PCR is SYBR Green I dye. SYBR Green I fluoresce when bound to double-stranded DNA (dsDNA), whereas it only binds

H. Nielsen (ed.), *RNA*, Methods in Molecular Biology 703,
DOI 10.1007/978-1-59745-248-9_14, © Springer Science+Business Media, LLC 2011

single-stranded DNA poorly. Unlike other detection chemistries used in real time PCR, SYBR Green I can present a specificity problem because the dye binds to any dsDNA formed during the real time PCR run. However, melting curve analysis can be used to discriminate the specific product from non-specific products such as primer-dimers, which may form in the reaction. Melting curve analysis determines the T_m of the PCR product, which is defined as the temperature where 50% of the DNA is single-stranded. T_m can form the basis for the inclusion of a 4th segment in the PCR amplification program if non-specific amplification is present, since it reduces the non-specific fluorescence signal generated in real time PCR (2). The 4th segment temperature is usually set at 3–5°C lower than the T_m for the specific product.

An important consideration for a successful real time PCR result is the precision of real time PCR cycler. Although replicates may improve the reliability of the results, in particular when quantifying targets with low copy number, surprisingly little attention is usually devoted to the fact that different real time PCR cyclers operate with very variable precision from virtually no variability to more than 100% variation (*see* **Note 1**). Equally important is the primer design. Software is available on the web which will optimise factors like length and G/C content of the primers as well as prevent secondary structures and overlapping sequences. Likewise, for reliable and accurate real time data it is imperative to use high-quality RNA (3), optimised PCR master mix and reagent cocktails as well as paying attention to mixing all reagent components thoroughly during the set-up of the experiments. Finally, any comparison of gene expression between different samples invariably also involves quantification of a reference gene in all samples (4). These general requirements for successful real time PCR procedures and possible pitfalls have been detailed in a recent article by Nolan et al. (5). Further information can be found on the authors' website (**www.realtimepcr.dk**).

All of the above-mentioned factors can influence real time PCR efficiency which is defined as the proportion of target amplicon, which is amplified from one cycle to the next. To determine the efficiency (E) of the reaction, C_T is plotted against the logarithm of the dilution of the template (6). In this graph, 100% efficiency corresponds to a slope of –3.3, and 90% or 110% to –3.6 or –3.1, respectively, since $E = 10^{-1/\text{slope}}$. Calculation of the real time PCR efficiency is of particular importance when the delta/delta method is employed to compare samples; such data is only valid when the amplification efficiencies of the target and reference genes are equal (5).

Single-stranded antisense RNAs have been demonstrated to inhibit effectively gene expression in mammalian systems (7). The production of single-stranded RNA is initiated by the action of dicer, a RNase III class enzyme acting upon endogenous double-stranded RNAs (dsRNAs) or exogenously introduced short dsRNAs, and finalised in the RNA-induced silencing complex (RISC), within which the RNA sense strand is degraded, whereas the antisense RNA strand can act as a guide sequence to mediate specific cleavage of targeted mRNA, resulting in RNAi. The length of duplex siRNA molecules is usually limited to less than 30 nucleotides (8) as non-specific effects such as cellular interferon activation may otherwise occur. In addition, vector-encoded RNAi, shRNAs (short-hairpin RNAs), have also been developed for long-term gene silencing experiments.

The success of any RNAi experiment depends on effective design of siRNAs or shRNAs, efficient delivery to the cells, an appropriate phenotypic assay to assess the RNAi effect and equally important: an assay for the target gene silencing. As RNAi exerts its effects at the mRNA level, the simplest, quickest and most sensitive assay for RNAi validation is real time PCR. Here, we demonstrate the use of real time PCR to validate the silencing of luciferase gene expression by shRNA. Hela cells were transfected with either luciferase shRNA construct or with a negative control shRNA. This was followed by isolation of total RNA, first strand cDNA synthesis and real time PCR by LightCycler. The result showed that luciferase mRNA was reduced fourfold in cells transfected with the luciferase shRNA construct compared with cells transfected with negative control shRNA.

2. Materials

2.1. Tissue Culture

1. Hela cells (American Tissue Culture Collection).
2. RPMI 1640 and Trypsin/EDTA (Lonza).
3. Foetal bovine serum, OptiMEM and Lipofectamine 2000 (Invitrogen).
4. 90% RMI1640 + 10% foetal bovine serum
5. T75 tissue culture flasks and 12-well tissue culture plates (Costar).
6. Plasmids pShag+ which encodes shRNA against luciferase and pShag-, a negative control, were supplied by Christer Einvik, Tromsø, Norway (9). pGL3-control encoding luciferase and pGL3 basic, a negative control (Promega).

2.2. RNA Isolation (see Note 2)

1. Trizol (Invitrogen).
2. Chloroform, isopropanol and ethanol (Merck Chemicals).
3. TE buffer: 1 mM Tris-HCl, pH 7.0, 10 mM EDTA.
4. UV spectrophotometer (BioPhotometer: Eppendorf).

2.3. Reverse Transcription

1. Reverse transcriptase kit (iScript from Bio-Rad) (*see* **Note 3**).

2.4. Real Time PCR

1. Real time PCR mix (FastStart^{plus} DNA Master SYBR Green I; Roche) (*see* **Note 4**).
2. PCR Primers (Sigma Genosys, England); purification: DST (*see* **Note 5**). Luciferase (fragment sized 201 bp), forward: ATG AAC GTG AAT TGC TCA ACA G, reverse: TAA AAC CGG GAG GTA GAT GAG A. Reference gene: Ribosomal protein L13 (RPL13, 227 bp product), forward: CCA CCC TAT GAC AAG AAA AAG C, reverse: ACA TTC TTT TCT GCC TGT TTC C.
3. Real time PCR cycler (LightCycler version 1.5 with LightCycler Software Version 3.5.3; Roche Molecular Biochemicals).
4. Reaction capillaries (Roche) (*see* **Note 6**).
5. TBE (Qiagen) and agarose (Invitrogen).
6. TOPO TA cloning kit and One Shot Competent *E. coli* (Invitrogen) for subcloning of PCR fragments (*see* **Note 7**).

3. Methods

3.1. Tissue Culture

1. Grow Hela cells in 90% RMI1640 + 10% foetal bovine serum in T75 bottles with medium replacements every other day (*see* **Note 8**).
2. Split cells when they reach approximately 70% confluence: remove the medium, rinse the cells briefly in 2 mL trypsin/EDTA, remove it and add 2 mL of fresh trypsin/EDTA. Leave it on the cells until they detach when the bottle is tilted. Add 8 mL of medium and triturate the cells (pipette the medium in and out of a 13-mL pipette 8–10 times until most of the cells are suspended as single cells). Transfer to 2–3 new T75 bottles and add medium to approximately 13 mL per bottle.
3. For transfection, split cells as above, but transfer the cells to 6-well plates to obtain a confluence of approximately 60% after 24 h of incubation.
4. For each transfection, mix 1.8 μg plasmid with 250 μL serum-free DME in a sterile Eppendorf tube and 3 μL Lipofectamine 2000 and 250 μL OptiMEM in another. Combine the content of the two tubes and leave for 20 min to form complexes. You need three transfections: pGL3

basic + pShag(−), pGL3 control + pShag- and pGL3 control + pShag+.

5. Replace medium on the cells (1.5 mL) and add the transfection mixtures to the cells.
6. Harvest the cells 48 h later.

3.2. RNA Isolation

1. Remove medium, and add 1 ml of cold (on ice) Trizol per well. Shake gently for 5 min at room temperature.
2. Transfer lysate to Eppendorf tubes and add 200 μL of chloroform.
3. Shake vigorously, and leave on ice for 20 min. Shake occasionally during this time.
4. Centrifuge for 30 min at 15,000×*g*.
5. Recover top (aqueous) phase.
6. Add 0.5 mL isopropanol. Leave at −20°C for 30 min.
7. Centrifuge for 30 min at 15,000×*g* and remove supernatant. Watch the pellet (*see* **Note 9**). Add 1 mL 75% ethanol and remove it again. Spin the tube for a few seconds and remove the remaining drops of ethanol.
8. Air dry for 5–10 min with open lid.
9. Redissolve the RNA in 20 μL RNAse-free water or 5 mM EDTA.
10. To quantify the RNA, dilute 1 μL of RNA in 99 μL of TE and measure the absorption at 260 and 280 nm. The 260/280 should be around 1.8–2 for high-quality RNA, lower with increasing amounts of protein contamination (aromatic amino acids absorbs at 260 nm). Calculate your stock RNA concentration in μg/μL as OD_{260} × 40 × 0.1 m/1 μL (light path = 0.1 cm).

3.3. First Strand cDNA Synthesis

1. Prepare master-mix for multiples samples, and add RNA last. Reaction can be scaled up when large amounts of input RNA is used (30 μL for 1.5 μg).

Component	Volume per reaction
5X iScript buffer	3 μL
iScript reverse transcriptase	0.75 μL
Nuclease-free water	1.25 μL
RNA template (up to 0.75 μg total RNA) (*see* **Note 10**)	10 μL
	15 μL

2. Incubate reaction mix in a thermocycler:

25°C	5 min
42°C	30 min
85°C	5 min
4°C	hold

3.4. Real Time PCR

3.4.1. Setting Up the Reaction

1. Cool rotor and centrifuge bucket at –20°C.
2. Prepare stock of master reaction mix on ice. Protect from light until needed. For each reaction you will need

Water	5 μL
Primers (*see* **Note 11**) (stock with each 10 μM)	1 μL
cDNA	2 μL
5× FastStart^{plus} DNA Master SYBR Green I (Roche)	2 μL

   Prepare one sample per cDNA to be analysed (*see* **Note 12**), eight samples for the standard curve (if you run consecutive runs, you only need one standard in each of the following rotors), and two negative controls: one (cDNA-) negative control to which no cDNA is added and one (RNA-) negative control where no RNA was added to the reverse transcription reaction.
3. Insert rotor into centrifuge bucket and put the bucket on ice.
4. Insert the capillaries into the rotor.
5. Pipette 8 μL of the reaction mix into each capillary (*see* **Note 13**).
6. Pipette 2 μL of cDNA into each capillary.
7. Cap the capillaries.
8. Spin the rotor in the special centrifuge which comes with the LightCycler.
9. Start execution of the program using the following parameters:
   slope (°C/s)

Pre -incubation			
Segment 1	Denaturation	95°C, 600 s	20
Amplification (40 cycles)			
Segment 1	Denaturation	95°C, 10 s	20
Segment 2	Anneal	60°C, 5 s	20
Segment 3	Extension	72°C, 15 s	20
Acquisition	(Single mode)		
Melting curve			
Segment 1	95°C	0 s	20
Segment 2	65°C	15 s	20
Segment 3	95°C	0 s	0.1
Cooling			
Segment 1	40°C	30 s	20

10. Repeat 1–9 using the housekeeping primer pair (*see* **Note 14**).

3.4.2. Preparation of the Real Time PCR Standard

1. PCR amplify a 150- to 250-bp fragment of your control gene using your preferred method.
2. Verify that the fragment has the correct size by agarose gel electrophoresis.
3. Subclone fragment:

PCR product (fresh ideally)	3 μL
Water	1 μL
Salt solution	1 μL
TOPO vector	1 μL
Total volume	6 μL

   Mix gently in tube using the pipette tip; incubate up to 30 min at room temperature (5 min recommended by manufacturer).
4. Place reaction tube on ice.
5. Thaw One Shot Competent cells (stored at –80°C) on ice.
6. Add 2 μL of the TOPO reaction to 40 μL of bacteria, and mix gently using pipette tip or flicking.
7. Incubate on ice for 15–30 min.
8. Heat-shock the cells for 45 s at 42°C, and place tube(s) on ice immediately.

9. To each vial (tube), add 250 μL room temperature SOC medium and cap and shake the tube at 37°C for 1 h (using Eppendorf Thermal Mixer).
10. Spread 10–50 μL from each transformation onto LB agar plate (imMedia AMP Blue from Fluka).
11. Incubate overnight at 37°C.
12. Pick white (or light blue) colonies, culture them over night in LB medium using Eppendorf LidBac, with Eppendorf Thermal Mixer at 37°C, 1,200 rpm and isolate plasmid DNA by a method which produces DNA suitable for sequencing (usually by affinity chromatography as for instance the Wizard kit from Promega or the FastPlasmid Mini from Eppendorf).
13. Linearise the plasmids:

Plasmid	2 μL
Restriction buffer	2 μL
EcoRI	1 μL
Water	5 μL

Incubate at 37°C for 1 h.

14. Analyse by 3% agarose gel electrophoresis at 50 V for 20 min (minigel). If the fragments are of the expected size, sequence the plasmid to verify that the expected DNA sequence has been introduced into the plasmid.
15. Upon verification of sequence, prepare a larger amount of linearised plasmid (as above, using 10 – 10 – 2 – 78 μL, respectively).
16. Check for completion of the reaction by analysing 5 μL of the reaction mixture by 3% agarose gel electrophoresis. If the reaction is not complete, extend the incubation and/or add additional restriction enzyme.
17. Inactivate the enzyme by incubating for 10 min at 65°C.
18. Store the digestion at –20°C for future use.
19. Prepare dilution standard:
    - Label eight tubes 1–8. Add 999 μL of water to tube 1 and 900 μL of water to tubes 2–8.
    - Transfer 1 μL of digestion reaction to tube 1. Mix thoroughly using a vortex mixer.
    - Transfer 100 μL to successive tubes, each time ensuring thorough mixing before transfer.

- Use the dilutions as a real time PCR standard: Substitute the 2 μL of cDNA normally used in the real time PCR reaction for 2 μL of the content of tubes 1–8.
- You may use the dilutions for the day, but otherwise make up fresh dilutions each time you want to use the standard.

3.4.3. Analysing the Results

1. For any new amplification, analyse a representative number of reactions by 3% agarose TBE gel electrophoresis. Verify that the resulting band is of the expected size and that other products are not present.
2. Check the melting curve analysis. For a 150–250 bp gene, you would expect a melting temperature of approximately 80–85°C. Curves showing melting temperatures around 60–65°C are almost certainly unspecific products like primer-dimer amplifications.
3. Check your (cDNA−) negative control. There should be no amplification, or the C_T should be much higher than reactions containing cDNA. If you obtain a C_T from the (cDNA−) negative control, consider if the amplification stems from non-specific amplification, and try to repeat the reaction with an additional step (4th segment) after the extension step. The temperature should be above the melting temperature of primer-dimers and below the melting temperature of your specific product.
4. Check your (RNA−) negative control. There should be no amplification. If there is, your chemicals might be contaminated, or your RNA might contain genomic DNA and would need DNAse treatment before being used in the real time PCR reaction.
5. Calculate the expression of your specific mRNA relative to the expression of your housekeeping gene to normalise for variation in the amount of cDNA transferred to the real time PCR reaction or variable efficiencies in reverse transcription.

3.4.4. Example of Results

Using the above method, cDNA samples from Hela cells transfected with pGL3 control plasmid encoding luciferase and either pSHAG− (control) or pSHAG+ (coding for shRNA against the luciferase gene) were analysed by the absolute quantification (*see* **Fig. 14.1**). Representative results are shown: A: standard curve. B: samples in duplicate. The three sets of curves represent (left to right): cells transfected with pGL3 control, pGL3 control/pSHAG+ and pGL3 basic, respectively. C: relative luciferase gene expression normalised against RPL13 which was used as a housekeeping gene. The shRNA induced a fourfold knockdown of luciferase gene expression.

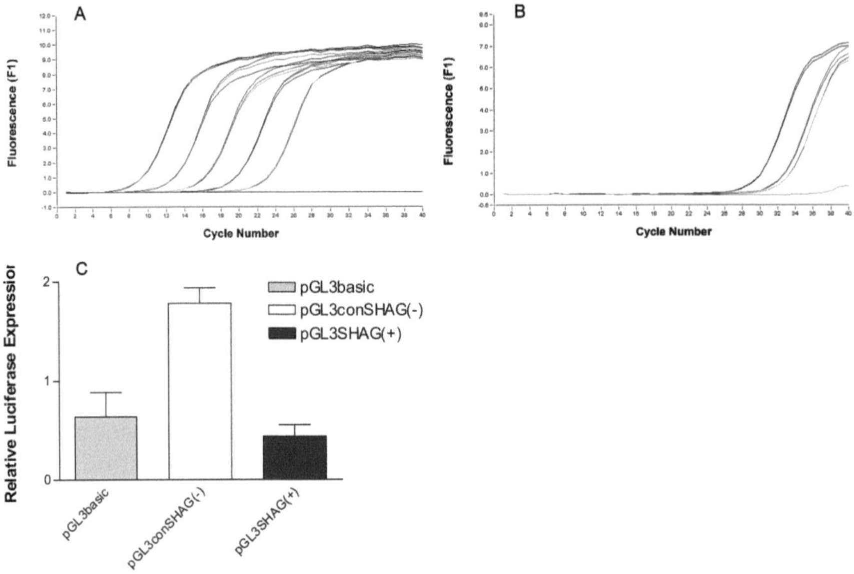

Fig. 14.1. Demonstration of shRNA downregulation of luciferase by real time PCR.

4. Notes

1. Real time PCR cyclers are either air based or block based. The former generate temperature variations in a reaction chamber by venting it by hot and cold air, whereas block cyclers are equipped with a semiconductor Peltier element. The air cyclers ramp the temperature extremely rapidly (20°C/min) which far exceeds the block cyclers due to the high heat capacity of metal; therefore a run is completed in about 40 min in an air cycler, whereas it takes around 2.5 h in a block cycler.

   A much more important feature, however, is the temperature homogeneity of the systems. Large temperature differences of 2–3 degrees are often seen across a thermal block, the periphery usually being cooler than the central area (10), and over time spots of temperature differences may also develop. Further, in some block cyclers, multiple sensors or light paths are used to read the fluorescence from the individual wells. According to the literature and equipment specifications, the sample to sample difference in block cyclers is at the best 0.2 cycles/20 cycles, whereas it is typically 10 times less in the (air based) LightCycler.

   These differences are absolutely crucial for measuring C_T of the real time reaction. To illustrate the performance of the two designs, we have performed identical amplifications in a block cycler (Bio-Rad Opticon II) and in a LightCycler ver. 1.5. The result showed that the sample to sample C_T variation was 5.4% using the block cycler (1.2 cycles in 22.4 cycles, $N = 27$), while the error was less than 0.3% in the Lightcycler (0.06 cycle in 23.2 cycles $N = 17$). Corresponding values for variations in C_T and mRNA quantification have been calculated for selected values in **Table 14.1**, assuming a C_T of 20. Since C_T is transformed by the exponential function in calculating the amount of mRNA, a one cycle difference per 20 cycles in the block cycler will not allow you to detect a twofold difference (100% variation in determining the amount of mRNA), while an inaccuracy typical for the air cycler only introduces a variation of 1.4% at the mRNA level. The numbers indicated were verified for several cyclers of each make.

Table 14.1
Corresponding values of variations in C_T and mRNA assuming 100% amplification efficiency

Variation (cycles)				Variation (mRNA level)			
Min	Max	Difference	%	Min	Max	Difference	%
19.99	20.01	0.02	0.1	94.7	96.0	1.3	1.4
19.9	20.1	0.2	1	89.0	102.2	13.2	14.9
19.5	20.5	1	5	67.4	134.9	67.4	100.0

In our experience, the precision indicated in the technical specifications of real time cyclers might be far from reality. It is therefore imperative that any real time PCR cycler be tested for sample to sample variation prior to its use. The test consists in preparing a master mix (including cDNA) sufficient for 10–20 identical samples, running the samples in various positions in the cycler and comparing the variation. The acceptable variation depends on the intended use, but since an induction or downregulation of a specific mRNA by a factor of 2 is often considered biologically significant, block-based cyclers might not suffice for this purpose. Precision also influence the economy of the real time PCR facility. The lesser the precision, the more replicates are needed to demonstrate a statistically significant difference. In our routine work, we do not run sample replicates on the LightCycler, since the intra-assay variation is essentially non-existent. We do, however, analyse multiple samples from parallel experimental setups. The variation in the mRNA quantifications therefore reflects biological rather than experimental variation.

Precision finally influences the capacity of a real time PCR system. When accuracy is low, multiple replicates must be analysed and therefore occupy more sample space. A 32-sample capacity in a precise instrument can therefore outmatch a 96-well instrument which is less precise, especially if the former operates by air, which completes runs faster than block based instruments.

2. As an alternative, RNeasy kit (Qiagen) has been used.
3. Superscript III (Invitrogen) works well too.
4. TaKaRa SYBR Premix Ex Taq (Lonza) also works very well and is considerably less expensive. Combining the TaKaRa mix with plastic capillaries in 10 μL reaction volume reduces the running cost by half.
5. Standard purification is usually sufficient.
6. Polycarbonate capillaries (Genaxxon) work equally well and are considerably less expensive than glass capillaries.
7. The GeneJet PCR cloning kit from Fermentas works equally well and is far less expensive.
8. We do not use antibiotics in our cell culture. It does not increase the infection rate, and it is convenient since some transfection protocols prescribe the use of antibiotic-free conditions. We change medium every other day as a standard, but cells grow faster if changed daily. Adjust your frequency to your purpose and type of cells you are growing.
9. Nucleic acid pellets are white in front light and reddish in backlight.
10. We routinely synthesise cDNA from 0.75 μg RNA.
11. Primer3: **http://frodo.wi.mit.edu**.
12. Initially you might want to convince yourself of the high precision of the LightCycler by running duplicates or triplicates, but in our experience replicates do not add extra information.
13. Glass capillaries are fragile and break easily. Moreover, they are expensive. We have therefore tested plastic capillaries (Roche and Genaxxon) and found that there is no difference in performance. Plastic capillaries require a specifically manufactured rotor.
14. As housekeeping gene you should choose a gene which shows a constant cellular expression level throughout your experiment. If such a gene has not already been described in the literature for your cell type and for the particular stimulatory conditions you are using, you will need to verify it yourself. Often genes like actin, GAP3DH (glyceralde-

hydephosphate 3-dehydrogenase), porphobilinogen deaminase or cyclophilin are used.

References

1. Paddison, P. J., Caudy, A. A., Bernstein, E., Hannon, G. J., Conklin, D. S. (2002) Short hairpin RNAs (shRNAs) induce sequence-specific silencing in mammalian cells. *Genes Dev* 16, 948–958.
2. Otsuka, Y., Ito, M., Yamaguchi, M., et al. (2002) Enhancement of lipopolysaccharide-stimulated cyclooxygenase-2 mRNA expression and prostaglandin E2 production in gingival fibroblasts from individuals with Down syndrome. *Mech Ageing Dev* 123, 663–674.
3. Fleige, S., Pfaffl, M. W. (2006) RNA integrity and the effect on the real-time qRT-PCR performance. *Mol Aspects Med* 27, 126–139.
4. Gilsbach, R., Kouta, M., Bonisch, H., Bruss, M. (2006) Comparison of in vitro and in vivo reference genes for internal standardization of real-time PCR data. *BioTechniques* 40, 173–177.
5. Nolan, T., Hands, R. E., Bustin, S. A. (2006) Quantification of mRNA using real-time RT-PCR. *Nat Protoc* 1, 1559–1582.
6. Pfaffl, M. W., Georgieva, T. M., Georgiev, I. P., Ontsouka, E., Hageleit, M., Blum, J. W. (2002) Real-time RT-PCR quantification of insulin-like growth factor (IGF)-1, IGF-1 receptor, IGF-2, IGF-2 receptor, insulin receptor, growth hormone receptor, IGF-binding proteins 1, 2 and 3 in the bovine species. *Domest Anim Endocrinol* 22, 91–102.
7. McManus, M. T., Sharp, P. A. (2002) Gene silencing in mammals by small interfering RNAs. *Nat Rev Genet* 3, 737–747.
8. Elbashir, S. M., Harborth, J., Lendeckel, W., Yalcin, A., Weber, K., Tuschl, T. (2001) Duplexes of 21-nucleotide RNAs mediate RNA interference in cultured mammalian cells. *Nature* 411, 494–498.
9. Henriksen, J. R., Lokke, C., Hammero, M., et al. (2007) Comparison of RNAi efficiency mediated by tetracycline-responsive H1 and U6 promoter variants in mammalian cell lines. *Nucleic Acids Res* 35, e67.
10. Editorial, July 2009. **www.Laboratorytalk.com/News/ida/idal09.html**

Chapter 15

Profiling RNA Polymerase II Using the Fast Chromatin Immunoprecipitation Method

Joel Nelson, Oleg Denisenko, and Karol Bomsztyk

Abstract

The traditional method for determining the transcription rate of a gene, nuclear run-on, is time consuming, laborious, and involves the use of high levels of radio-labeled nucleotides. When combined with measurements of mRNA levels, RNA polymerase II (Pol II) chromatin immunoprecipitation (ChIP) is a simpler alternative to determine the transcription rate of genes. Moreover, this approach provides more information about the transcriptional regulation of a gene than nuclear run-on. The power of the ChIP assay is that it gives a researcher the ability to not only detect a specific protein–DNA interaction in vivo, for instance with Pol II, but also to determine the relative density of factors along genes or the entire genome. Though powerful, the conventional ChIP assay is time consuming (involving 2 days or more) and involves labor intensive steps. With Fast ChIP we simplified the assay to greatly reduce the time and labor involved. The improved assay is especially useful for studies which involve many samples, including the probing of multiple transcriptionally related factors simultaneously and/or looking at transcription events over several time points. Using Fast ChIP, 24 sheared chromatin samples can be processed to yield PCR ready DNA in 5 h.

Key words: Chromatin immunoprecipitation, RNA polymerase II, ChIP-chip, tissue ChIP, transcription.

1. Introduction

Microarray and deep sequencing technologies have allowed researchers to determine RNA levels on a genome-wide level. These RNA levels are often used as surrogates to estimate the level of transcription of genes but they also reflect the turnover rate of transcripts and are not accurate indicators of the transcription rate. Traditionally, nuclear run-on has been used to directly

H. Nielsen (ed.), *RNA*, Methods in Molecular Biology 703,
DOI 10.1007/978-1-59745-248-9_15, © Springer Science+Business Media, LLC 2011

interrogate the transcription rate of a gene in vitro, by measuring the level of radioactive nucleotides incorporated into a particular RNA by RNA polymerase II (Pol II) (1, 2). Nuclear run-on is time consuming, uses very large numbers of cells and high levels of radioactivity, and requires highly and rapidly purified nuclei (2). Using chromatin immunoprecipitation (ChIP) to detect Pol II offers a simpler, less time consuming, and less laborious alternative for estimating transcription rates.

Chromatin immunoprecipitation (ChIP) is a powerful method used to study the interactions of proteins (or specific modified forms of proteins) with DNA in vivo (3, 4). ChIP can be used not only to detect an interaction of a protein with a specific region of the genome but also to estimate a relative density of this interaction. The ChIP assay represents a major advancement in the study of chromatin processes and its use has increased dramatically over the last few years.

ChIP to detect Pol II along genes has been used to estimate transcription rates with similar accuracy as nuclear run-on (5). Essentially, Pol II density determined by ChIP, within the body of a gene where it should be undergoing elongation, is proportional to the amount of transcription taking place. For example, the kinetics of induction of the short half-life egr-1 transcript in rat mesangial cells treated with the 12-O-tetradecanoylphorbol-13-acetate, TPA, are very similar to the kinetics of Pol II density at the end of the gene (*see* **Fig. 15.1**).

For the Pol II density to be proportional to transcription it is important to avoid detection of Pol II that may be stalled. For this reason, when determining the transcription rate, the region directly downstream of the transcription start site (within a few hundred basepairs, depending on the resolution of the assay) should be avoided as Pol II is more likely to be stalled there. Alternatively, the density of Pol II which is elongating can be determined using ChIP. The C-terminal domain (CTD) of the largest subunit of elongation competent Pol II is phosphorylated on the YSPTSPS repeat at serine 2 (ser2 CTD). Detection of phospho-ser2 CTD using ChIP, may avoid the detection of stalled transcription and can be used to estimate the transcription rate.

In addition to estimating the transcription level of a gene, Pol II ChIP may provide information about the way it is regulated. Though regulation of transcription has traditionally been thought to occur primarily at the level of promoter recruitment of Pol II, it has become apparent that, for many genes regulation occurs later, at the transition from promoter proximal pausing to productive elongation (6, 7). For instance, the Pol II level at the promoter of several *Drosophila* genes is highly enriched but does not correlate with transcript levels whereas Pol II in the downstream regions does (8). Thus, by comparing the level of Pol II at the transcription start site (TSS) and within the body of a gene

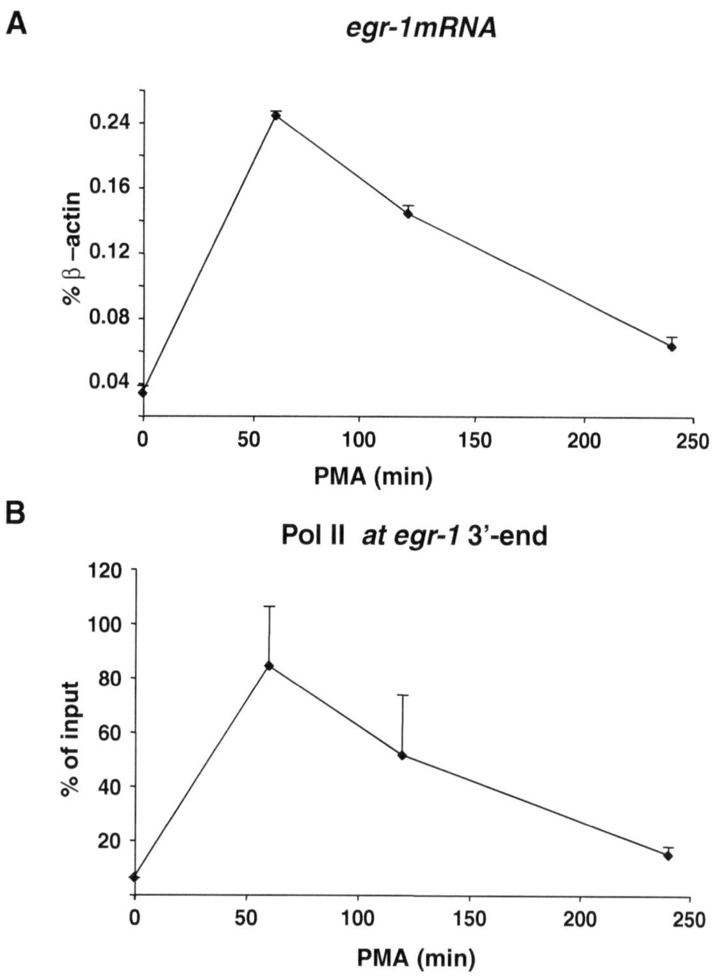

Fig. 15.1. Kinetics of mRNA and Pol II levels for *egr-1*. **a** Trizol extracted total RNA from PMA treated (10^{-7} M) rat mesangial cells (RMCs) was used in RT-qPCR reactions with random hexamers and primers specific to *egr-1* or β-*actin*. Data are expressed as the percentage of the β-*actin* PCR signal and represents the mean of three independent experiments plus or minus SEM. **b** Chromatin from PMA treated (10^{-7} M) RMCs was used in ChIP assay with antibodies to the N-terminus of the Rbp1 subunit of Pol II (Santa Cruz) or non-immune total IgG (mock). ChIPed DNA was used in qPCR with primers to the second exon of *egr-1*. Data are expressed as a ratio of the PCR signal for Pol II over mock and represents the mean of three independent experiments plus or minus SEM.

to the transcript level a researcher can gain insight into the contribution of Pol II recruitment and post recruitment processes to the transcription level of a gene.

ChIP for specific covalent modifications of histone proteins in the nucleosomes along a gene can also provide transcriptionally relevant information. For example, the presence of di- and tri-methylation of lysine 4 of histone H3 and lack of tri-methylation of lysine 27 of H3 at the 5′- ends of genes indicate genes that are transcribed or permissive to transcription (9). Also, the presence of tri-methylation of lysine 36 of histone H3 in the body of a gene is indicative of transcription elongation (10, 11). Thus, using ChIP for probing several factors and their modifications along a gene gives much more information on the transcriptional state of the gene than just the rate of transcription. We developed Fast ChIP to facilitate the tracking of several proteins simultaneously over multiple time points.

The ChIP assay begins with the crosslinking of protein–DNA complexes by the fixation of cells/tissues with formaldehyde (3, 4, 12). After lysis of the fixed cells, the nuclei are disrupted and the chromatin is sheared either by sonication (3, 4) or by digestion with micrococcal nuclease (13). The chromatin fragments, typically between 500 and 1,000 bp in length, are immunoprecipitated using an antibody specific to the protein of interest (3, 4). After reversal of the cross-links, the DNA is isolated and used in one of several detection methods including dot/slot blot (14), PCR or qPCR (15), hybridization to a DNA microarray (ChIP-chip) (16), or sequenced using deep sequencing technology (ChIP-seq) (17). Enrichment of a particular DNA region over other sites where the factor is not expected to bind indicates that the protein interacts with this region.

The traditional ChIP assay, though it has proved to be powerful, is burdensome for studies involving many samples. The slowest step of the traditional ChIP assay is the several hour long reversal of cross-linking (12) and the most laborious step is the DNA cleanup, which involves phenol/chloroform extractions and ethanol precipitation (3). In Fast ChIP, cross-links are reversed during a 10 min incubation at 100°C in the presence of Chelex 100. In addition, since Fast ChIP does not require the addition of sodium bicarbonate/SDS buffer to elute the chromatin from the beads (the high temperature is sufficient), the DNA cleanup step is not necessary. After incubation at 100°C, the DNA-containing supernatant is directly used in PCR (15). Thus, several hours and a great deal of labor in the traditional assay are replaced with 10-min incubation in Fast ChIP.

Another improvement in Fast ChIP is the use of an ultrasonic bath to increase the rate of antibody–chromatin interaction (18). In the traditional ChIP assay, the antibody–chromatin incubation can take anywhere from 1 h to overnight (3). With the use of the ultrasonic bath this incubation is decreased to 15 min (15). The combination of these two improvements in Fast ChIP not only allows the assay to be easily completed in 1 day, starting with sonicated chromatin extracts, but also gives enough time for the products to be analyzed by qPCR in the same day (*see* **Fig. 15.2** for an outline of the method).

Beginning with sheared chromatin, 24 ChIP samples can be easily processed using Fast ChIP to yield PCR ready DNA in 5 h.

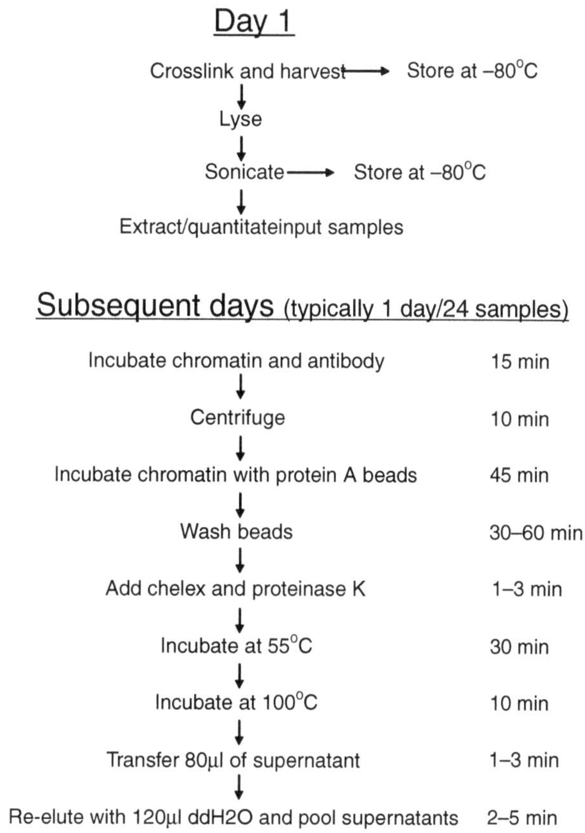

Fig. 15.2. A suggested outline for Fast ChIP. This outline assumes that sonication conditions have not been optimized for the cell/tissue type used and/or that tissue is being used as the chromatin input, requiring quantification of the input DNA to adjust the input chromatin (*see* **Section 3.4**, **Steps 1–14**). If neither of these cases applies, the method can be condensed into 1 day.

We have used Fast ChIP with chromatin from tissue culture (19), mammalian tissues (20), and yeast cultures (21) and it is likely that it is compatible with most other sources of chromatin. Though we have designed Fast ChIP for analysis by PCR or qPCR it has also been used, with the addition of a column cleanup step, in ChIP-chip studies (16). Thus, it is likely that Fast ChIP may be used for most ChIP applications, including ChIP-seq.

2. Materials

2.1. Reagents

1. Chelex-100 (BioRad).
2. Proteinase K.
3. Protein A–Sepharose (Amersham).

4. Formaldehyde 37% (w/v). Very toxic by inhalation, ingestion, and absorption through skin. Use in a fume hood.
5. Phosphate-Buffered Saline (PBS) (Sigma).
6. NP-40 (IGEPAL CA-630) (MP Biomedicals).
7. Triton X-100 (MP Biomedicals).
8. Phenyl methyl sulfonylfluoride (PMSF) (Sigma). Toxic by ingestion, contact, or inhalation of dust; powder reacts with water to form an explosive gas.
9. Leupeptin (Sigma).
10. SYBR Green PCR Master Mix (ABI Biotechnology). OPTIONAL:
11. Sodium molybdate dihydrate ($Na_2MoO_4 \bullet 2H_2O$) (Sigma).
12. β-Glycerophosphate (Sigma).
13. Sodium fluoride (NaF) (Sigma). Very toxic if ingested or if dust is inhaled.
14. Sodium orthovanidate (Na_3VO_4) (Sigma).
15. *p*-Nitrophenylphosphate di(tris) salt (Sigma).

2.2. Buffers and Solutions

1. 1 M Glycine.
2. IP buffer: 150 mM NaCl, 50 mM Tris-HCl (pH 7.8), 5 mM EDTA (pH 8.0), 0.5% (v/v) NP-40, 1% (v/v) Triton X-100.
3. Lysis/sonication buffer: Make fresh before each use. Per 1 mL of IP buffer add the following protease inhibitors: 5 μL PMSF (0.1 M in isopropanol; stored at –20°C; redissolve at RT before pipetting) and 1 μL leupeptin (10 μg/μL; aliquoted and stored at –20°C). Keep on ice. In addition, the following phosphatase inhibitors may be added if required for ChIP with phosphospecific antibodies: 10 μL β-glycerophosphate (1 M; stored at 4°C), 10 μL sodium fluoride (1 M; stored at 4°C; resuspend before pipetting), 10 μL sodium molybdate dihydrate (10 mM; stored at 4°C), 1 μL sodium orthovanadate (100 mM; stored at –20°C) and 13.84 mg *p*-nitrophenylphosphate (stored at 4°C).
4. 10% Chelex-100 in ddH_2O.
5. 20 μg/μL proteinase K in ddH_2O.
6. TE pH 9.0: 10 mM Tris, 1 mM EDTA, bring to pH 9 with 5 M NaOH.

2.3. Equipment

1. Sonicator with microtip (e.g., Misonix Sonicator 3000 with microtip).
2. Refrigerated microcentrifuge.

3. Heat blocks and hot plate (for 55°C incubation and boiling water incubation).
4. Tube rotator or tumbler at 4°C.
5. Set up for quantitative PCR (e.g., ABI 7900 real time PCR system).

   OPTIONAL:
6. Ultrasonic bath (e.g., Branson).

3. Methods

The steps which make Fast ChIP unique from other ChIP methods are the immunoprecipitation and the preparation of PCR ready DNA. Therefore, the following methods for cross-linking, lysis, and sonication are based on what has worked in our lab but are certainly not the only methods compatible with Fast ChIP. If a researcher has previously established his/her own chromatin preparation method for ChIP, they should continue to use this method with Fast ChIP.

To ensure equal loading of different chromatin samples, especially necessary when tissue fragments are used, we suggest extracting total DNA from each chromatin sample (*see* **Section 3.4**, **Steps 1–14**) and measuring the amount of DNA for each by using qPCR. If the samples differ by more than 25%, the amount of chromatin loaded (*see* **Section 3.5**, **Step 1**) should be adjusted based on this measurement. If the amount of chromatin is adjusted remember to use an average of the input samples in calculating the percent of input (*see* **Section 3.6**).

If extracting the input DNA for quantification to adjust chromatin loading for ChIP or if analyzing the chromatin fragmentation when optimizing the sonication conditions, we suggest doing the cross-linking, lysis, and sonication steps on a separate day. If using cells from tissue culture where the sonication conditions have already been optimized, the entire assay can be completed in 1 day with the extraction of input DNA and the ChIP being processed simultaneously.

3.1. Cross-Linking

3.1.1. Tissue Culture

1. Keep in mind that approximately $4 \times 10^5 - 1 \times 10^6$ cells are required per IP sample.
2. Add 40 μL of 37% formaldehyde per mL of tissue culture medium directly to the dish/flask (1.42% final concentration), swirl, and incubate at room temperature for 15 min (*see* **Note 1**).

3. Quench formaldehyde by adding 141 μL of 1 M glycine per mL of medium (125 mM final concentration) and incubate for 5 min at room temperature.
4. Harvest the cells by scraping and centrifuging at 2,000×*g* for 5 min (4°C).
5. Keep cells on ice and wash twice with ice-cold IP buffer. After aspirating the PBS, the cell pellet can be stored at –80°C for at least a year. Otherwise, proceed to the lysis step (*see* **Section 3.2**)

3.1.2. Fresh or Frozen Tissue

This method has been used in our lab for ChIP on both kidney and liver tissue and is likely to be effective in other tissues which have similar numbers of cells per volume of tissue.

1. Place approximately 0.1 cm^2 piece of fresh or frozen (–80°C) tissue in 1 mL of PBS containing 1% formaldehyde at room temperature and quickly mince with forceps into 1–2 mm^2 fragments.
2. Incubate tissue fragments at room temperature for 20 min (*see* **Note 1**).
3. Centrifuge at 2,000–3,000×*g* for 1 min (4°C), and discard the supernatant.
4. Suspend pellet in 1 mL PBS with 125 mM glycine and incubate for 5 min at room temperature.
5. Centrifuge tissue fragments and discard the supernatant.
6. Wash twice with PBS and place on ice for the lysis/sonication step (*see* **Section 3.2**, **Step 4**).

3.1.3. Yeast Culture

For both crosslinking and lysis of yeast cells we use the method described by Kuo and Allis (3) up to the point where whole cell lysate is obtained (*see* **Note 1**). At this point, Fast ChIP can be used, beginning at the sonication steps (*see* **Section 3.3**).

3.2. Lysis

1. Lyse approximately 1 × 10^7 cells by resuspension in 1 mL ice cold lysis/sonication buffer (*see* **Note 2**) and pipetting up and down several times.
2. Collect the insoluble material, which includes the nuclei, by centrifugation at 12,000×*g* for 1 min (4°C), and aspirate the supernatant.
3. Resuspend the pellet once more in 1 mL lysis/sonication buffer. Collect the pellet by centrifugation, and aspirate the supernatant. This washes away residual soluble proteins from the pellet leaving insoluble chromatin, nuclear matrix, and associated cytoskeleton.

4. For tissues, resuspend cross-linked fragments (*see* **Section 3.1.2**, **Step 6**) in 1 mL lysis buffer and proceed to the sonication step (*see* **Section 3.3**).

3.3. Sonication

1. Resuspend pellet again before splitting into two 500-μL fractions. At this point both fractions should be in 1.5-mL microcentrifuge tubes. Both the volume of buffer and the geometry of the tube used for sonication affect fragmentation efficiency, with volumes of 500 μL or less and 1.5-mL microcentrifuge tubes (for tissues 1 mL buffer and 2.0-mL tubes) being optimal.
2. The protocol used for sonication can vary widely and must be optimized for each cell type/tissue and sonicator set-up (*see* **Note 3** *for suggestions*). Optimal fragment sizes are typically between 0.5 and 1 kb as determined by running sonicated chromatin on 1% agarose after DNA extraction and reversal of cross-links (*see* **Section 3.4**, **Steps 1–14**).
3. After sonication, the chromatin should be cleared by centrifugation at 12,000×*g* for 10 min (4°C).
4. Transfer the supernatant to a new tube and aliquot for storage at –80°C (*see* **Note 4**). Save one aliquot of 10 μL for extracting total DNA for the "input" sample.

3.4. Isolation of Total DNA (Input Sample)

Unless otherwise stated, Steps 1–14 can be performed at room temperature.

1. Precipitate DNA from the 10-μL aliquot from **Section 3.3**, **Step 4** for 10 min at room temperature with 30 μL of absolute ethanol.
2. Pellet the DNA by centrifugation at 12,000×*g* for 3 min (4°C).
3. Aspirate or decant the supernatant and add 50 μL of 75% ethanol.
4. Centrifuge at 12,000×*g* for 1 min (4°C) and remove as much of the supernatant as possible.
5. Dry the pellets to completion. They should become transparent after drying.
6. Add 100 μL of 10% Chelex-100 slurry to the dried pellets (*see* **Note 5**).
7. Boil for 10 min and cool by centrifugation for 1 min (4°C).
8. Add 1 μl of 20 μg/μL proteinase K to each tube and vortex. Briefly centrifuge to bring contents to the bottom of the tube.
9. Incubate at 55°C for 30 min, gently resuspending the Chelex once or twice by flicking the tube during the incubation.

10. Boil for 10 min and centrifuge the condensate to the bottom of the tube at 10,000×*g* for 1 min (4°C).
11. Transfer 80 μL of supernatant to a new tube.
12. Add 120 μL of ddH_2O to each tube containing Chelex slurry, vortex, and centrifuge the contents to the bottom of the tube.
13. Remove 120 μL of the supernatant and pool with the 80 μL supernatant from step 11 (*see* **Note 6**).
14. The DNA can be run undiluted on 1% agarose gels. For PCR, use no less than a 1:20 dilution in TE, since some of the remaining contaminants can be inhibitory to PCR.

3.5. Immunoprecipitation

1. For each IP sample dilute the equivalent of 1×10^6 cells of chromatin to 200 μL with ice-cold lysis/sonication buffer (*see* **Note** 7).
2. Add specific or mock antibodies to each sample and mix by inverting (*see* **Notes 8** and **9**).
3. Turn the ultrasonic bath on and float samples in bath for 15 min at 4°C (*see* **Note 10**).
4. Clear the solution by centrifugation at 12,000×*g* for 10 min (4°C). *This step is essential* to remove non-specific insoluble chromatin aggregates which may contaminate the final product.
5. While the chromatin and antibodies are incubating transfer approximately 20 μL per IP sample of protein A agarose slurry to a clean tube (*see* **Notes 5** and **11**). Wash 1–3 times with IP buffer to remove ethanol.
6. Resuspend the beads in 180 μL IP buffer for every 20 μL of beads (*see* **Note 12**). Dispense 200 μL of the diluted slurry to new tubes, one tube for each IP sample (*see* **Note 5**). Centrifuge and aspirate the buffer. Visually inspect the tubes to make sure that each one has the same amount of beads.
7. Transfer no more than the top 90% of each cleared chromatin sample from Step 4 (avoiding the pellet at the bottom of the tube) to the tubes with the beads.
8. Rotate tubes at 4°C for 45 min with a rotating platform or tumbler. The rotation should be fast enough to keep the beads suspended.
9. Centrifuge the tubes at 10,000×*g* for 1 min (4°C) and aspirate the supernatant.
10. Wash the beads (resuspend with buffer, centrifuge, and aspirate the supernatant) five times with 1 mL ice-cold IP

buffer. After the last wash remove as much supernatant as possible without removing beads.

11. Add 100 μL of 10% Chelex-100 slurry to the washed beads (*see* **Note 5**).
12. Add 1 μL of 20 μg/μL proteinase K to each tube and vortex. Briefly centrifuge contents to the bottom of the tube.
13. Incubate at 55°C for 30 min. Gently resuspend beads and Chelex once or twice during the incubation.
14. Boil samples for 10 min.
15. Centrifuge samples at 10,000×*g* for 1 min (4°C) to cool samples and bring condensate to the bottom of the tube.
16. Transfer 80 μL of supernatant to new tubes.
17. Add 120 μL ddH_2O to each tube containing Chelex/protein A beads slurry, vortex, and centrifuge contents to the bottom of the tube (*see* **Note 6**).
18. Remove 120 μL of the supernatant and pool with the 80 μL supernatant from Step 16.
19. The PCR ready DNA can be stored at −20°C and repeatedly thawed and frozen over several months without loss of PCR signal.

3.6. PCR and Calculation of Enrichment

We use 2.35 μL of IP DNA or diluted input DNA in 5 μL reactions with 0.15 μL of primer pair (each primer at 10 μM), and 2.5 μL of master mix (SensiMix containing SYBR green and ROX). The reactions are run in triplicate in 384-well PCR plates on the ABI 7900 for 40 cycles with the default two-step method. Data are acquired and analyzed using SDS 2.2.1 software. The threshold is set manually and Ct-values are imported to Excel for calculations.

We express enrichment of the immunoprecipitated region of the genome as the percent of input DNA. To eliminate the differences in amplification efficiencies of different primers, relative amounts of DNA for the IP, mock, and input samples are calculated for each primer using a standard curve. The standard curve consists of serial dilutions of total DNA from the same cell type or tissue used in the experiment and is run each time a primer pair is used. We suggest making up a large amount of each dilution in TE buffer and aliquoting them for multiple uses so that the standard curve can be run repeatedly without error due to degradation of the DNA.

PCR-primer efficiency curves are fit to the natural log of concentration versus Ct for each dilution using an r-squared best fit. The relative amount for each ChIP and input DNA sample is calculated from their respective averaged CT values using the formula

$$[DNA] = \frac{b \times e^{m \times AvgCT}}{Dilution} \qquad [1]$$

where b and m are the curve fit parameters from the primer calibration curve that is generated for each PCR experiment. *Dilution* is the cumulative dilution of ChIP DNA as compared to the input DNA sample. Final results are expressed as either a fraction or percent of input using the following equation.

$$\% \text{ of input} = \frac{[DNA_{sample}] - [DNA_{mock}]}{[DNA_{input}]}.100 \qquad [2]$$

where DNA concentrations were computed from [1], DNA_{sample}, ChIP DNA sample; DNA_{mock}, IgG mock IP control, and DNA_{input}; input DNA used in ChIP. Remember that if the chromatin amount used in ChIP was adjusted based on measurement of the input samples (*see* **Section 3.4, Step 14**), then DNA_{input} should be an average of the input for all the samples.

4. Analysis

The enrichment (percent of input) determined using the above calculations is, in itself, not a meaningful number. To determine the significance of the enrichment at a region of interest, this region must be compared to another region where the factor of interest is not expected to bind (the negative control region). The enrichment at the negative control region gives a baseline which is assumed to represent zero binding and the significance of the enrichment at the region of interest depends on the signal at this region being significantly above the baseline.

Another means of determining the significance of enrichment of a factor at a particular locus is to compare that enrichment in cells where the factor is present to those where the factor has been knocked out. The enrichment in the knockout cells represents the zero binding baseline and enrichment is significant at the region of interest only if it is above this baseline.

5. Notes

1. The cross-linking times and formaldehyde concentrations used here are suggestions and may need to be optimized depending on the cell/tissue type used as well as on the

factor being immunoprecipitated. Longer cross-linking times or higher formaldehyde concentrations can improve the immunoprecipitation of some factors by increasing the number of cross-links between the factor and the DNA. Conversely, longer cross-linking times can be detrimental for pull down of some factors because epitopes in the factor may be masked by the cross-linking. At the upper range of fixing, tissue or cells may become resistant to shearing of the chromatin by sonication.

2. Both PMSF and leupeptin have short half-lives in aqueous solutions at room temperature. It is important to prepare the lysis/sonication buffer fresh and keep it on ice before use.

3. Suggestions for sonication:
   a. Sonication can cause heating of the sample so the tube should be immersed in an ice-water bath during sonication.
   b. Foaming can occur if the microtip gets too close to the surface of the sample during sonication. The tip should remain no more than a few millimeters from the bottom of the tube during sonication. If foaming does occur, stop sonication and wait till the majority of bubbles rise to the surface before continuing sonication.
   c. The two variables to optimize are the total amount of sonication time and the power output of the sonicator.
   d. To avoid excessive heating, the total sonication time should be broken up into rounds of 10–20 s each, with at least 2 min of rest on ice between each round. In addition, sonication is more efficient if each round is broken up into ~1 s pulses rather than continuous sonication, since the power of sonication decreases gradually after the beginning of each pulse.
   e. The higher the power output of the sonicator the faster the fragmentation of the chromatin and the more heating the sample is exposed to. Start with a power output 50% or less of the total power output for the sonicator and increase as needed such that the samples are not overheated by the end of each round of sonication but the amount of time required for sonication is not prohibitive considering the number of samples to be sonicated.
   f. Other factors which affect sonication efficiency are the cell concentration and extent of cross-linking of the chromatin. Diluting the chromatin and/or reducing the

cross-linking time or concentration of formaldehyde can increase sonication efficiency.

4. The chromatin preparations can be stored at –80°C for months without loss of pull down efficiency; however, repeated thawing and freezing can reduce this efficiency. To avoid frequent thawing of chromatin, make aliquots just large enough for each experiment you are planning.
5. Keep the slurry in suspension while pipetting and use a tip with the end cut off to avoid clogging.
6. 17 mM Tris, 1.7 mM EDTA (final pH 9.0) may be substituted here to improve DNA stability over time. Check to make sure that PCR amplification is not negatively affected by the use of this buffer.
7. The amount of chromatin used here is a suggested starting point. In our experience in some cases, using smaller amounts of chromatin can increase the difference between the IP and mock signals by decreasing the background without significantly decreasing the signal from the specific pull down.
8. For some antibodies the amount required may need to be determined empirically, however 1–2 μg per sample is sufficient for many antibodies. For a mock IP (control for non-specific binding) either the same antibody blocked with saturating amounts of an epitope-specific peptide, a pre-immune IgG, or no antibody can be used.
9. Those antibodies that have worked well for us in Fast ChIP include those for the Pol II CTD (4H8), which can be purchased from Abcam (ab5408) or Santa Cruz (sc-47701), the Pol II n-terminus from Santa Cruz (sc-899), trimethyl lysine 4 of histone H3 from Abcam (ab8580) or Novus Biologicals (NB500-173), trimethyl lysine 27 of H3 from Abcam (ab6002), and trimethyl lysine 36 from Abcam (ab9050). Antibodies for the serine 2 phosphorylated CTD of Pol II have also been used in ChIP, including clone H5 from Covance (MMS-129R) and an antibody generated in David Bentley's lab (22); however, we have not used these with Fast ChIP.
10. If an ultrasonic bath is not available, samples may need to be incubated for 1–2 h at 4°C depending on the antibody (some antibodies may require longer times up to overnight incubations; this should be determined empirically).
11. Non-specific binding of the chromatin to the protein A beads accounts for the majority of the mock signal. Therefore, reducing the amount of beads used may reduce the mock signal (improving the IP – mock difference). The 20 μL suggested here is far above what is necessary to bind

the antibodies and this amount is only used as it is convenient to visualize the pellet while aspirating the washes.

12. To reduce non-specific binding to the protein A beads, blocking reagents may be used both to block the beads prior to the IP and during the IP. To make blocking buffers, add 5% BSA (fraction V) and 100 μg/mL sheared salmon sperm DNA (ssDNA) to aliquots of lysis/sonication buffer and IP buffer. The chromatin should be diluted in lysis sonication buffer with BSA and ssDNA before incubating with antibody. Also the beads should be pre-incubated with 200 μL of IP buffer with BSA and ssDNA while resuspended on a rotating platform for 0.5 h. This buffer should be aspirated off the beads before transferring the chromatin/antibody mix.

Acknowledgment

We thank members of KB lab for valuable discussions of the method. This work was supported by NIH DK45978 and GM45134 (K.B.).

References

1. Gariglio, P., Bellard, M., Chambon, P. (1981) Clustering of RNA polymerase-B molecules in the 5' moiety of the adult beta-globin gene of hen erythrocytes. *Nucleic Acids Res* 9, 2589–2598.
2. Srivastava, R. A., Schonfeld, G. (1998) Measurements of rate of transcription in isolated nuclei by nuclear "run-off" assay. *Methods Mol Biol* 86, 201–207.
3. Kuo, M. H., Allis, C. D. (1999) In vivo cross-linking and immunoprecipitation for studying dynamic Protein:DNA associations in a chromatin environment. *Methods* 19, 425–433.
4. Orlando, V., Strutt, H., Paro, R. (1997) Analysis of chromatin structure by in vivo formaldehyde cross-linking. *Methods* 11, 205–214.
5. Sandoval, J., Rodriguez, J. L., Tur, G., Serviddio, G., Pereda, J., Boukaba, A., Sastre, J., Torres, L., Franco, L., Lopez-Rodas, G. (2004) RNAPol-ChIP: a novel application of chromatin immunoprecipitation to the analysis of real-time gene transcription. *Nucleic Acids Res* 32, e88.
6. Core, L. J., Lis, J. T. (2008) Transcription regulation through promoter-proximal pausing of RNA polymerase II. *Science* 319, 1791–1792.
7. Saunders, A., Core, L. J., Lis, J. T. (2006) Breaking barriers to transcription elongation. *Nat Rev Mol Cell Biol* 7, 557–567.
8. Muse, G. W., Gilchrist, D. A., Nechaev, S., Shah, R., Parker, J. S., Grissom, S. F., Zeitlinger, J., Adelman, K. (2007) RNA polymerase is poised for activation across the genome. *Nat Genet* 39, 1507–1511.
9. Barski, A., Cuddapah, S., Cui, K., Roh, T. Y., Schones, D. E., Wang, Z., Wei, G., Chepelev, I., Zhao, K. (2007) High-resolution profiling of histone methylations in the human genome. *Cell* 129, 823–837.
10. Kouzarides, T. (2007) Chromatin modifications and their function. *Cell* 128, 693–705.
11. Shilatifard, A. (2004) Transcriptional elongation control by RNA polymerase II: a new frontier. *Biochim Biophys Acta* 1677, 79–86.
12. Solomon, M. J., Varshavsky, A. (1985) Formaldehyde-mediated DNA-protein crosslinking: a probe for in vivo chromatin structures. *Proc Natl Acad Sci USA* 82, 6470–6474.

13. Thorne, A. W., Myers, F. A., Hebbes, T. R. (2004) Native chromatin immunoprecipitation. *Methods Mol Biol* 287, 21–44.
14. Solomon, M. J., Larsen, P. L., Varshavsky, A. (1988) Mapping protein-DNA interactions in vivo with formaldehyde: evidence that histone H4 is retained on a highly transcribed gene. *Cell* 53, 937–947.
15. Nelson, J. D., Denisenko, O., Sova, P., Bomsztyk, K. (2006) Fast chromatin immunoprecipitation assay. *Nucleic Acids Res* 34, e2.
16. Huebert, D. J., Kamal, M., O'Donovan, A., Bernstein, B. E. (2006) Genome-wide analysis of histone modifications by ChIP-on-chip. *Methods* 40, 365–369.
17. Johnson, D. S., Mortazavi, A., Myers, R. M., Wold, B. (2007) Genome-wide mapping of in vivo protein-DNA interactions. *Science* 316, 1497–1502.
18. Chen, R., Weng, L., Sizto, N. C., Osorio, B., Hsu, C. J., Rodgers, R., Litman, D. J. (1984) Ultrasound-accelerated immunoassay, as exemplified by enzyme immunoassay of choriogonadotropin. *Clin Chem* 30, 1446–1451.
19. Nelson, J. D., Flanagin, S., Kawata, Y., Denisenko, O., Bomsztyk, K. (2008) Transcription of laminin {gamma}1 chain gene in rat mesangial cells: constitutive and inducible RNA polymerase II recruitment and chromatin states. *Am J Physiol Renal Physiol* 294, F525–F533.
20. Zager, R. A., Johnson, A. C., Naito, M., Bomsztyk, K. (2008) Maleate nephrotoxicity: mechanisms of injury and correlates with ischemic/hypoxic tubular cell death. *Am J Physiol Renal Physiol* 294, F187–F197.
21. Denisenko, O., Bomsztyk, K. (2008) Epistatic interaction between the K-homology domain protein HEK2 and SIR1 at HMR and telomeres in yeast. *J Mol Biol* 375, 1178–1187.
22. Glover-Cutter, K., et al. (2008) RNA polymerase II pauses and associates with pre-mRNA processing factors at both ends of genes. *Nat Struct Mol Biol* 15(1), 71.

Part II

RNP Methods

Chapter 16

The Post-transcriptional Operon

Scott A. Tenenbaum, Jan Christiansen, and Henrik Nielsen

Abstract

A post-transcriptional operon is a set of monocistronic mRNAs encoding functionally related proteins that are co-regulated by a group of RNA-binding proteins and/or small non-coding RNAs so that protein expression is coordinated at the post-transcriptional level. The post-transcriptional operon model (PTO) is used to describe data from an assortment of methods (e.g. RIP-Chip, CLIP-Chip, miRNA profiling, ribosome profiling) that globally address the functionality of mRNA. Several examples of post-transcriptional operons have been documented in the literature and demonstrate the usefulness of the model in identifying new participants in cellular pathways as well as in deepening our understanding of cellular responses.

Key words: mRNA, RNA-binding proteins, miRNA, coordination, regulation.

1. Introduction

Gene expression is regulated at multiple levels. This chapter explores a layer of regulation based on a model called the post-transcriptional operon, a ribonucleoprotein infrastructure in which multiple mRNAs are coordinately regulated by RNA-binding proteins and/or small non-coding RNAs. Post-transcriptional operons are analogous to classical prokaryotic, polycistronic DNA operons in that they coordinate the expression of a subset of functionally related genes. However, they are comprised of combinatorial RNA and protein complexes operating at the post-transcriptional level. The post-transcriptional operon model or PTO was originally proposed by Jack Keene and Scott Tenenbaum (1–3) and has more recently been reviewed (4, 5). In this chapter, we initially provide an outline of many of the ways in which coordination of gene expression can be exerted. We then present the post-transcriptional operon model

H. Nielsen (ed.), *RNA*, Methods in Molecular Biology 703,
DOI 10.1007/978-1-59745-248-9_16,

and describe key observations that emphasize the importance of post-transcriptional regulation, especially in multicellular organisms. Finally, we introduce many of the methods that are typically being used to study this level of regulation and present several examples of data in support of the PTO model.

2. Coordination of Gene Expression

Coordination of gene expression is a fundamental feature of cellular life. It is a requirement for cells to grow proportionally, divide, maintain cellular homeostasis, and respond adequately to external stimuli. Coordination is obtained in multiple ways and at all levels of gene expression. All known mechanisms involved in the processing of genetic information require some regulation of at least a fraction of the regulated genes, which will be affected to differing amounts, leading to coordination in gene expression. The first level that plays a role in the coordination of gene expression is the physical linkage of some genes dictated by the architecture of the genome. Here, the expression of some protein coding sequences is coordinated by being transcribed from a single promoter into a polycistronic transcript. This is characteristic of the prokaryotic operon organization of genes but is infrequent in eukaryotes due to the cap-dependent translation of the mRNA. However, in eukaryotes, the structure of the genome may contribute to coordination by transcriptional interference between transcription units. This is known from yeast in which cryptic unstable transcripts regulate the expression of several genes (6). Transcriptional interference may also be relevant in the many sense/antisense pairs of transcripts that have been observed in analyses of the human transcriptome (7). Other examples of coordination mediated by the physical linkage of genes include the globin gene clusters and certain transcription factor genes (Hox genes) that include a homeobox DNA-binding sequence. The latter are found in a cluster in which the order of the genes is directly correlated to the regions in the body plan they affect as well as the timing of expression.

The chromatin structure of the genome provides coordination of gene expression at least at two levels. DNA found in the heterochromatin fraction is largely transcriptionally inactive and the formation and recession of facultative heterochromatin will coordinately affect neighbouring genes. Within the euchromatin fraction, the detailed chromatin structure governs the access of transcription factors and thus the transcription of the genes. Here too, the physical linkage of genes may influence their expression. There are several examples of "gene expression neighbourhoods" where genes are expressed in similar patterns due to their physical linkage (8).

Transcription initiation is one of the key regulatory steps in gene expression and cell- and tissue-specific transcription factors allow the coordinated expression of large sets of genes. Transcription factors recognize short sequences and act in a combinatorial fashion. Furthermore, these factors are frequently modified by post-translational modifications allowing modulation and fine-tuning of their regulatory function. More recent discoveries of promoter-associated short RNA transcripts in both the sense and antisense orientation at RNA pol II promoters may provide an additional aspect to this step of regulation of gene expression (9, 10).

The elongating polymerase is associated with numerous factors that participate in or regulate co- or post-transcriptional processing of the transcript. Splicing and polyadenylation are highly regulated and coordinated events that are controlled by some of these factors (11, 12). Thus, transcription-associated processes provide another level at which coordination of expression can take place. The use of alternative upstream promoters as well as alternative splicing and polyadenylation leads to formation of mRNA isoforms. Widespread use of alternative transcription start sites located >100 kb upstream was revealed by the ENCODE pilot project that systematically analysed 1% of the human genome (13). These transcripts result in the incorporation of exons derived from transcription units that were previously annotated as independent as well as exons derived from what was thought to be intergenic regions. Conventional alternative splicing has been estimated to occur in primary transcripts of almost all human genes with an average of approximately seven alternative splicing events per multiexon gene (14, 15). This is one of the main contributions to diversity of gene expression in humans. On top of this, alternative polyadenylation is increasingly being observed, in particular during development. In general, mouse genes tend to be expressed with longer 3′ UTRs as embryonic development progresses by a mechanism that involves suppression of short type polyA sites (16). Conversely, short type alternative polyA sites are favoured at late stages of differentiation following stimulation of resting T cells (17).

Once formed, the fate of an mRNA is governed by the proteins with which it is associated (18). Proteins associate with the mRNA during transcription and are exchanged during the life of the mRNA. A prominent example of this is the exchange of nuclear versions of cap- and polyA tail-binding proteins with the corresponding cytoplasmic versions upon export to the cytoplasm. Following export many, if not all, mRNAs are localized within the cytoplasm. mRNAs that encode membrane-associated proteins or proteins that are secreted are specifically localized to the cytoplasmic side of the endoplasmic reticulum where they are translated. Fractionation studies have shown that 44% of mRNAs in human cells belong to this class (19). Around 100 examples of individual mRNA species that are specifically located within the cytoplasm are known in diverse organisms (20) and more than 400 mRNAs have been identified in mammalian neuronal dendrites (21).

The overall level of translation is coordinately regulated, at least in part, by the availability of initiation factor eIF2 and by the trimeric eIF4F discriminating between capped and non-capped mRNAs. The non-capped mRNAs are characterized by the presence of internal ribosomal entry sites (IRES) or other 5′ UTR elements (22) that mediate alternative translation initiation in situations of cellular stress or during mitosis when the cap-binding eIF-4E subunit is absent. Besides regulation of the overall level of translation, regulation of individual mRNAs and groups of mRNAs take place by association of protein factors and microRNAs. Some protein factors operate in an all or none fashion but others are combinatorial. MicroRNAs regulate more than 30% of human genes at the translational level. Individual microRNAs have many targets and conversely, each mRNA exhibits target sequences for several microRNAs. Thereby a combinatorial type of coordinated regulation that serves to fine-tune gene expression is established. Recent advances in ribosome profiling using high-throughput sequencing (23) are likely to give a more detailed picture of coordinated regulation than previous methods. Decay of the mRNA is the ultimate alternative to translation and is not only an important step in the regulation of individual mRNAs but also one that displays many examples of coordinated regulation, in particular of mRNA decay initiated by Argonaute-2 induced endonucleolytic cleavage in the 3′ UTR or AU-rich element dependent de-adenylation.

Finally, at the protein level there are several crucial mechanisms for coordinated regulation of sets of proteins with related functions. Proteins may be co-localized because they share a common sorting signal. They may be regulated in a concerted fashion because they are substrates of the same kinases and/or phosphatases or other modification enzymes that influence their biological properties. Eventually, they may be degraded simultaneously by being targeted by the same ubiquitin ligase that marks the proteins for degradation by the proteasome.

Clearly, coordination of gene expression operates at several levels that are intimately related. This interrelatedness follows from simple observations, including that every transcription regulator must be translated and every translation factor must be transcribed. It is also clear that biological processes are too complex to be studied exhaustively by focusing at single or individual levels of regulation. For example, UV radiation of eukaryotic cells activate pathways that include signal transduction, DNA synthesis, transcriptional regulation of both initiation and elongation rate, alternative splicing, translational repression, and protein turnover. Some of these pathways are independent, while others are interdependent, and presently there are no means of studying the response in its entirety. Thus, the key question is whether new concepts, such as the PTO model provides a simplifying principle that can be applied to understand some of the observations made by experiments and to provide testable predictions.

3. The Post-transcriptional Operon Model of Co-regulated Gene Expression

The post-transcriptional operon model is based on the existence of distinct mRNP (messenger ribonucleoprotein) particles as a central regulatory entity of the cell. These are relatively stable in their protein content but are dynamic and can change upon cellular perturbation. A post-transcriptional *operon* has been described as a set of monocistronic mRNAs that encode functionally related proteins (analogous to the classical polycistronic operon of bacteria) that are regulated by a group of RNA-binding proteins and small non-coding RNAs so that protein expression is coordinated at the post-transcriptional level. Thus, an RNA operon comprises many different mRNA species, any of which can be distributed between multiple post-transcriptional operons. The term *RNA regulon* is sometimes used as a further extension of the parallel to regulation at the DNA level in bacteria. A DNA regulon is a set of operons that are co-regulated by a protein. One example is the catabolite activator protein (CAP; also known as cAMP receptor protein, CRP) that co-regulates the transcription of several polycistronic mRNAs. By analogy, an RNA regulon represents sets of monocistronic mRNAs in eukaryotes that are co-regulated by a protein factor or a family of RNA-binding proteins.

The post-transcriptional operon model is applied to events that take place both during transcription (e.g. splicing and polyadenylation) and after the formation of the mature mRNA. These events are clearly separable and the inclusion of both may seem confusing. However, this is justified by the facts that co- and post-transcriptional events in many cases are coordinated and that several RNA-binding proteins, that are involved in processing (e.g. splicing factors such as U2AF, SR, and PTB), are also found to be preferentially associated with different species of mature mRNAs (4).

A term that is often used in conjunction with the post-transcriptional operon model is the *ribonome*. This term refers to the total cellular complement of regulated RNAs and their regulatory factors functioning dynamically in time and space within RNP complexes. The ribonome is highly dynamic and responsive to cellular perturbations and is distinct from the transcriptome, which is comprised of the steady-state RNA complement of the expressed genes from the genome.

Several methods have been specifically developed to facilitate the study of post-transcriptional gene expression including RIP-Chip (*R*NA-binding protein *I*mmuno*p*recipitation-micro*chip* profiling, also called *ribonomics*), in which an antibody specific to the targeted RBP is used to immunoprecipitate endogenously formed mRNP complexes followed by quantitative analysis by microarrays (24, 25). In variations of the technique, SAGE (serial analysis of gene expression) or high-throughput sequencing (also known as "deep sequencing" or RNA-seq) is used to characterize the mRNAs. The related CLIP-Chip (cross-linking and immunoprecipitation) uses in vivo cross-linking prior to the immunoprecipitation step in order to faithfully represent interactions taking place within cells. A related experimental approach uses tagged mRNAs to purify mRNPs followed by characterization of the mRNA as well as the protein complement of the isolated material. Given that microRNAs function in a manner equivalent to RNA-binding proteins in the post-transcriptional operon hypothesis, microRNA profiling and target characterization are frequently used with the assortment of methods that are relevant in this context. Finally, ribosome profiling and other methods that globally address functionality of mRNAs can be used together with RIP-Chip and related methods to obtain a more detailed description of the post-transcriptional operon in question.

3.1. Basic Observations in Support of the Importance of Post-transcriptional Regulation and the Post-transcriptional Operon Model

There are a number of basic observations that indirectly support the idea of the post-transcriptional operon. First, the transcriptome and the proteome are poorly correlated. This has been shown in several studies of the expression of the yeast genome, and is probably even more pronounced in the expression of the mammalian genome. In a study of two mouse haematopoietic cell lines representing distinct stages of myeloid differentiation, as well as experimentally perturbed livers, it was observed that the differential expression of mRNA (up or down) can only account for 40% at most of the variation of protein expression (26). These studies are supported by numerous observations of the expression of individual genes that all point to an important role of post-transcriptional regulation. Second, pre-mRNAs are generally processed into several isoforms by alternative splicing and other mechanisms. Differential expression of these must necessarily be based on post-transcriptional events. In many cases isoforms are produced by alternative polyadenylation to encode identical proteins expressed from mRNAs that differ in their UTRs resulting in differential regulation by miRNAs and RNA-binding proteins targeting the 3′-UTR (16, 17). Third, RNA-binding proteins are very diverse and function in a modular and cooperative fashion that promotes target binding and facilitates a diversity of biological functions (27). The number of RNA interacting proteins in eukaryotes has been suggested to be approximately 2,500, almost twice the number of DNA-binding proteins (2, 28). Fourth, there are many examples of RNA-binding proteins that regulate their own expression as well as the expression of other RNA-binding proteins (4, 5). This is a property that would be expected if post-transcriptional operons constitute a major regulatory layer of gene expression. Taken together, post-transcriptional regulation is mandatory for proper gene expression. Regulation of mRNAs by RNA-binding proteins and small non-coding RNAs is the most obvious way to obtain this, but other models may also be relevant.

3.2. Examples of Post-transcriptional Operons

Recent reviews discuss numerous examples (4, 5, 20) of data in support of post-transcriptional operons based predominantly on RIP-Chip and CLIP-Chip experiments and a list comprising more than 30 examples of RIP-Chip analysis of specific RNA-binding proteins can be found in (20). These examples include RNA operons encoding mRNAs involved in mRNA splicing, polyadenylation, export, translation, and decay. Moreover, additional examples include studies based on tagged RNA-binding proteins. The current view is that many RNA operons operate predominantly at the level of translation and mRNA stability and several examples focusing on these functions will be detailed below.

A clear cut example of a post-transcriptional operon derives from a study of Pumilio RNA-binding proteins (PUFs) in yeast (28). These proteins repress translation of their target mRNAs by binding to elements located in the 3′ UTR. Five different members were each tagged and the mRNAs associated with individual PUFs were characterized. Distinct groups of 40–220 different mRNAs with striking similarities in terms of functions and subcellular localization of the proteins they encode were found associated with each PUF member. Among them, Puf3p bound to a subset of 154 mRNAs, 135 of which could be assigned to mitochondrial proteins. The involvement of Puf3p in mitochondrial biogenesis has subsequently been the focus of a more detailed study that found even more targets for Puf3p and corroborated the physical localization of mRNAs in the vicinity of mitochondria (29). Thus, it can be concluded that Puf3p defines a post-transcriptional operon involved in synthesis and motility of mitochondria.

Several examples of post-transcriptional operons are related to mRNA turnover. Half-lives of mRNAs in humans vary tremendously and decay of mRNA is an important regulatory step. An intuitive example is that of the non-polyadenylated mRNAs encoding histones. Most histone mRNAs are required for synthesis of the major histones (H1, H2A, H2B, H3, and H4) during the S-phase of the cell cycle. The mRNAs are regulated at several steps, including decay, by the attachment of a stem-loop binding protein (SLBP) to a conserved hairpin in the 3′ UTR of the mRNAs. RIP-Chip analysis of SLBP targets identified approximately 30 such mRNAs that might function as an RNA operon for coordination of histone production during replication (30). A broader example in the same vein is the regulation of mRNAs that exhibit coordinated decay by the presence of AU-rich elements (AREs) in their 3′ UTRs. These elements of 50–150 nt are found in 5–8% of all mRNAs and are typically associated with mRNAs with a short half-life that has to be tightly regulated as part of a transient cellular response (31). AREs are bound by different kinds of ARE-binding proteins. As an example, it has been shown that distinct classes of mRNAs that encode regulators of the immune response, such as cytokines and chemokines, have similar half-lives that can be altered in a concerted manner. RIP-Chip and other types of analyses have shown that this can be mediated by the binding of ARE-binding proteins such as ELAV/Hu (1) and tristetraproline (TTP). RNA operons that are involved in decay seem to be relatively common and examples include the decay of mRNAs involved in ribosome biogenesis and iron metabolism (20).

4. Conclusions

The post-transcriptional operon model appears to be valuable at our current conceptual understanding of molecular biology. Importantly, it is associated with specific methods that provide a new type of information that can be used to reveal unexpected participants in cellular pathways and deepen our understanding of cellular responses. Furthermore, recent advances in high-throughput sequencing will make these methods even more attractive and contribute to the generation of a knowledge platform on post-transcriptional regulation that will be helpful in the discipline of systems biology.

References

1. Tenenbaum, S. A., Carson, C. C., Lager, P. J., Keene, J. D. (2000) Identifying mRNA subsets in messenger ribonucleoprotein complexes by using cDNA arrays. *Proc Natl Acad Sci USA* 97, 14085–14090.
2. Keene, J. D. (2001) Ribonucleoprotein infrastructure regulating the flow of genetic information between the genome and the proteome. *Proc Natl Acad Sci USA* 98, 7018–7024.
3. Keene, J. D., Tenenbaum, S. A. (2002) Eukaryotic mRNPs may represent post-transcriptional operons. *Mol Cell* 9, 1161–1167.
4. Keene, J. D. (2007) RNA regulons: coordination of post-transcriptional events. *Nat Rev Genet* 8, 533–543.
5. Mansfield, K. D., Keene, J. D. (2009) The ribonome: a dominant force in co-ordinating gene expression. *Biol Cell* 101, 169–181.
6. Neil, H., Malabat, C., ubenton-Carafa, Y., Xu, Z., Steinmetz, L. M., Jacquier, A. (2009) Widespread bidirectional promoters are the

major source of cryptic transcripts in yeast. *Nature* 457, 1038–1042.
7. He, Y., Vogelstein, B., Velculescu, V. E., Papadopoulos, N., Kinzler, K. W. (2008) The antisense transcriptomes of human cells. *Science* 322, 1855–1857.
8. Oliver, B., Parisi, M., Clark, D. (2002) Gene expression neighborhoods. *J Biol* 1, 4.
9. Carninci, P. (2009) Molecular biology: the long and short of RNAs. *Nature* 457, 974–975.
10. Seila, A. C., Calabrese, J. M., Levine, S. S., Yeo, G. W., Rahl, P. B., Flynn, R. A., Young, R. A., Sharp, P. A. (2008) Divergent transcription from active promoters. *Science* 322, 1849–1851.
11. Maniatis, T., Reed, R. (2002) An extensive network of coupling among gene expression machines. *Nature* 416, 499–506.
12. Moore, M. J., Proudfoot, N. J. (2009) Pre-mRNA processing reaches back to transcription and ahead to translation. *Cell* 136, 688–700.
13. Gerstein, M. B., Bruce, C., Rozowsky, J. S., Zheng, D., Du, J., Korbel, J. O., Emanuelsson, O., Zhang, Z. D., Weissman, S., Snyder, M. (2007) What is a gene, post-ENCODE? History and updated definition. *Genome Res* 17, 669–681.
14. Pan, Q., Shai, O., Lee, L. J., Frey, B. J., Blencowe, B. J. (2008) Deep surveying of alternative splicing complexity in the human transcriptome by high-throughput sequencing. *Nat Genet* 40, 1413–1415.
15. Wang, E. T., Sandberg, R., Luo, S., Khrebtukova, I., Zhang, L., Mayr, C., Kingsmore, S. F., Schroth, G. P., Burge, C. B. (2008) Alternative isoform regulation in human tissue transcriptomes. *Nature* 456, 470–476.
16. Ji, Z., Lee, J. Y., Pan, Z., Jiang, B., Tian, B. (2009) Progressive lengthening of 3′ untranslated regions of mRNAs by alternative polyadenylation during mouse embryonic development. *Proc Natl Acad Sci USA* 106, 7028–7033.
17. Sandberg, R., Neilson, J. R., Sarma, A., Sharp, P. A., Burge, C. B. (2008) Proliferating cells express mRNAs with shortened 3′ untranslated regions and fewer microRNA target sites. *Science* 320, 1643–1647.
18. Moore, M. J. (2005) From birth to death: the complex lives of eukaryotic mRNAs. *Science* 309, 1514–1518.
19. Diehn, M., Bhattacharya, R., Botstein, D., Brown, P. O. (2006) Genome-scale identification of membrane-associated human mRNAs. *PLoS Genet* 2, e11.
20. Halbeisen, R. E., Galgano, A., Scherrer, T., Gerber, A. P. (2008) Post-transcriptional gene regulation: from genome-wide studies to principles. *Cell Mol Life Sci* 65, 798–813.
21. St Johnston, D. (2005) Moving messages: the intracellular localization of mRNAs. *Nat Rev Mol Cell Biol* 6, 363–375.
22. Gilbert, W. V., Zhou, K., Butler, T. K., Doudna, J. A. (2007) Cap-independent translation is required for starvation-induced differentiation in yeast. *Science* 317, 1224–1227.
23. Ingolia, N. T., Ghaemmaghami, S., Newman, J. R., Weissman, J. S. (2009) Genome-wide analysis in vivo of translation with nucleotide resolution using ribosome profiling. *Science* 324, 218–223.
24. Keene, J. D., Komisarow, J. M., Friedersdorf, M. B. (2006) RIP-Chip: the isolation and identification of mRNAs, microRNAs and protein components of ribonucleoprotein complexes from cell extracts. *Nat Protoc* 1, 302–307.
25. Baroni, T. E., Chittur, S. V., George, A. D., Tenenbaum, S. A. (2008) Advances in RIP-chip analysis: RNA-binding protein immunoprecipitation-microarray profiling. *Methods Mol Biol* 419, 93–108.
26. Tian, Q., Stepaniants, S. B., Mao, M., Weng, L., Feetham, M. C., Doyle, M. J., Yi, E. C., Dai, H., Thorsson, V., Eng, J., et al. (2004) Integrated genomic and proteomic analyses of gene expression in Mammalian cells. *Mol Cell Proteomics* 3, 960–969.
27. Lunde, B. M., Moore, C., Varani, G. (2007) RNA-binding proteins: modular design for efficient function. *Nat Rev Mol Cell Biol* 8, 479–490.
28. Gerber, A. P., Herschlag, D., Brown, P. O. (2004) Extensive association of functionally and cytotopically related mRNAs with Puf family RNA-binding proteins in yeast. *PLoS Biol* 2, E79.
29. Saint-Georges, Y., Garcia, M., Delaveau, T., Jourdren, L., Le, C. S., Lemoine, S., Tanty, V., Devaux, F., Jacq, C. (2008) Yeast mitochondrial biogenesis: a role for the PUF RNA-binding protein Puf3p in mRNA localization. *PLoS One* 3, e2293.
30. Townley-Tilson, W. H., Pendergrass, S. A., Marzluff, W. F., Whitfield, M. L. (2006) Genome-wide analysis of mRNAs bound to the histone stem-loop binding protein. *RNA* 12, 1853–1867.
31. Barreau, C., Paillard, L., Osborne, H. B. (2005) AU-rich elements and associated factors: are there unifying principles? Nucleic Acids Res 33, 7138–7150.

Chapter 17

RIP-Chip Analysis: *R*NA-Binding Protein *I*mmuno*p*recipitation-Microarray (*Chip*) Profiling

Ritu Jain, Tiffany Devine, Ajish D. George, Sridar V. Chittur, Timothy E. Baroni, Luiz O. Penalva, and Scott A. Tenenbaum

Abstract

Post-transcriptional regulation of gene expression plays an important role in complex cellular processes. Just like transcription factors regulate gene expression through combinatorial binding to multiple, physically dispersed cis elements, mRNA binding proteins can regulate the translation of functionally related gene products by coordinately binding to subsets of mRNAs. The networks of mRNA binding proteins that facilitate this fine-tuning of gene expression are poorly understood. By combining genomic technologies with standard molecular biology tools, we have helped pioneer the development of high-throughput technologies for the global analysis of subsets of mRNAs bound to RNA-binding proteins. This technique is termed RIP-Chip and stands for *R*NA-Binding Protein *I*mmuno*p*recipitation-Microarray (*Chip*) Profiling. This approach is also referred to as "ribonomic profiling" and has revealed valuable information about the workings of mRNP networks in the cell and the regulation of gene expression. In this chapter, we describe the latest advances that we have made in the RIP-CHIP technology.

Key words: Post-transcriptional gene regulation, Ribonomics, RIP-Chip, RNA-binding Protein (RBP), Immunoprecipitation (IP), microarray, microarray expression profiling, array, systems biology.

1. Introduction

In eukaryotic organisms, a tight regulation of gene expression is required for complex processes such as development, differentiation, cell specification, and response of each cell to environmental signals. Any deviation from this regulation may result in disease and other abnormal conditions. The multistep process of translating the information contained in the genetic code of

H. Nielsen (ed.), *RNA*, Methods in Molecular Biology 703,
DOI 10.1007/978-1-59745-248-9_17,

every cell into proteins that ultimately control the function of the cell is regulated at every step (for a review, *see* 1). Thus far, scientists have extensively studied the beginning (transcription), middle (splicing), and the end (translation and post-translational modifications) of this process. However, it has been well documented that there is a discrepancy between mRNA transcript levels and the final amount of protein produced. There is increasing evidence that RNA does not exist as a naked entity in the cell but is always bound by proteins that guide it through every step. All events beginning with the transcription of mRNA to processing, transport, localization, stability, degradation, and translation are regulated by various RNA-Binding Proteins (RBPs), and contribute significantly to the final amount of protein product produced (for a review, *see* 2–4). These RNA binding proteins are required to accomplish several complicated tasks such as localization of mRNAs during embryonic patterning (5), development of synaptic plasticity (6, 7), maintaining iron homeostasis in the cell (8), response to growth signals (9) and regulation of cellular growth (10). The complex of mRNA and its associated set of proteins and small non-coding RNAs is commonly referred to as an mRNP (messenger ribonucleoprotein particle). The set of proteins associated with mRNA may change dynamically and all these processes are coordinated to generate a gene expression profile specific for that cell under a particular condition.

In eukaryotes, the process of transcription is physically separated from translation by the nuclear membrane. In addition the functionally related genes are dispersed on the chromosome. However, coordinated expression of functionally related genes can be achieved through RBPs, which can sequester functionally related mRNAs together. These "post-transcriptional operons" may allow the cell to generate a rapid response to an incoming signal by regulating the stability and translation of specific mRNAs (11). The post-transcriptional operon model can be used to explain the discordance that is often observed between the levels of mRNA and the protein of any given gene. Just because a gene is transcribed does not necessarily mean that it is translated. It may be kept on hold, or degraded by various RBPs. Experimental evidence from many different labs in recent years has demonstrated that mRNAs which either participate in a common function or subcellular localization tend to be part of the same RNP complex, thus providing further evidence for the existence of post-transcriptional operons. Often, these RBPs bind to mRNAs through elements that contain sequence and structural specificity and are present in their 5′ or 3′ untranslated regions. Each mRNA often contains binding-sites for more than one RBP and each RBP has the potential to associate with more than one mRNA, giving rise to combinatorial, systems-level regulatory networks with the potential for tremendous complexity, yet elegant simplicity (12). This combinatorial post-transcriptional networking likely gives rise to much of the complexity of the human genome that has yet to be fully understood. RIP-Chip provides the fundamental technology to elucidate this information.

The complete human genome sequence has revealed the predicted presence of 1,000–2,000 RBPs. Each of these RBPs may bind subsets of 10–1,000 s of mRNAs at a time. Individual analysis of each of these mRNAs bound to their respective proteins one-at-a-time would be painstakingly slow and laborious. To facilitate this, we have helped pioneer the development of high-throughput technologies to analyze the entire subset of mRNAs associated with a particular RBP at a global level, using microarrays (12–15). This technique which employs immunoprecipitation of endogenously formed mRNP complexes using an antibody specific to the RBP in question, followed by purification of associated mRNAs and their global, quantitative analysis using a genomic read-out such as microarrays, SAGE analysis, or deep-sequencing has been termed RIP-Chip (*R*NA-Binding Protein *I*mmuno*p*recipitation-Microarray (*Chip*)), and is outlined in **Fig. 17.1**. Many refer to the approach of characterizing mRNA subsets associated with mRNA binding proteins as "ribonomics," and this approach has been widely used to identify unique mRNA regulatory networks targeted by various mRNA binding proteins (13–57). Since this method can also detect mRNAs that do not directly bind to a particular RBP but may form part of the larger mRNP complex, this method also provides a way to determine the mRNA infrastructure of a cell or tissue. Ribonomic profiling has provided several valuable insights into post-transcriptional gene networks, e.g., we were able to show that upon functional perturbation (stimulation of P19 embryonic carcinoma cells with Retinoic Acid, stress, differentiation, etc.), the subset of mRNAs associated with any particular RBP may change (13). Comparison of mRNAs associated with the Fragile-X Mental Retardation protein (FMRP) from normal and Fragile-X syndrome patients revealed the subset of mRNAs whose translation may change in the absence of FMRP (16). Also, upon phosphorylation, the localization of the La protein and its associated RNAs were altered (18).

In the last few years many groups have used RIP-Chip profiling to study the mRNA profiles associated with selected RBPs (12–58). Some have used different methods to co-immunoprecipitate mRNAs bound to a particular RBP to further analyze the mRNA targets (e.g., cloning and characterization, high-throughput sequencing, etc.). As a further modification of the technique, groups have ectopically expressed the RBP as a recombinant fusion protein (e.g., Flag-tag or TAP-tag), and performed RIP-Chip using antibodies against the tag. Several have employed UV

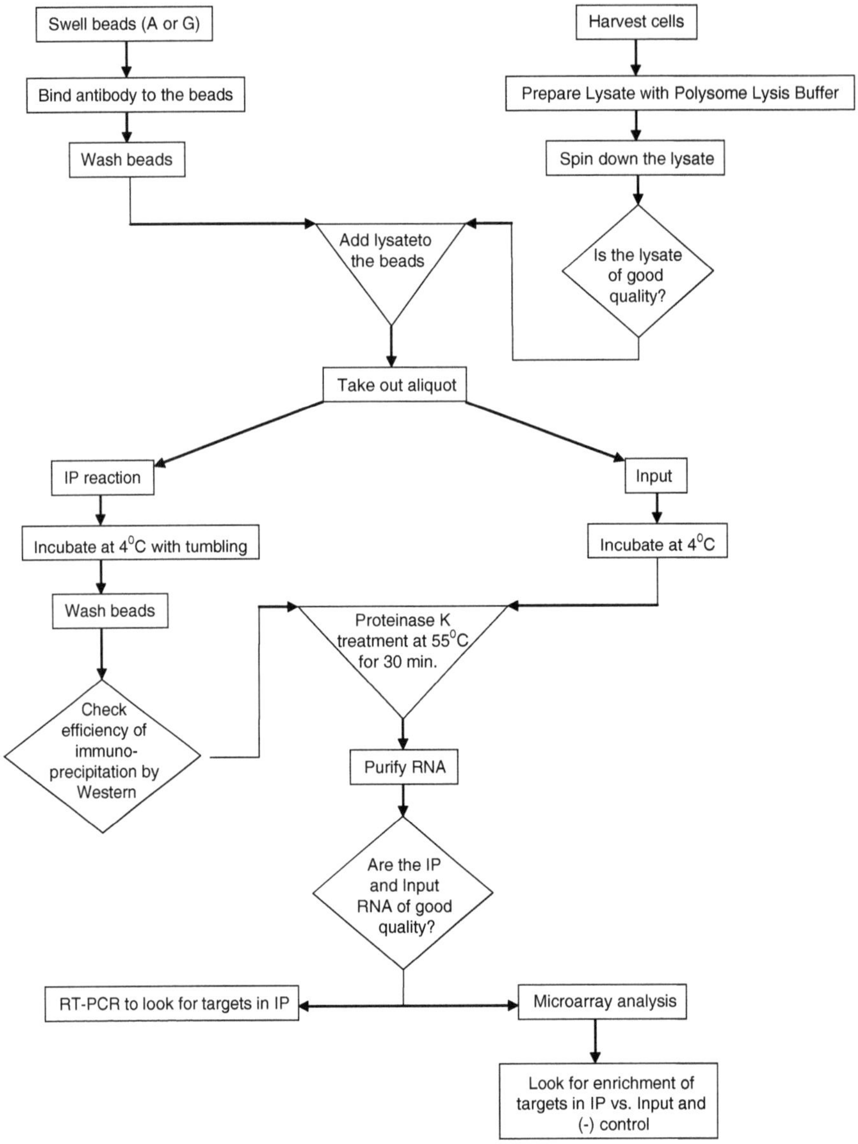

Fig. 17.1. Flow-chart for the RIP-CHIP technique.

cross-linking to irreversibly bind mRNA targets to RBP, a technique known as CLIP (40, 41). Collectively of these studies have provided valuable insights into the function of mRNPs. For example, a systematic analysis of mRNAs bound to 40 different RBPs of yeast has revealed that each binds a subset of mRNAs with common functional or subcellular localization themes (28). These subsets may overlap indicating a combinatorial network of regulation. Ribonomic profiling has been extended to include analysis of miRNA regulation of translation. Immunopurification of miRNP has revealed valuable information about miRNA and their mRNA targets (33–38). Interestingly, many well-known RBPs (e.g., HuR, FMRP, Staufen) were found to be associated with miRNP complexes providing evidence for extensive connections between RBP and miRNA regulatory networks (27, 38). Immunopurification-based methods for identifying miRNA targets has the potential to be more biologically relevant than computer prediction-based approaches because it allows the direct identification of miRNA targets with non-ideal sequence complementarity and without any prior knowledge of the rules of interaction between miRNA and its targets. Lastly, ribonomic profiling has also been used to identify mRNA targets of RBPs from microorganisms such as *Salmonella* and *Escherichia coli* (43, 44).

In this chapter, we describe the most current RIP-Chip techniques as we presently practice them and provide helpful tips and suggestions to present a comprehensive overview of the technology in hopes that it can be as widely used as possible.

2. Materials

We recommend that throughout this method, all standard precautions should be taken to minimize RNAse contamination (*see* **Notes 1** and **2**). All reagents, glassware, and plasticware should either be purchased RNAse-DNAse free or be treated with 0.1% DEPC to ensure that they are RNAse free. All cell culture should be done with cell-culture grade media.

1. Tissues and/or cells of interest.
2. Phosphate-buffered Saline (PBS).
3. Bovine Serum Albumin Fraction V.
4. Antibody to the RNA Binding Protein of interest.
5. Antibody binding matrix (e.g., Protein A Sepharose (Sigma) or Protein G Sepharose beads (Sigma), or the pre-swollen beads from GE Healthcare, or Magnetic beads (Dynabeads) from Invitrogen). Choice of beads depends on the antibody being used (*see* **Notes 3** and **4**).
6. 0.1 M sodium phosphate buffer (pH 8.1) for Dynabeads A or 0.1 M citrate-phosphate buffer (pH 5.0) for Dynabeads G.
7. 0.2 M triethanolamine (Sigma).
8. 20 mM DMP (dimethyl pimelimidate (Sigma)) in 0.2 M triethanolamine (freshly prepared).
9. 0.1 M DTT.

10. 0.5 M EDTA, pH 8.0.
11. 5 M ammonium acetate.
12. 7.5 M lithium chloride.
13. Glycogen, 5 mg/mL (Ambion).
14. Proteinase K, 20 mg/mL (Ambion).
15. Acid-phenol:chloroform, pH 4.5 (with isoamyl alcohol, 25:24:1) (Ambion).
16. Chloroform.
17. Absolute ethanol.
18. 80% ethanol.
19. RNAse OUT, 40 U/mL (Invitrogen) or RNAse Inhibitor (New England Biolabs) or Superase•IN^{TM} (Ambion, 20 U/μL).
20. Complete^{TM} proteinase inhibitor cocktail (Roche).
21. Nuclease-free water (Ambion).

We recommend using nuclease-free water to prepare the following buffers:

1. *Polysome Lysis Buffer (PLB):* 10 mM Hepes, pH 7.0, 100 mM KCl, 5 mM $MgCl_2$, 0.5% NP-40.

   Typically we prepare a 10× stock of this buffer from which a ready to use Polysome Lysis Buffer can be prepared. At the time of use, 1 mM DTT, 100 Units/mL RNAse OUT, and complete protease inhibitor cocktail should be added (*see* **Note 5**).
2. *NT-2 Buffer:* 50 mM Tris-HCl, pH 7.4, 150 mM NaCl, 1 mM $MgCl_2$, 0.05% NP-40. Typically we prepare a 5× stock of NT-2 Buffer, from which a ready to use buffer can be prepared as and when desired.
3. *NET-2 Buffer (Binding buffer for RNP to antibody):* NT-2 buffer supplemented with 20 mM EDTA, pH 8.0, 1 mM DTT, 100 U/mL RNAse OUT.
4. *Proteinase K digestion buffer:* NT-2 buffer supplemented with 1% SDS, 1.2 mg/mL Proteinase K.
5. *Antibody binding buffer:* 5% BSA in NT-2 buffer.

3. Methods

The method described here has been successfully used with mammalian cells and tissue samples, as well as *Caenorhabditis elegans* and *Drosophila melanogaster* cells. However, we assume that this method could be adapted or modified to samples from other organisms. A flow-chart explaining the various steps involved in RIP-Chip is shown in **Fig. 17.1**. We have successfully immunoprecipitated RNA Binding Protein-mRNA complexes (RIP) using both Sepharose and magnetic beads. Both methods are detailed here. Typically immunoprecipitations are performed in a 1 mL volume, this allows for reproducible kinetics, reduced mRNA re-assortment potential, and ease of calculations.

3.1. Preparation of PLB Lysates

1. Grow desired cells to about 70–80% confluence, on 150- or 100-mm plates.
2. Wash twice with ice-cold PBS.
3. Harvest cells by scraping in ice-cold PBS and transfer to a centrifuge tube. If desired, Trypan Blue assay can be used to check the viability of the cells at this point.
4. Collect cells by centrifugation at 2,500×*g* for 5 min.
5. Aspirate PBS from the cell pellet and add equal pellet volume of PLB (containing protease and RNAse inhibitors) to the cell pellet. Mix by pipetting up and down until the mixture looks homogeneous. Keep on ice for 5 min to allow the hypotonic PLB buffer to swell the cells. Dispense into 1-mL aliquots and store at –80°C. This method depends on repeated freeze–thaws to gently lyse the cells. The cell lysates are ready to proceed to RBP immunoprecipitation (*see* **Note 6**).

3.2. Coating of Bead Matrix with Antibody

3.2.1. Coating Sepharose Beads

1. Swell the desired amount of protein G or protein A agarose beads in NT2 buffer containing 5% BSA. Typically we add 30 mL of buffer to 1 g of beads. This yields 20% slurry (1 mL of slurry yields about 200 μL of packed bead volume). Alternatively, pre-swollen beads equilibrated with NT2 buffer can be used.
2. Add the immunoprecipitation antibody or serum and incubate for at least 1 h at room temperature on a rotating device or overnight (or longer) at 4°C (*see* **Notes** 7 and **8**). Typically for each immunoprecipitation reaction we use 300 μL of slurry, dilute it with 1 mL of NT2 buffer containing 5% BSA, and add 5 μg of commercially purchased antibody.

3.2.2. Coating Magnetic Beads

For both Dynabeads protein A and protein G, the method is essentially as recommended by the manufacturer.

1. For each immunoprecipitation reaction use 50 μL of Dynabeads A suspension, wash twice with 0.5 mL of 0.1 M sodium phosphate buffer (pH 8.1) using the magnet Dynal. For Dynabeads G use citrate-phosphate buffer

(pH 5.0) instead of 0.1 M sodium phosphate buffer (pH 8.1).

2. Re-suspend the beads in 50 μL of the same buffer and add 5 μg of the antibody.
3. Incubate with rotation for 10 min at room temperature.
4. Wash three times with 0.5 mL of 0.1 M sodium phosphate buffer (for Dynabeads A) or citrate-phosphate buffer (pH 5.0) (for Dynabeads G).
5. For cross-linking the antibody to the beads (optional), resuspend the beads in 1.0 mL of 0.2 M triethanolamine. Wash twice with 1.0 mL of 0.2 M triethanolamine.
6. Add 1.0 mL of freshly prepared 20 mM dimethyl pimelimidate in 0.2 M triethanolamine and incubate with rotational mixing for 30 min at room temperature.
7. Wash three times with NT-2 buffer. The beads are now ready to proceed with RNP-binding. We have not attempted to store antibody-coated magnetic beads for extended periods.

3.3. Immunoprecipitation of RNA Binding Protein-mRNA Complex (RIP)

The method for immunoprecipitation on Sepharose versus Dynabeads is almost identical with minor variations.

1. Wash the Sepharose beads (*see* **Section 3.2.1, Step 2**), six times with 1.0 mL of cold NT-2 buffer.
2. Resuspend the antibody-coated beads (Sepharose or Dynabeads) in 900 μL of NET-2 buffer. Typically for each immunoprecipitation we mix: 850 μL NT-2, 35 μL 0.5 M EDTA, 10 μL 0.1 M DTT, 5 μL RNAse OUT to obtain 900 μL of NET-2 buffer.
3. Thaw the PLB lysate quickly by holding between your fingers and centrifuge at 14,000 × *g* for 10 min at 4°C. Remove 100 μL of the supernatant and add to the beads in NET-2. The final volume of the immunoprecipitation reaction will now be 1.0 mL.
4. Mix and briefly centrifuge to bring the beads down. Remove 100 μL of the supernatant. This represents the starting material or "input" which will be processed alongside the immunoprecipitation to monitor any RNAse contamination, and to compare with immunoprecipitated mRNAs at the end.
5. Tumble the reactions end-over-end for 3 h to overnight at 4°C (1 h at room temperature for Dynabeads) (*see* **Notes 9, 10** and **11**).
6. It is a good idea to also include a negative control for the immunoprecipitation using an unrelated antibody, e.g., T7-tag antibody, generic IgG, or pre-immune serum to monitor the non-specific binding to the beads.

7. After incubation is complete, wash the beads six times with cold NT-2 buffer and resuspend in 150 μL of proteinase K buffer (*see* **Notes 12** and **13**). Typically for each immunoprecipitation we mix 124 μL NT-2 buffer, 15 μL of 10% SDS, 9 μL of 20 mg/mL proteinase K to obtain 150 μL of proteinase K digestion buffer. Also add the same amount of SDS and proteinase K to the tubes labeled "input" and bring up the volume to 150 μL with NT-2. Incubate all tubes at 55°C for 30 min to digest the protein and release the RNA from RNP complex.

4. Purification of RNA

RNA is purified from the supernatants using standard phenol–chloroform extraction (*see* **Note 14**).

1. Add 150 μL phenol:chloroform:isoamyl alcohol to the tubes containing beads in proteinase K buffer. For Dynabeads, remove the supernatant using the magnet and place it in a separate tube. Add phenol–chloroform only to the supernatant. Vortex to mix and centrifuge at 14,000×*g* for 10 min to separate the phases.
2. Remove the aqueous phase carefully and place it in a new tube. Add 150 μL of chloroform and repeat the extraction.
3. To each tube add 50 μL of 5 M ammonium acetate, 15 μL 7.5 M lithium chloride, 5 μL of 5 mg/mL glycogen, and 850 μL absolute ethanol. Mix and keep at –80°C from 1 h to overnight to precipitate the RNA.
4. To collect the RNA centrifuge at 14,000×*g* for 30 min at 4°C, pour off the supernatant, wash the pellet once with 80% ethanol, centrifuge again at 14,000×*g* for 30 min at 4°C. Pour off the supernatant, dry the pellet, and resuspend in 20 μL of RNAse-free water.
5. If desired, the RNA can be further purified using Qiagen RNEasy Micro clean-up, as per manufacturer's instructions. However, we have not found it to be necessary.

4.1. Assessment of RNA Quality

1. Typically, optical absorbance of the RNA can be measured using a Nanodrop® spectrophotometer. Ideally we expect both the A_{260}/A_{280} and A_{260}/A_{230} ratios to be close to 2.0, implying the purity of RNA and the absence of any contaminating proteins or chemicals. If this ratio is less than 1.8, there may be problems with further downstream applications.

2. The molecular weight profile of the subset of RNAs immunoprecipitated can be analyzed on a nanochip using Agilent's BioAnalyzer. Nanochip is a convenient alternative to using formaldehyde-agarose gels. For the total RNA (or "input") it is helpful to evaluate the 28S/18S rRNA ratio. A ratio between 1.6 and 2.0 indicates RNA of good integrity. However for the immunoprecipitated material such ratio can not be obtained unless rRNAs are known to be targets of the RNA Binding Protein in question. And so, a bioanalyzer profile of immunoprecipitated RNA only serves to provide information about any extensive degradation of RNA. Usually, if the total RNA shows a good 28S/18S rRNA ratio, we can safely assume the immunoprecipitated RNA to be of good integrity.

4.2. Analysis of Immunoprecipitated RNA

1. If the mRNAs binding to the RBP in question are known, RT-PCR using gene-specific primers is the method of choice due to cost and convenience. An example of RT-PCR to analyze the mRNAs associated with the histone Stem-loop binding protein and Poly-A binding protein is shown in **Fig. 17.2**.

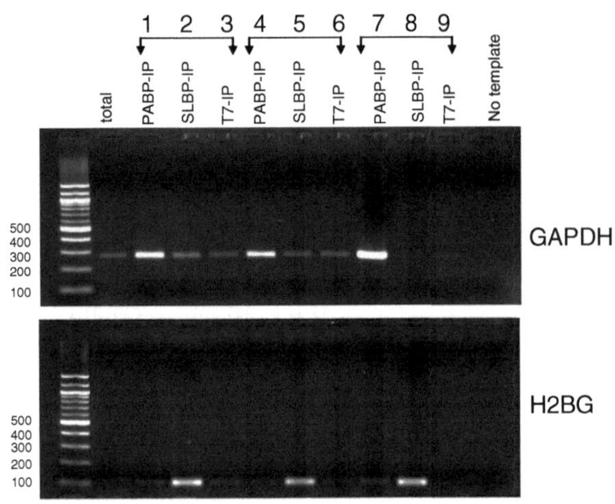

Fig. 17.2. Example of RT-PCR analysis of mRNA targets recovered after RIP. This figure shows a comparison of immunoprecipitation on agarose versus magnetic beads using antibodies against poly-A binding protein (PABP), histone stem-loop binding protein (SLBP) and T7-tag (negative control). Lanes 1, 2, and 3 show immunoprecipitation from protein-G Sepharose beads. Lanes 4, 5, and 6 show immunoprecipitation on protein-G Sepharose beads performed with pre-cleared lysate (lysate was incubated with beads only for 30 min at 4°C). Lanes 7, 8, and 9 show immunoprecipitation on protein-G magnetic beads. The PCR was performed using gene specific primers for GAPDH, HIST1H2BG, and HIST1H4B. GAPDH mRNA has a poly-A tail and so is expected to be present in PABP-IP and not SLBP-IP, whereas the histone H2BG has a stem-loop in its 3′ UTR and so is expected to be present in SLBP-IP and not PABP-IP. The conditions for PCR were as follows. GAPDH was amplified for 18 cycles using forward primer: GGCTCTCCAGAACATCATCCCTGC and reverse primer: GGGTGTCGCTGTTGAAGTCAGAGG H2BG was amplified for 25 cycles using forward primer: ACAAGCGCTCGACCATTACCT and reverse primer: TGGTGACAGCCTTGGTACCTTC.

2. However, if the entire subset of mRNAs binding to a particular RBP needs to be analyzed, microarray analysis is the perfect high-throughput method of choice. We have used Affymetrix Exon arrays, Agilent expression arrays, and Nimblegen Tiling arrays.
3. Multiprobe-based RNase protection assays (PharMingen) are an ideal alternative for the optimization and high-throughput analysis of mRNP immunoprecipitations.
4. We have identified many RBP-associated mRNAs using cDNA/genomic arrays and have found the Affymetrix GeneChip® array platform and the BD Atlas Nylon cDNA Expression Array platform are excellent for conducting RIP-Chip analysis. However, other array platforms have worked with varying success.
5. If gene expression analysis is performed using glass arrays that utilize Cy3 and Cy5 labeling or on Affymetrix arrays, we typically increase the amount of extract by 2–3 times that required for Atlas Nylon Arrays.

4.3. Synthesis of Labeled cRNA and Microarray Hybridization

1. We have successfully used 10–100 ng of immunoprecipitated RNA for microarray studies using the Affymetrix GeneChip® platform.
2. We recommend using 50 ng of immunoprecipitated material if available. The RNA is first converted to T7-oligo(dT) primed double stranded cDNA using the Affymetrix two-cycle cDNA synthesis kit as per the manufacturer's protocol.
3. This is then converted to amplified RNA (aRNA) by in vitro transcription using MEGAscript® T7 polymerase (Ambion).
4. The aRNA is subsequently processed to double stranded cDNA in a manner similar to the first strand cDNA, but uses random primers.
5. Then, biotinylated UTPs are incorporated in the second-IVT step (Affymetrix IVT Labeling kit) resulting in labeled, amplified cRNA.
6. The cRNA (15 μg) is fragmented using a metal induced hydrolysis step to obtain 25–200 bp fragments that are then hybridized to the GeneChip® arrays as per manufacturer's protocol.

4.4. Statistical Analysis of Ribonomic Data from Microarrays

1. The difference in signal intensities obtained from hybridizing cRNA from immunoprecipitated samples and from total (a.k.a. input) RNA confounds traditional microarray analysis methods. The problem resides primarily in the normalization techniques used to distribute the signal intensities on the array.

2. We have developed a strategy to overcome this problem and have successfully identified targets of RBPs through ribonomic profiling of GeneChip® microarrays. To obtain a robustly confident list of genes associated with a given RBP, we use both MAS5 as well as GCRMA algorithms in data analysis to determine the subset of genes that stands out, regardless of the probe intensity normalization method.
3. While using MAS5 for analysis, we use a filter based on the requirement that the target must be called present in the input RNA.
4. Fold change based filtering of these lists (we prefer a 4-fold or higher level of enrichment between input [total RNA] and IP RNA in all replicate sets) results in a final list of targets with a very high level of confidence.
5. We strongly recommend the use of replicates in the experiments using microarray technology for ribonomic profiling.
6. Additionally, we recommend the use of one or more control immunoprecipitation (i.e., IgG, T7, HA, FLAG, etc.) to estimate the degree of target "background" binding to proteins and/or agarose beads.
7. The output of the Affymetrix protocols results in a high level of signal enrichment over traditional methods like northern blots. Consequently, the protocol creates the impression that a large number of targets are being immunoprecipitated by the "negative control."
8. In our analysis, we subtract the signal of those "negative" targets present in order to decrease the complexity of the resulting data and increase the confidence level.
9. In our analysis, we use the "negative" targets as an estimate of our background noise, subtracting their signals from the IP signals to adjust the confidence of high-background probes.

5. Notes

1. For general precautions on working with RNA, see Current protocols in Molecular Biology (Chap 4.1.1). All instruments, glassware and plasticware that touch cells or cell lysates should be certified DNase-free and RNase-free or should be pre-washed with RNase Zap (Ambion) or RNase Away (Molecular BioProducts) followed by DEPC water and allowed to air dry.
2. Generally, solutions which are certified DNase-free and RNase-free from the manufacturer will make for easier

solution preparation and allow for faster troubleshooting if they are handled properly. Ambion's buffer kit contains concentrated solutions of Tris (pH 7 and pH 8), EDTA, sodium chloride, magnesium chloride, potassium chloride, ammonium acetate, and DEPC-treated water.

3. The type of beads used will depend on the binding affinity of the particular antibody to the beads. Most mouse monoclonal antibodies bind strongly to protein G beads, and rabbit polyclonal antibodies bind strongly to both protein A and G beads. Consult the binding chart for the beads from the manufacturer to determine the best choice of beads for the particular isotype of antibody being used.
4. Pre-swollen beads are more expensive; however, they can save time and give more consistent results.
5. One complete protease inhibitor tablet is sufficient for 10 mL of Polysome Lysis Buffer (PLB). Typically we prepare 10 mL of ready to use solution at a time, which can be dispensed into 1-mL aliquots and stored at –80°C for extended periods of time without any deleterious effects (more than 6 months).
6. The volume of cell lysate obtained depends on the cytoplasmic volume of the cell; typically we obtain 400–500 μL lysate from six 15-cm plates. The objective is to get an extremely concentrated lysate ranging in concentration from 20 to 50 mg/mL of total protein. On average, $1–5 \times 10^6$ cells will generate enough lysate for a typical immunoprecipitation reaction.
7. Typically 2–20 μL sera or 5 μg of commercially available antibody per immunoprecipitation reaction is used, depending on the affinity of the antibody. We have found that using excess antibodies when possible greatly reduces non-specific binding of proteins to Sepharose beads, possibly by reducing the number of available binding sites on the beads.
8. Antibody-coated beads can be prepared in bulk and stored at 4°C with 0.02% sodium azide.
9. To minimize re-assortment potential, the final volume of re-suspended beads in NET2 buffer should correspond to approximately ten times the original volume of the RNP lysate being used (a 1:10 fold dilution of lysate). Since we typically perform our reactions in 1 mL, we use 100 μL lysate. Performing the immunoprecipitation reactions in larger volumes can help decrease background problems.
10. Provided there is no RNA degradation or RNAse problems, longer incubations will result in better recovery of RNA.

11. A concern when isolating mRNP complexes is the possibility of exchange of proteins and mRNAs. In principle, cross-linking agents, such as formaldehyde, could prevent this (58). However, we have found mRNA exchange to occur at a minimal level using these methods described here, and crosslinking, therefore, to be unnecessary. In some cases, formaldehyde actually can interfere with subsequent mRNA detection methods and increase the background (15).
12. Several additional washes with NT2 buffer supplemented with 1–3 M urea can increase specificity and reduce nonspecific binding. However, it is important to first determine whether urea disrupts binding of the antibody to the target protein and/or the RBP-mRNA interaction.
13. It is a good idea to remove an aliquot during the last wash, to test the efficiency of immunoprecipitation by western blotting. The proteins can be eluted off the beads by resuspending the beads in 1× SDS-PAGE loading buffer followed by heating at 95°C. The beads can then be centrifuged down and the supernatant directly applied on SDS-PAGE. This is helpful when using a new antibody for the first time. Many commercially available antibodies are not good quality and in our experience, sometimes we had to try as many as five different antibodies before we were able to successfully immunoprecipitate the target RBP.
14. We have tried to use the Qiagen RNEasy kit to avoid the phenol–chloroform extraction step. After the proteinase K digestion, the supernatant can be removed, mixed with buffer RLT/β-ME, and RNA can be purified from the spin column according to manufacturer's directions. Although this process offers ease of use and saves time, we found the recovery of RNA to be at least tenfold less compared to phenol–chloroform extraction.

Acknowledgments

We would like to acknowledge the expert technical help from David Frank and Marcy Kuentzel of the microarray core, Center for Functional Genomics, University at Albany-SUNY and input from the other Tenenbaum Lab members. This work was supported in part by NIH grant U01HG004571 to SAT from the NHGRI.

References

1. Orphanides, G., Reinberg, D. (2002) A unified theory of gene expression. *Cell* 108, 439–451.
2. Moore, M. J. (2005) From birth to death: the complex lives of eukaryotic mRNAs. *Science* 309, 1514–1518.
3. Keene, J. D. (2001) Ribonucleoprotein infrastructure regulating the flow of genetic information between the genome and the proteome. *Proc Natl Acad Sci USA* 98, 7018–7024.
4. Sanchez-Diaz, P., Penalva, L. O. (2006) Post-transcription meets post-genomic: the saga of RNA binding proteins in a new era. *RNA Biol* 3, 101–109, Review.
5. King, M. L., Messitt, T. J., Mowry, K. L. (2005) Putting RNAs in the right place at the right time: RNA localization in the frog oocyte. *Biol Cell* 97, 19–33.
6. Feng, Y., Absher, D., Eberhardt, D. E., Brown, V., Malter, H. E., Warren, S. T. (1997) FMRP associates with polyribosomes as an mRNP and the I304N mutation of severe fragile-X-syndrome abolishes this association. *Mol Cell* 1, 109–118.
7. Bassell, G. J., Warren, S. T. (2008) Fragile X Syndrome: loss of local mRNA regulation alters synaptic development and function. *Neuron* 60, 201–214.
8. Klausner, R. D., Rouault, T. A., Harford, J. B. (1993) Regulating the fate of mRNA: the control of cellular iron metabolism. *Cell* 72, 19–26.
9. López de Silanes, I., Lal, A., Gorospe, M. (2005) HuR: Post-transcriptional paths to malignancy. *RNA Biol* 2, 11–13.
10. Kato, T., Hayama, S., Yamabuki, T., Ishikawa, N., Miyamoto, M., Ito, T., Tsuchiya, E., Kondo, S., Nakamura, Y., Daigo, Y. (2007) Increased expression of insulin-like growth factor-II messenger RNA binding protein-1 is associated with tumor progression in patients with lung cancer. *Clin Cancer Res* 13, 434–442.
11. Keene, J. D., Tenenbaum, S. A. (2002) Eukaryotic mRNPs may represent posttranscriptional operons. *Mol Cell* 9, 1161–1167.
12. Hieronymus, H., Silver, P. (2004) A systems view of mRNP biology. *Genes Dev* 18, 2845–2860.
13. Tenenbaum, S. A., Carson, C. C., Lager, P. J., Keene, J. D. (2000) Identifying mRNA subsets in messenger ribonucleoprotein complexes by using cDNA arrays. *Proc Natl Acad Sci USA* 97, 14085–14090.
14. Tenenbaum, S. A., Lager, P. J., Carson, C. C., Keene, J. D. (2002) Ribonomics: identifying mRNA subsets in mRNP complexes using antibodies to RNA-binding proteins and genomic arrays. *Methods* 26, 191–198.
15. Penalva, L. O., Tenenbaum, S. A., Keene, J. D. (2004) Gene Expression Analysis of Messenger RNP Complexes. *Methods Mol Biol* 257, 125–134.
16. Brown, V., Jin, P., Ceman, S., Darnell, J. C., O'Donnell, W. T., Tenenbaum, S. A., Jin, X., Feng, Y., Wilkinson, K. D., Keene, J. D., Darnell, R. B., Warren, S. T. (2001) Microarray identification of FMRP-associated brain mRNAs and altered mRNA translational profiles in fragile X syndrome. *Cell* 107, 477–487.
17. Eystathioy, T., Chan, E. K., Tenenbaum, S. A., Keene, J. D., Griffith, K., Fritzler, M. J. (2002) A phosphorylated cytoplasmic autoantigen, GW182, associates with a unique population of human mRNAs within novel cytoplasmic speckles. *Mol Biol Cell* 13, 1338–1351.
18. Intine, R. V., Tenenbaum, S. A., Sakulich, A. L., Keene, J. D., Mariah, R. J. (2003) Differential phosphorylation and subcellular localization of La RNPs associated with precursor tRNAs and translation-related mRNAs. *Mol Cell* 12, 1301–1307.
19. Tenenbaum, S. A., Carson, C. C., Atasoy, U., Keene, J. D. (2003) Genome-wide regulatory analysis using en masse nuclear run-ons and ribonomic profiling with autoimmune sera. *Gene* 317, 79–87.
20. Stoecklin, G., Tenenbaum, S. A., Mayo, T., Chittur, S. V., George, A. D., Baroni, T. E., Blackshear, P. J., Anderson, P. (2008) Genome-wide analysis identifies interleukin-10 mRNA as target of tristetraprolin. *J Biol Chem* 283, 11689–11699.
21. Sanchez-diaz, P. C., Burton, T. L., Burns, S. C., Hung, J. Y., Penalva, L. O. (2008) Musashi modulates cell proliferation genes in the medulloblastoma cell line Daoy. *BMC Cancer* 8, 280.
22. Lopez-de-Silanes, I., Fan, J., Yang, X., Zonderman, A. B., Potapova, O., Pizer, E. S., Gorospe, M. (2003) Role of the RNA-binding protein HuR in colon carcinogenesis. *Oncogene* 22, 7146–7154.
23. Mazan-Mamczarz, K., Patrick, R. H., Dai, B., Wood, W. H., Zhang, Y., Becker, K. G., Liu, Z., Gartenhaus, R. B. (2008) Identification of transformation-related pathways in a breast epithelial cell model using a ribonomics approach. *Cancer Res* 68, 7730–7735.

24. Mazan-Mamczarz, K., Hagner, P. R., Corl, S., Srikantan, S., Wood, W. H., Becker, K. G., Gorospe, M., Keene, J. D., Levenson, A. S., Gartenhaus, R. B. (2008) Post-transcriptional gene regulation by HuR promotes a more tumorigenic phenotype. *Oncogene* 27, 6151–6163.
25. Gerber, A. P., Herschlag, D., Brown, P. O. (2004) Extensive association of functionally and cytotopically related mRNAs with Puf family RNA-binding proteins in yeast. *PLoS Biol* 2, E79.
26. Gerber, A. P., Luschnig, S., Krasnow, M. A., Brown, P. O., Herschlag, D. (2006) Genome-wide identification of mRNAs associated with the translational regulator PUMILIO in *Drosophila melanogaster*. *Proc Natl Acad Sci USA* 103, 4487–4492.
27. Galgano, A., Forrer, M., Jaskiewicz, L., Kanitz, A., Zavolan, M., Gerber, A. P. (2008) Comparative analysis of mRNA targets of the human PUF-family proteins suggests an extensive interaction with the miRNA regulatory system. *PLoS One* 3, e3164.
28. Hogan, D. J., Riordan, D. P., Gerber, A. P., Herschlag, D., Brown, P. O. (2008) Diverse RNA-binding proteins interact with functionally related sets of RNAs, suggesting an extensive regulatory system. *PLoS Biol* 6, e255.
29. Hieronymus, H., Silver, P. A. (2003) Genome-wide analysis of RNA-protein interactions illustrates specificity of the mRNA export machinery. *Nat Genet* 33, 155–161.
30. Furic, L., Maher-Laporte, M., Desgroseillers, L. (2008) A genome-wide approach identifies distinct but overlapping subsets of cellular mRNAs associated with Staufen1- and Staufen2-containing ribonucleoprotein complexes. *RNA* 14, 324–335.
31. Townley-Tilson, W. H., Pendergrass, S. A., Marzluff, W. F., Whitfield, M. L. (2006) Genome-wide analysis of mRNAs bound to the histone stem-loop binding protein. *RNA* 12, 1853–1867.
32. Duttagupta, R., Tian, B., Wilusz, C. J., Khounh, T. J., Soteropoulos, P., Ouyang, M., Dougherty, J. P., Peltz, S. W. (2005) Global analysis of Pub1p targets reveals a coordinate control of gene expression through modulation of binding and stability. *Mol Cell Biol* 25, 5499–5513.
33. Duan, R., Jin, P. (2006) Identification of messenger RNAs and microRNAs associated with fragile-X-mental-retardation protein. *Methods Mol Biol* 342, 267–276.
34. Easow, G., Teleman, A. A., Cohen, S. M. (2007) Isolation of microRNA targets by miRNP immunopurification. *RNA* 13, 1198–1204.
35. Zhang, L., Ding, L., Cheung, T. H., Dong, M. Q., Chen, J., Sewell, A. K., Liu, X., Yates, J. R., 3rd, and Han, M. (2007) Systematic identification of *C. elegans* miRISC proteins, miRNAs and mRNA targets by their interactions with GW182 proteins AIN-1 and AIN-2. *Mol Cell* 28, 598–613.
36. Karginov, F. V., Conaco, C., Xuan, Z., Schmidt, B. H., Parker, J. S., Mandel, G., Hannon, G. J. (2007) A biochemical approach to identifying microRNA targets. *Proc Natl Acad Sci USA* 104, 19291–19296.
37. Hammell, M., Long, D., Zhang, L., Lee, A., Carmack, C. S., Han, M., Ding, Y., Ambros, V. (2008) mirWIP: microRNA target prediction based on microRNA-containing-ribonucleoprotein-enriched transcripts. *Nat Methods* 5, 813–819.
38. Landthaler, M., Gaidatzis, D., Rothballer, A., Chen, P. Y., Soll, S. J., Dinic, L., Ojo, T., Hafner, M., Zavolon, M., Tuschl, T. (2008) Molecular characterization of human argonaute-containing ribonucleoprotein complexes and their bound target mRNAs. *RNA* 14, 2580–2596.
39. Sanford, J. R., Coutinho, P., Hackett, J. A., Wang, X., Ranahan, W, Caceres, J. F. (2008) Identification of nuclear and cytoplasmic mRNA targets for the shuttling protein SF2/ASF. *PLoS One* 3, e3369.
40. Cho, Y. S., Iguchi, N., Yang, J., Handel, M. A., Hecht, N. B. (2005) Meiotic messenger RNAs and non-coding RNA targets of the protein Translin (TSN) in mouse testis. *Biol Reprod* 73, 840–847.
41. Ule, J., Jensen, K. B., Ruggiu, M., Mele, A., Ule, A., Darnell, R. B. (2003) CLIP identifies Nova-regulated RNA networks in the brain. *Science* 302, 1212–1215.
42. Noe, G., De Gaudenzi, J. G., Frasch, A. C. (2008) Functionally related transcripts have common RNA motifs for specific RNA-binding proteins in trypanosomes. *BMC Mol Biol* 9, 107.
43. Zhang, A., Wassarman, K. M., Rosenow, C., Tjagden, B. C., Storz, G., Gottesman, S. (2003) Global analysis of small RNAs and mRNA targets of *Hfq*. *Mol Microbiol* 50, 1111–1124.
44. Sittka, A., Lucchini, S., Papenfort, K., Sharma, C. M., Rolle, K., Binnewies, T. T., Hinton, J. C., Vogel, J. (2008) Deep sequencing analysis of small non-coding RNA and mRNA targets of the global post-transcriptional regulator *Hfq*. *PLoS Genet* 4, e1000163.

45. He, Y., Rothnagel, J. A., Epis, M. R., Leedman, P. J., Smith, R. (2009) Downstream targets of heterogeneous nuclear ribonucleoprotein A2 mediate cell proliferation. *Mol Carcinog* 48, 167–179.
46. Kim Guisbert, K., Duncan, K., Li, H., Guthrie, C. (2005) Functional specificity of shuttling hnRNPs revealed by genome-wide analysis of their RNA binding profiles. *RNA* 11, 383–393.
47. Mazan-Mamczarz, K., Kuwano, Y., Zhan, M., White, E. J., Martindale, J. L., Lal, A., Gorospe, M. (2008) Identification of a signature motif in target mRNAs of RNA binding protein AUF1. *Nucleic Acids Res* 25 Nov 2008, Epub ahead of print. 2009 Jan; 37(1):204–14.
48. Banihashemi, L., Wilson, G. M., Das, N., Brewer, G. (2006) Upf1/Upf2 regulation of 3′ untranslated region splice variants of AUF1 links nonsense-mediated and A+U rich element-mediated decay. *Mol Cell Biol* 26, 8743–8754.
49. Graindorge, A., Le Tonqueze, O., Thuret, R., Pollet, N., Osborne, H. B., Audic, Y. (2008) Identification of CUG-BP1/EDEN-BP target mRNAs in Xenopus tropicalis. *Nucleic Acids Res* 36, 1861–1870.
50. Kim, H. S., Kuwano, Y., Zhan, M., Pullman, R., Jr., Mazan-mamczarz, K., Li, H., Kedersha, N., Anderson, P., Wilce, M. C., Gorospe, M., Wice, J. A. (2007) Elucidation of a C-rich signature motif in target mRNAs of RNA-binding protein TIAR. *Mol Cell Biol* 27, 6806–6817.
51. Tremblay, G. A., Richard, S. (2006) mRNAs associated with the Sam68 RNA binding protein. *RNA Biol* 3, 90–93.
52. Johnson, E. M., Kinoshita, Y., Weinrub, D. B., Wortman, M. J., Simon, R., Khalili, K., Winckler, B., Gordon, J. (2006) Role of pur alpha in targeting mRNA to sites of translation in hippocampal neuronal dendrites. *J Neurosci Res* 83, 929–943.
53. Liao, B., Hu, Y., Herrick, D. J., Brewer, G. (2005) The RNA – binding protein IMP-3 is a translational activator of insulin-like growth factor II leader-3 mRNA during proliferation of human K562 leukemia cells. *J Biol Chem* 280, 18517–18524.
54. Lee, M. H., Schedl, T. (2001) Identification of in vivo mRNA targets of GLD-1, a maxi-KH motif containing protein required for *C. elegans* germ cell development. *Genes Dev* 15, 2408–2420.
55. Yang, J., Chennathukuzhi, V., Miki, K., O'Brien, D. A., Hecht, N. B. (2003) Mouse testis brain RNA-binding protein/translin selectively binds to the messenger RNA of the fibrous sheath protein glycerldehyde-3-phosphate-dehydrogenase-S and suppresses its translation in vitro. *Biol Reprod* 68, 853–859.
56. Squires, J. E., Stoytchev, I., Forry, E. P., Berry, M. J. (2007) SBP2 binding affinity is a mjor determinant in differential selenoprotein mRNA translation and sensitivity to non-sense mediated decay. *Mol Cell Biol* 27, 7848–7855.
57. Huang, Y. S., Richter, J. D. (2007) Analysis of mRNA translation in cultured hippocampal neurons. *Methods Enzymol* 431, 143–162.
58. Niranjanakumari, S., Lasda, E., Brazas, R., Garcia-Blanco, M. A. (2002) Reversible cross-linking combined with immunoprecipitation to study RNA-protein interactions in vivo. *Methods* 26, 182–190.

Chapter 18

Isolation of RNP Granules

Lars Jønson, Finn Cilius Nielsen, and Jan Christiansen

Abstract

The post-transcriptional operon provides a means of synexpression of mRNAs encoding interrelated proteins. The coordination of gene expression may be achieved by a *trans*-acting RNA-binding protein attaching to similar *cis*-elements in different, yet functionally clustered, mRNAs. The RNP granule can be regarded as a supramolecular assembly of RNA and protein, probably representing several overlapping post-transcriptional operons. The present protocol describes how RNP granules may be isolated by the transgenic expression of a 3X FLAG version of an RNA-binding protein under tetracycline control via the tetracycline receptor/operator complex. In this way, inclusion of an appropriate tetracycline concentration ensures expression of the tagged version at the endogenous level, and the 3X FLAG tag is a convenient "handle" for the subsequent immunoprecipitation by immobilized anti-FLAG antibody.

Key words: 3X FLAG tag, post-transcriptional operon, RNA-binding protein, RNP granule, stable transfection, tetracycline induction.

1. Introduction

Coordination of gene expression in eukaryotes relies heavily on post-transcriptional processes, especially during development of multicelluar organisms. Therefore, the concept of the post-transcriptional operon has emerged, stating that the expression of different mRNAs, encompassing a functional entity, is coordinated by the attachment of an RNA-binding protein to a common *cis*-element in the mRNAs (1). A convincing case of the concept is seen with the *Saccharomyces cerevisiae* Puf protein family, where the five members interact preferentially with different clusters of mRNAs, e.g. Puf3p binds nearly exclusively to

H. Nielsen (ed.), *RNA*, Methods in Molecular Biology 703,
DOI 10.1007/978-1-59745-248-9_18, © Springer Science+Business Media, LLC 2011

cytoplasmic mRNAs that encode mitochondrial proteins (2). The synexpression in terms of RNA stability, localization or translatability does not necessarily have to be mediated by an RNA-binding protein, since the concept can be extended to a miRNA binding to different, yet co-regulated, mRNAs.

The hub of the post-transcriptional operon is the ribonucleoprotein (RNP) particle, which may appear as a granule in the μm range (3). RNP granules have mainly been studied in neurons, where they provide an appealing way of transporting genetic information in a silent manner for later deployment at dendritic spines. However, it is likely that most cell types contain compartmentalized genetic information packaged in these "organelles."

RNP granules have been isolated from various sources with or without a "handle," but often involving an affinity purification step (4–6), and major concerns are non-specific interactions occurring from rearrangements during cell lysis and indirect associations via cytoskeletal components during immunoprecipitations (7). Here, we present a procedure based on chromosomal integration of a transgenic 3X FLAG version of the oncofetal RNA-binding protein IMP1 under the control of a tetracycline repressor/operator complex. The crucial aspect of this approach is that the expression of the transgenic 3X FLAG version provides a covenient "handle" for the RNP granule isolation, that can be followed on SDS-PAGE and Western blots, but avoids overexpression artefacts, since the expression of 3X FLAG-IMP1 can be gauged at the same level as the endogenous form of IMP1 by an appropriate concentration of tetracycline (*see* **Fig. 18.1**). Following cell lysis, particles containing the "handle" are isolated by affinity purification on FLAG-coated anti-FLAG antibody beads (*see* **Fig. 18.2**), and the structural integrity of granules may be visualized, e.g. by atomic force microscopy (*see* **Fig. 18.3**). We have excluded the description of downstream proteomics and ribonomics analyses from this chapter, since services are now commercially available.

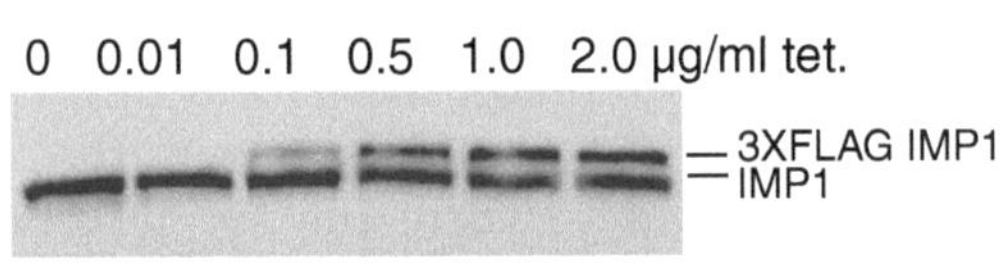

Fig. 18.1. Induction of FLAG-tagged RNA-binding protein by tetracycline. Western blot analysis with anti-IMP1 antibody of cell lysates from HEK293 cells expressing a transgenic version of 3X FLAG-IMP1 under tetracycline control. The use of a 3X FLAG tag instead of mono-FLAG facilitates the electrophoretic separation of the RNA-binding protein with a "handle" from the endogenous protein. In a typical induction experiment, 0.5 μg/mL tetracycline was used.

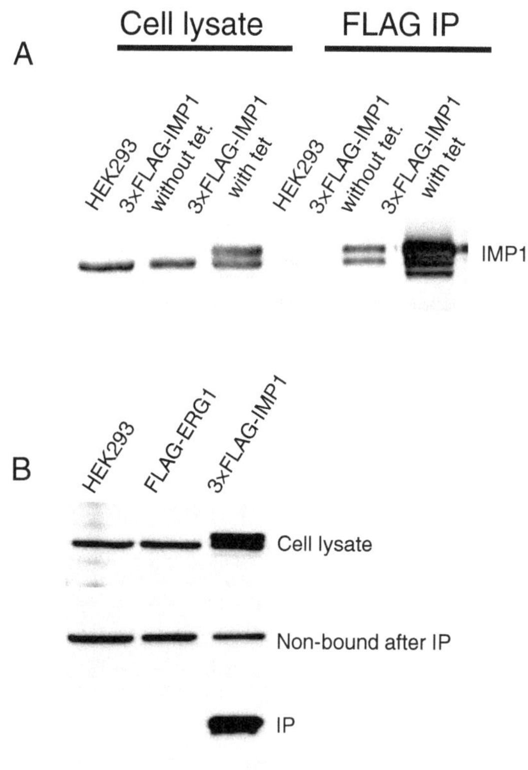

Fig. 18.2. Immunoprecipitation of FLAG-tagged RNA-binding protein by anti-FLAG antibody. ***A*** Western blot analysis with anti-IMP1 antibody of cell lysates and immunoprecipitates from HEK293 cells expressing a transgenic version of 3X FLAG-IMP1 under tetracycline control. Although 3X FLAG-IMP1 is absent in *lane 2*, a small amount is produced due to promoter leakage (see *lane 5*). Both *lanes 5 and 6* reveal that 3X FLAG-IMP1 is able to associate with endogenous IMP1, either via direct dimerization or by being present in the same RNP granule. ***B*** Western blot analysis with anti-IMP1 antibody of cell lysates (*upper panel*), non-bound fractions (*middle panel*) and immunoprecipitates (*lower panel*) from HEK293 cells expressing transgenic versions of FLAG-ERG1 and 3X FLAG-IMP1. Note that 3X FLAG-IMP1 is absent from the non-bound fraction, indicating complete immunoprecipitation.

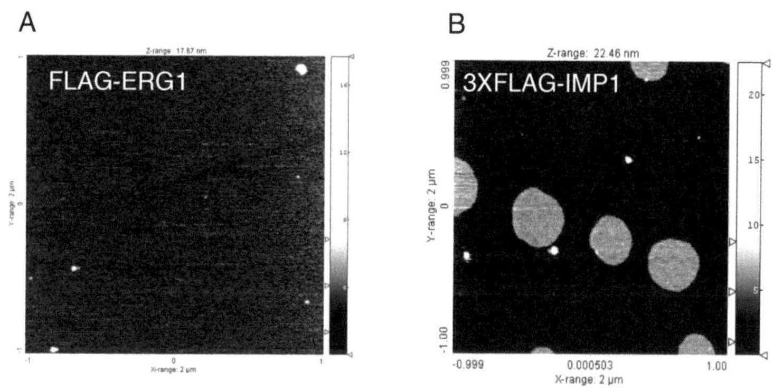

Fig. 18.3. Atomic Force Microscopy of immunoprecipitated RNPs. Anti-FLAG antibody did not precipitate granules from cells expressing FLAG-ERG1 (*panel A*), whereas granules with a diameter of about 0.3 μm are apparent from cells expressing 3X FLAG-IMP1. The samples have been diluted 1,000 times in TBS buffer prior to AFM.

2. Materials

2.1. Generation of a Cell-Line Expressing 3X FLAG-Tagged RNA-Binding Protein Under Tetracycline Control

1. Plasmids (all purchased from Invitrogen: pFRT/*lac*Zeo, pcDNA6/TR, pcDNA5/FRT/TO and pOG44).
2. GIBCO™ Phosphate-buffered saline – PBS (Invitrogen).
3. Hygromycin B (50 mg/mL) in PBS (Invitrogen).
4. Blasticidin S hydrochloride (Invitrogen).
5. Zeocin (Invitrogen).
6. Cycloheximide (Sigma-Aldrich).
7. Trypsin (0.25%) in PBS (Gibco).
8. Tetracycline (Invitrogen).
9. FuGENE 6 transfection reagent (Roche).
10. Scienceware® cloning cylinders, polystyrene (Sigma-Aldrich).
11. 2× SDS load buffer: 100 mM Tris-HCl (pH 6.8), 20% glycerol, 4% SDS, 200 mM DTT, 0.2% bromophenol blue.

2.2. Cell Lysis and Capturing of Granules Containing 3X FLAG-Tagged RNA-Binding Protein

1. Lysis buffer: 50 mM Tris-HCl, pH 7.4, 150 mM KCl, 1 mM EDTA, 1% Triton X-100, 0.5 mM PMSF (Sigma-Aldrich), 5 μL/mL protease inhibitor cocktail (Sigma-Aldrich).
2. Tris-buffered saline (TBS): 50 mM Tris-HCl, pH 7.4, 150 mM KCl, 0.5 mM PMSF (Sigma-Aldrich), 5 μL/mL protease inhibitor cocktail (Sigma-Aldrich).
3. Precoated anti-FLAG antibody-conjugated agarose beads: Place 100 μL packed Ezview red anti-FLAG M2 affinity gel (Sigma-Aldrich) in a 2-mL Eppendorf tube and add 1.5 mL TBS. Invert the tube several times and centrifuge for 30 s at 8,200×*g*. Aspirate the supernatant and add 1 mL TBS for an additional wash. Resuspend the washed beads in 2 mL lysis buffer containing 5 μg FLAG peptide (Sigma-Aldrich) and 10 μg *S. cerevisiae* ribosomal RNA. Rotate for 1 h at 4°C. Pellet the beads at 8,200×*g* for 1 min at 4°C, and wash the precoated beads twice in TBS (*see* **Note 1**).
4. Synthetic 3X FLAG peptide for elution (Sigma-Aldrich).
5. Anti-FLAG M2 monoclonal antibody (Sigma-Aldrich).
6. RNase inhibitor (Roche).

3. Methods

3.1. Generation of a Cell-Line Expressing 3X FLAG-Tagged RNA-Binding Protein Under Tetracycline Control

The protocol presented in this section is based on the Flp-In^{TM} T-Rex^{TM} system developed by Invitrogen, and it is subdivided into three parts: **Section 3.1.1** describes the construction of a host cell-line encompassing both the gene for the tetracycline repressor (Blasticidin resistance) and the FRT recombination site (Zeocin resistance). **Section 3.1.2** provides the protocol for the Flp recombinase-catalysed intermolecular recombination between the FRT site in the host cell-line and the FRT site in the vector encoding the 3X FLAG-tagged RNA-binding protein (Hygromycin B resistance). **Section 3.1.3** describes how the expression of 3X FLAG RNA-binding protein is gauged at an appropriate level by tetracycline induction.

3.1.1. Construction of a Host-Cell Line

1. Grow the chosen cell-line in complete medium (*see* **Note 2**).
2. Split cells into dishes of 10 cm in diameter, each containing 10^6 cells.
3. Transfect cells with pcDNA6/TR expressing the Tet repressor from the *tetR* gene using FuGENE 6.
4. Change medium after 24 h to a medium containing 5 μg/mL Blasticidin.
5. Grow cells to 80% confluence and split into two 10-cm Petri dishes.
6. At 50% confluence, transfect one dish with pFRT/*lac*Zeo and one dish without DNA as a control using FuGENE 6 as transfection reagent.
7. Grow cells to 80–90% confluence and harvest the cells.
8. Seed the cells in two 10-cm Petri dishes in medium containing 100 μg/mL Zeocin.
9. Change medium after 24 h to remove dead cells.
10. Grow the cells for 10–12 generations until colonies are visible in a light microscope with a diameter of 3–4 mm. At this point nothing should be alive in the negative control.
11. Remove medium and wash once in PBS.
12. Place cloning cylinders around 3–4 individual colonies.
13. Add 100 μL trypsin into each cylinder, and place the cells in the incubator for 3–5 min.
14. Add 300 μL media into each cylinder and release cells by pipetting.
15. Transfer each clone into a 3.5-cm diameter Petri dish and grow to 80% confluence.

16. Expand growth via 10-cm-diameter Petri dishes into two 175-cm^2 bottles.
17. Harvest 10^6 cells for preparation of genomic DNA (*see* **Note 3**)
18. Make a Southern blot after digestion of the DNA with *Hind*III and *Sac*I. Use a probe against *lacZ* (*see* **Note 4**).
19. Select a clone with a single insertion of the FRT (*F*lp *R*ecombination *T*arget) site and make a freezer stock. These cells are equivalent to the cells available from Invitrogen named Flp-In^{TM} T-Rex^{TM} cells.

3.1.2. Insertion of the Gene of Interest

1. Amplify your gene of interest with a 3X FLAG sequence by PCR (*see* **Note 5**).
2. Insert the PCR product into pcDNA5/FRT/TO and store until **Step 5**.
3. Seed 10^6 cells of the selected clone with a single FRT site into a 10-cm Petri dish.
4. The next day – a few hours before transfection – the medium is changed to complete medium without Zeocin, but still containing 5 μg/mL Blasticidin.
5. Co-transfect with pcDNA5/FRT/TO containing your reading-frame of interest fused to the 3X FLAG, and pOG44 expressing the Flp-recombinase using FuGENE 6 as the transfection reagent. The ratio between FuGENE 6 and the DNA should be 2:1, and the ratio between the two plasmids (pcDNA5/FRT/TO:pOG44) should be 1:10. Include a negative control that contains DNA (*see* **Note 6**).
6. The next day the medium is changed.
7. Forty-eight hours after transfection, cells are harvested and placed in a new 10-cm Petri dish with complete medium containing 5 μg/mL Blasticidin and 100 μg/mL hygromycin B.
8. The next day the medium is changed to remove dead cells.
9. The clones are grown and selected as described above (*see* **Section 3.1.1, Steps 11–17**).

3.1.3. Adjustment of Gene Expression by Tetracycline Induction

1. Test the expression of the 3X FLAG-tagged RNA-binding protein in the tetracycline-inducible system by seeding 125,000 cells in 6-well cell culture multidishes.
2. Next day the medium should be changed and tetracycline included (final concentrations of 0, 0.01, 0.1, 0.5, 1 and 2 μg/mL).
3. Two days after the inclusion of tetracycline, cells should by harvested in 2× SDS-load buffer, and Western blotting

performed using either an antibody against the FLAG tag or a specific antibody against the FLAG-tagged protein (*see* **Fig. 18.1**).

4. Select an appropriate tetracycline concentration (e.g. 0.5 μg/mL) to obtain comparable levels of endogenous and 3X FLAG-tagged versions of your protein of interest and grow cells in 175-cm^2 bottles. Include tetracycline when the cells are 50% confluent and grow the cells for 48 h.
5. Harvest the cells gently by using trypsin, and wash the cells three times in cold PBS (*see* **Note 7**).
6. Count the cells and freeze the cells without PBS at –80°C.

3.2. Cell Lysis and Capturing of Granules Containing 3X FLAG-Tagged RNA-Binding Protein

1. Resuspend harvested cells (about 4×10^8) in 1.6 mL lysis buffer.
2. Homogenize by sonication 20 times of approximately 1 s at 15 μm amplitude on ice (*see* **Note 8**).
3. After lysis, centrifuge at 8,200×*g* for 10 min at 4°C (*see* **Note 9**).
4. Remove 200 μL supernatant to a new tube as a protein control for the cell lysate or for isolation of total RNA.
5. Transfer the remaining of the supernatant (about 1.4 mL) to a 2-mL NUNC tube.
6. Add 5 μL RNasin (40 units per μL) and 75 μL FLAG-coated and washed anti-FLAG antibody-conjugated agarose beads.
7. Rotate for 2 h at 4°C.
8. Pellet the beads by centrifugation at 8,200×*g* for 1 min at 4°C.
9. Save 200 μL of the non-bound fraction for protein or RNA analysis.
10. Resuspend the pelletted beads by inversion in 1.8 mL TBS and repeat **Step 8**.
11. Repeat this wash step twice (three washes in total).
12. After removal of the final wash solution (*see* **Note 10**), the immunoprecipitated material is eluted with 500 μL TBS containing 50 μg synthetic 3X FLAG peptide for 1 h by rotation at 4°C (*see* **Note 11**).
13. After elution, the beads are removed by centrifugation at 8,200×*g* for 1 min at 4°C, and the supernatant transferred to a new tube.
14. If proteins are to be visualized by Coomassie Brilliant Blue or silver staining, trichloroacetic acid precipitation of the proteins is necessary prior to SDS-PAGE (*see* **Note 12**).

4. Notes

1. It is recommended to coat the beads with FLAG peptide to avoid non-specific binding of β-actin, some heat-shock proteins, and α-actinin-4 to the anti-FLAG antibody-conjugated agarose beads (8). If the 3X FLAG-tagged protein is expressed at very low levels or from a limited number of cells due to transient transfection, the pretreatment with synthetic FLAG peptide should be omitted to ensure complete immunoprecipitation of 3X FLAG-tagged protein, but bearing in mind the likelihood of non-specific binding of the above-mentioned proteins.
2. In our studies, we have used HEK293, HeLa and HT-1080 host cell-lines. The former can be purchased from Invitrogen.
3. For preparation of genomic DNA and subsequent Southern analysis, follow standard protocols as described in, e.g. *Molecular Cloning 3rd edition* by Sambrook and Russell (CSHL Press, 2001).
4. In addition to the Southern blot and the resistance to Zeocin, it is possible to verify the protein expression of the fusion protein between β-galactosidase and the Zeocin resistance marker by a β-galactosidase assay.
5. Provided an N-terminal 3X FLAG "handle" is to be inserted, the oligodeoxynucleotide should be synthesized so that the endogenous translational initiation codon is replaced by the cDNA encoding the 3X FLAG: 5′-ATG GACTACAAAGACCATGACGGTGATTATAAAGATCAT GACATCGATTACAAGGATGACGATGACAAG-3′. If N-terminal signal sequences are to be preserved, the cDNA encoding the 3X FLAG is inserted immediately before the stop-codon.
6. It is recommended to include a negative control, which encodes a FLAG version of a different protein than your protein of interest, to exclude putative protein interaction partners, which derive from non-specific binding to the agarose beads (*see* **Fig. 18.2b**).
7. Ribosome run-off during lysis may be inhibited by the inclusion of 12.5 μg/mL cycloheximide in the cell medium 20 min prior to harvest.
8. Cells can also be lysed by a Dounce homogenizer.
9. To examine if reassociations between RNA-binding proteins and mRNAs during cell lysis is a problem leading to artefacts, cells from the chosen expression system and

Drosophila S2 cells may be mixed prior to cell lysis. If none of the *Drosophila* transcripts, as monitored by microarray analysis on a *Drosophila* chip, are enriched after immunoprecipitation by anti-FLAG antibody, artefactual reassociations are unlikely to have taken place.

10. Isolated granules may be flash-frozen for storage at −80°C.
11. Total granular RNA can be isolated directly from the beads by addition of 1 mL Trizol (Invitrogen) and 5 μg yeast carrier RNA (tRNA or rRNA) or glycogen.
12. Precipitation of the proteins can be performed by the addition of trichloroacetic acid to a final concentration of 10%. Place the sample on ice for 10 min followed by centrifugation at 16,000 ×*g* for 10 min at 4°C. Wash the pellet in 500 μL ice-cold acetone and spin 16,000 ×*g* for 5 min at 4°C. Repeat the wash step twice. Dry the pellet after the final wash at 95°C for 5–10 min. Resuspend the dried pellet in SDS load buffer.

Acknowledgements

This work was supported by the Danish Natural Science and Medical Research Councils, and the Lundbeck Foundation.

References

1 Keene, J. D. (2007) RNA regulons: coordination of post-transcriptional events. *Nat Rev Genet* 8, 533–543.

2 Gerber, A. P., Herschlag, D., Brown, P. O. (2004) Extensive association of functionally and cytotopically related mRNAs with Puf family RNA-binding proteins in yeast. *PLoS Biol* 2, E79.

3 Krichevsky, A. M., Kosik, K. S. (2001) Neuronal RNA granules: a link between RNA localization and stimulation-dependent translation. *Neuron* 32, 683–696.

4 Kanai, Y., Dohmae, N., Hirokawa, N. (2004) Kinesin transports RNA: isolation and characterization of an RNA-transporting granule. *Neuron* 43, 513–525.

5 Villace, P., Marion, R. M., Ortin, J. (2004) The composition of Staufen-containing RNA granules from human cells indicates their role in the regulated transport and translation of messenger RNAs. *Nucleic Acids Res* 32, 2411–2420.

6 Elvira, G., Wasiak, S., Blandford, V., Tong, X. K., Serrano, A., Fan, X., del Rayo Sanchez-Carbente, M., Servant, F., Bell, A. W., Boismenu, D., Lacaille, J. C., McPherson, P. S., DesGroseillers, L., Sossin, W. S. (2006) Characterization of an RNA granule from developing brain. *Mol Cell Proteomics* 5, 635–651.

7 Mili, S., Steitz, J. A. (2004) Evidence for reassociation of RNA-binding proteins after cell lysis: implications for the interpretation of immunoprecipitation analyses. *RNA* 10, 1692–1694.

8 Jonson, L., Vikesaa, J., Krogh, A., Nielsen, L. K., Hansen, T., Borup, R., Johnsen, A. H., Christiansen, J., Nielsen, F. C. (2007) Molecular composition of IMP1 ribonucleoprotein granules. *Mol Cell Proteomics* 6, 798–811.

Chapter 19

Electrophoretic Mobility Shift Assay for Characterizing RNA–Protein Interaction

Keith T. Gagnon and E. Stuart Maxwell

Abstract

Electrophoretic mobility shift assay, or EMSA, is a well-established technique for separating macromolecules under native conditions based on a combination of shape, size, and charge. The use of EMSA can provide both general and specific information concerning the interaction between two macromolecules such as RNA and protein. Here we present a protocol for the practical use of EMSA to assess protein-RNA interactions and ribonucleoprotein (RNP) assembly. The conceptual framework of the assay is discussed along with a step-by-step procedure for the binding of archaeal ribosomal protein L7Ae to a box C/D sRNA. Potential pitfalls and common mistakes to avoid are emphasized with technical tips and a notes section. This protocol provides a starting point for the design and implementation of EMSA in studying a wide variety of RNP complexes.

Key words: EMSA, gel-shift, RNA–protein interaction, RNP assembly, radiolabeled RNA.

1. Introduction

During the course of research, it often becomes necessary to characterize the interaction between a protein and an RNA. Many methods are available for the analysis of protein–RNA interactions. Each approach depends upon the particular question being asked. Electrophoretic mobility shift assays (EMSA), commonly referred to as gel shift or band shift assays, provide a sensitive, straightforward, and low cost analysis of protein–RNA interactions. Here we will focus on using gel-shifts to observe the interaction between an in vitro synthesized box C/D sRNA and

H. Nielsen (ed.), *RNA*, Methods in Molecular Biology 703,
DOI 10.1007/978-1-59745-248-9_19, © Springer Science+Business Media, LLC 2011

recombinant ribosomal protein L7Ae from *Methanocaldococcus jannaschii*.

Gel electrophoresis is based upon the principle that charged biological molecules will migrate through a gel or porous matrix in an electric field toward the opposite charge (1, 2). Polyacrylamide gels are the standard matrix for EMSA, giving a good balance between band resolution and broad separation ranges. Because EMSA is gel electrophoresis under native, non-denaturing conditions with a buffer of near neutral pH and low ionic strength, macromolecules are separated based not only on their size and charge but also on their shape. For example, an elongated or odd shaped protein or RNA will typically run slower than a more compact, globular protein or RNA with otherwise identical molecular weight and charge. For this reason it is not possible to use molecular weight standards in native gel electrophoresis to accurately estimate protein, RNA, or ribonucleoprotein (RNP) size. Native conditions are necessary to maintain stable non-covalent interactions between protein and RNA in an electric field. The RNA, being uniformly negatively charged, will migrate toward the cathode. RNAs bound by protein will typically migrate slower through the gel due to the increased size of the RNP complex, thus causing a "shift" in the RNA band observed on the gel.

The benefits of gel-shifts over other techniques for analyzing RNA–protein interactions include sensitivity, simple setup, relatively low cost in time and materials, and a limited requirement for knowledge of the RNA–protein interaction under investigation (3). The assay only requires knowing, with some degree of precision, what the DNA or RNA is that the protein binds to and having a relatively pure form of the nucleic acid. The protein can be recombinant or a purified fraction from an extract, but an extract itself may suffice, especially if antibodies are available for the protein of interest. Only minute amounts of radiolabeled RNA and small quantities of the protein are required since the RNA is usually limiting in the reaction and the reaction volume must be small enough to load on a gel. In general, the size and absolute purity of the nucleic acid or protein is not a concern, unlike in other methods, as long as their interaction causes an observable shift in the migration of the RNA or DNA through the gel (3). Furthermore, once the basic gel-running apparatus has been setup and reagents have been prepared, multiple gel shifts can be run simultaneously and the results easily known within the day of the experiment.

EMSA is a technique often used early in characterizing RNA–protein interaction, providing the information necessary to move on to more specific experiments. On the other hand, it can be used to ask very specific questions about an RNA–protein interaction, such as through systematic mutation of the RNA or protein followed by a series of gel-shifts to assay binding. Combined with other biochemical, biophysical, or genetic approaches, EMSA is an exceptionally useful and informative tool. Although gel-shifts are simple in concept, they can sometimes pose difficult technical problems or generate puzzling results. In this chapter, we walk through an established experimental protocol showing real results and their interpretation, noting common mistakes to watch out for and tips to ensure high-quality data. A special notes section takes much of the guesswork and troubleshooting out of the method. The protocol shown here involves three parts: (1) preparation of radioactively labeled RNA, (2) a binding reaction that combines radiolabeled RNA and protein, and (3) separation of unbound RNA from protein-bound RNA by native polyacrylamide gel electrophoresis (PAGE).

The binding of ribosomal protein L7Ae to the sR8 RNA, a box C/D sRNA containing two K-turn motifs, is well-characterized and commonly used in our laboratory to train new students in the art of EMSA. Both protein and RNA genes have been cloned in our laboratory from the archaeal thermophile *M. jannaschii* (4, 5). Purification of L7Ae as a recombinant His(6X)-tagged protein is straightforward and the sR8 RNA can be quickly synthesized using an in vitro T7 RNA polymerase transcription kit (6). While the specific binding of L7Ae to a K-turn RNA has now been extensively studied with biophysical techniques, such as X-ray crystallography, fluorescence resonance energy transfer (FRET), and circular dichroism (7–11), it was originally characterized and continues to be investigated by EMSA (5, 11–14). In Archaea, L7Ae specifically recognizes a K-turn motif in the large ribosomal subunit as well as k-turns of the box C/D and box H/ACA sRNAs. For the box C/D sRNAs, L7Ae binds a terminal K-turn motif, called the box C/D, and an internal K-turn motif called the box C′/D′ (5). The box C/D sRNAs direct 2′-O-methylation of specific nucleotides through complementary base-pairing with target RNA substrates (*see* **Fig. 19.1a**). The initial in vitro binding of L7Ae is required for the subsequent binding of two other core proteins, Nop56/58 and fibrillarin, to generate an enzymatically active box C/D sRNP (5, 13) (*see* **Fig. 19.1b**). The bound core proteins are the catalytic engine of the RNP and are guided to the correct target RNA substrates by the RNA guide sequence.

2. Materials

2.1. General Methods

1. Redistilled phenol equilibrated in Tris-HCl, pH 8.0.
2. Chloroform:isoamyl alcohol (24:1).

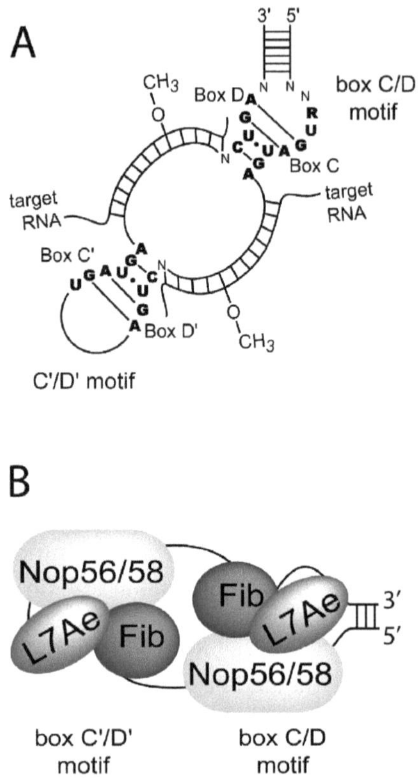

Fig. 19.1. Structure and function of the archaeal box C/D sRNP. **a** Secondary structure of archaeal box C/D sRNA base-paired to target RNA substrates. The conserved box C/D and box C′/D′ motif sequences are indicated. Guide regions base-pair with complementary target RNA substrates to guide site-specific 2′-O-methylation. **b** Three core proteins bind the archaeal box C/D sRNP to assemble in vitro an enzymatically active RNP. L7Ae initiates assembly by specifically recognizing and binding the terminal box C/D motif and internal box C′/D′ motif. Nop56/58 and fibrillarin core proteins then bind at each RNP.

3. RNase-free distilled/deionized water (ddH_2O).
4. 3 M sodium acetate solution, pH 5.2.
5. 100% ethanol.
6. 70% ethanol.

2.2. Preparation of Radiolabeled RNA

1. Calf intestinal phosphatase (CIP) and 10× CIP buffer: 0.5 M Tris-HCl pH 9.0, 100 mM $MgCl_2$, 10 mM $ZnCl_2$, 0.1 M spermidine-HCl.
2. Polynucleotide kinase (PNK) and 10× PNK buffer: 0.5 M Tris-HCl pH 7.6, 70 mM $MgCl_2$, 50 mM dithiothreitol (DTT).
3. [γ-^{32}P] adenosine triphosphate (ATP).
4. G-25 sephadex and minispin columns (Amersham Pharmacia).

5. TE buffer:10 mM Tris-HCl, pH 7.5, 1 mM EDTA.
6. 19:1 acrylamide:bisacrylamide.
7. 10× TBE: 0.89 M Tris base, 0.89 M boric acid, 20 mM EDTA.
8. Urea, molecular biology grade.
9. 10% ammonium persulfate (APS), prepared fresh.
10. N,N,N,N'-Tetramethyl-ethylenediamine (TEMED).
11. Gel loading buffer: 80% formamide, 1× TBE, 10 mM EDTA.
12. Bromophenol blue and xylene cyanol dyes.
13. Clear plastic wrap (Saran^{TM} wrap).
14. Black India ink.
15. RNA elution buffer: 0.3 M sodium acetate, 5 mM EDTA, 10 mM Tris-HCl, pH 7.4, 0.1% SDS.
16. Phosphorimager cassette or X-ray film (for visualizing radioactivity).
17. 0.45-μm syringe filter.

2.3. EMSA to Characterize RNA–Protein Interaction

1. Buffer D: 20 mM HEPES, pH 7.0, 0.1 M NaCl, 3 mM $MgCl_2$, 0.4 mM EDTA, 1 mM DTT, 20% glycerol.
2. 10× binding buffer: 0.1 M HEPES, pH 7.0, 1 M NaCl.
3. 10× phosphate dye: 25 mM potassium phosphate, pH 7.0, 25% sucrose, 0.1 mg/mL bromophenol blue.
4. 10× phosphate buffer: 0.25 M potassium phosphate, pH 7.0.
5. Glycerol, molecular biology grade.
6. 3MM Whatman filter paper.

3. Methods

3.1. General Methods

3.1.1. RNase-Free Technique

1. Use baked glassware and certified RNase-free or DEPC-treated plastic ware.
2. Wear gloves at all times. RNases from skin are the most common form of contamination.
3. Never reuse tips or tubes. Discard if you are unsure whether it has been contaminated.
4. Keep work surfaces clean and free of dust. Clean automatic pipettors regularly.

5. All reagents and buffers should be certified RNase-free from the manufacturer or prepared with RNase free chemicals and RNase-free water (ddH_2O). Everything that will touch the RNA must be free of RNases, especially protein solutions.
6. Solutions of RNA should be handled appropriately. Store dry or aqueous stocks at –20°C or colder. Do not expose RNA solutions to high concentrations of divalent metal ions, high pH (>9.0), or elevated temperatures for extended periods of time.

3.1.2. Phenol/Chloroform Extraction of RNA Solutions (Removal of Protein)

1. To an RNA solution, add 1 volume of phenol (*see* **Note 1**). Mix vigorously.
2. Separate aqueous and phenol layers by centrifugation at 10,000×*g* for 3 min.
3. Carefully transfer the top aqueous phase into a fresh tube with a pipette (*see* **Note 2**).
4. Add 1 volume of water to the phenol layer and repeat mixing and centrifugation.
5. Pool the first and second aqueous layers and add 1 volume of chloroform. Mix vigorously. Centrifuge at 10,000×*g* for 3 min.
6. Carefully transfer the top aqueous phase into a fresh tube with a pipette.
7. Precipitate the aqueous RNA solution.

3.1.3. Precipitation of RNA Solutions

1. To an aqueous RNA solution, add 1/10 volume of 3 M sodium acetate, pH 5.2.
2. Add ice-cold 100% ethanol to a final volume of 70% (a general rule of two volumes is sufficient), invert to mix, and incubate at –20°C for > 1 h (*see* **Note 3**).
3. Pellet precipitated RNA by centrifugation at >10,000×*g* for 20 min at room temperature. Carefully aspirate the ethanol solution.
4. Wash the pellet with one volume of ice-cold 70% ethanol by inverting tube several times. Immediately centrifuge at >10,000×*g* for 5 min. Carefully aspirate the ethanol.
5. Dry the pellet by lyophilization (using a "speed-vac") or laying the tube on its side in a hood.
6. Resuspend the pellet in ddH_2O and quantitate by absorbance at 260 nm (*see* **Note 4**).

3.2. Preparation of Radiolabeled RNA

RNA is most commonly "body-labeled," where the RNA transcript contains radioactive nucleotides within its sequence, or "end-labeled," where a radioactive nucleotide or phosphate is placed at the end of the RNA sequence. Here we use 5′-end labeling, which requires that the RNA does not have a 5′-phosphate (*see* **Note 5**).

3.2.1. Dephosphorylation of RNA with Calf Intestinal Phosphatase (CIP)

1. Mix the reaction components below in a 1.5-mL microfuge tube:

   20 μg RNA

   20 μL 10× CIP buffer

   10 μL CIP (1 U/μL)

   ddH_2O to 200 μL

2. Incubate at 37°C for 45 min. Phenol/chloroform extract the reaction and precipitate the RNA.

4. Resuspend the dried pellet in 30 μL ddH_2O and quantitate by absorbance at 260 nm.

3.2.2. 5′-End Labeling with T4 Polynucleotide Kinase (PNK)

1. Mix the reaction components below in a 1.5-mL microfuge tube:

   50–80 pmol CIP-treated RNA (1–2 μg)

   2.5 μL 10× PNK buffer

   8–10 μL [γ-^{32}P] ATP (1 μCi/μL)

   1 μL PNK (20 U/μL)

   ddH_2O to 25 μL

2. Incubate at 37°C for 1.5 h. Add 25 μL of ddH_2O then phenol/chloroform extract.

CAUTION: Work behind a shield and use proper technique when handling radioactivity.

3.2.3. Purification of 5′-End Labeled RNA

Two methods are available for purification of radiolabeled RNA. The phenol/chloroform-extracted RNA can be filtered through size exclusion resin to remove free radioactive nucleotides and salts or purified by denaturing gel electrophoresis. Although more time consuming, gel purification is recommended for gel shifts of the highest quality. Gel purification is desirable if the starting RNA was not initially purified or degradation occurs during the labeling process. Simply label twice as much RNA and scale up the labeling reaction proportionately if you plan to gel purify your RNA.

3.2.3.1. Removing Unincorporated [γ-^{32}P]ATP by Size Exclusion

1. Filter phenol/chloroform-extracted RNA (50 μL) by centrifugation through a 2-cm bed of G-25 size exclusion resin packed in a mini-spin column (Amersham Pharmacia) (*see* **Note 6**). Spin at low speed (<1,000×*g*) for 3 min.

2. Check radioactivity by Cerenkov counting in a scintillation counter (*see* **Note 7**). Do not use scintillation fluid. Place 1

μL of eluate in a 0.5-mL microfuge tube in a scintillation vial for counting.

3. Record date and radioactive counts and store at –20°C. The half-life of ^{32}P is 14.2 days.

3.2.3.2. Gel Purification of Radiolabeled RNA

1. Prepare a 40 mL solution containing 6% acrylamide (19:1 acrylamide:bisacrylamide), 1× TBE and 7 M urea (*see* **Note 8**).
2. Add 10% APS (8 μL/mL) and TEMED (1 μL/mL). Mix by inverting.
3. Pour into assembled gel apparatus (15 × 17 × 1.5 cm) and position a comb in the top of the gel.
4. Allow the gel to polymerize for 20 min. Remove the comb; rinse out the wells with 1× TBE, and pre-run the gel with 1× TBE running buffer for ~20 min at 40–45 mA (*see* **Note 9**).
5. Add 1/2 volume (25 μL) of gel loading buffer to the extracted and 5′-end labeled RNA.
6. Boil the RNA sample for 3–5 min. Cool to room temperature.
7. Load the sample. In a separate lane load 20 μL gel loading dye (gel loading buffer +0.1 mg/mL bromophenol blue and xylene cyanol) (*see* **Table 19.1** for dye migration distances).

Table 19.1
Migration of dyes in EMSA

Acrylamide (19:1) (%)	Bromophenol blue (lower) dye band	Xylene cyanol (upper) dye band
5	35 nucleotides	130 nucleotides
6	26 nucleotides	105 nucleotides
8	19 nucleotides	75 nucleotides
10	12 nucleotides	55 nucleotides

8. Run gel at 40–45 mA (*see* **Note 9**). Run time is from 1 to 2 h. Use the dye bands as an approximation for where the RNA's migration position is in the gel (*see* **Table 19.1**).
9. Separate the glass plates with a wedge so that the gel sticks to one plate.
10. Cover the exposed gel with clear plastic wrap (Saran™ wrap).
11. Place three small drops of radioactive dye (1 μL [γ-^{32}P]ATP in 30 μL black india ink) on the Saran™ wrap at three corners of the gel and let them air dry.

12. Cover the dried drops with clear tape and place a phosphorimager cassette on top.
13. Expose for 5–10 min. Scan cassette in phosphorimager and place print out of gel under the glass plate (*see* **Note 10**). Align the dots and cut out the radioactive RNA band with a heat-treated razor blade.
14. Crush the gel slice into a fine paste in a 1.5-mL microfuge tube using a 1-mL pipette tip that has been sealed at the tip with a heat source.
15. Add 500 μL of RNA elution buffer. Rock at room temperature for 45 min (*see* **Note 11**).
16. Recover the eluted RNA by spinning at 10,000×g for 2 min. Filter the elution (solution on top of the gel bits) through a 0.45 μm syringe filter into a new 1.5-mL microfuge tube.
17. Repeat the elution with 300 μL of RNA elution buffer. Spin again and filter through the same syringe to pool with the previous elution. Split the elution into two tubes (400 μL each) and ethanol precipitate (omit addition of sodium acetate).
18. Check the radioactivity and handle as in **Section 3.2.3.1** above.

3.3. EMSA to Characterize RNA–Protein Interaction

3.3.1. RNA–Protein Binding Reactions

1. Add the components indicated in **Table 19.2** to a microfuge tube at room temperature in the order shown:
   a. Mix 10× binding buffer with the tRNA (*see* **Note 12**) and ddH_2O.
   b. Add sR8 RNA (*see* **Note 13**).
   c. Add buffer D and L7Ae protein (*see* **Note 14**).
2. Mix the reaction gently and incubate at 70°C for 8 min (*see* **Note 15**). Cool to room temperature, spin down to remove any precipitation. Transfer the reaction to a new tube, add 2 μL of 10× phosphate dye, and mix gently.

3.3.2. Resolving RNA–Protein Complexes by Native Page

3.3.2.1. Preparing the Native Polyacrylamide Gel

1. Prepare a 40-mL solution containing 6% acrylamide (19:1 acrylamide:bisacrylamide), 1× phosphate buffer, and 2% glycerol (*see* **Note 16**).

Table 19.2
Reaction components and set up for titration of L7Ae with sR8 RNA

	[serial dilution]	1×	20 kcpm/μL 0.2 pmol/μL	10 mg/mL	10×			
Rxn #	L7Ae (μL)	Buffer D (μL)	sR8 (μL)	tRNA (μL)	Binding buffer (μL)	ddH_2O (μL)	Final volume (μL)	Final [L7](nM)
1	1	10	1	1	2	5	20	0
2	1	9	1	2	2	5	20	20
3	1	9	1	2	2	5	20	50
4	1	9	1	2	2	5	20	100
5	1	9	1	2	2	5	20	200
6	1	9	1	2	2	5	20	300
7	1	9	1	2	2	5	20	400
8	1	9	1	2	2	5	20	600
9	1	9	1	2	2	5	20	800
10	1	9	1	2	2	5	20	1,000

2. Add 10% APS (8 μL/mL) and TEMED (1 μL/mL) and mix by inverting.
3. Pour into an assembled gel apparatus (16 × 18 × 1.5 cm) and position a 15-well comb in the top of the gel (*see* **Note 17**).
4. Allow the gel to polymerize for 20 min. Remove the comb and place the gel in a Hoefer SE600 apparatus (or comparable apparatus) filled with 3.5 L of 1× phosphate puffer (*see* **Note 18**). Add 0.5 L of buffer to the top tank and rinse out the wells with a glass Pasteur pipette.

3.3.2.2. Running the Gel

1. Load samples into the wells. Turn on cold water and stir bar to circulate buffer (*see* **Note 19**).
2. Run the gel at 150 V until the bromophenol blue band has migrated about 2/3 of the way through the gel (1.5–2 h).

3.3.2.3. Visualizing the Gel

1. Separate the glass plates so that the gel sticks to one plate.
2. Cut two pieces of 3MM Whatman filter paper slightly larger than the gel. Press filter paper against the gel. Peel the paper off the plate. The gel will stick to the paper (*see* **Note 20**).
3. Cover the exposed side of the gel with clear plastic wrap and dry in a gel dryer under vacuum at 80°C for 1 h.
4. When dry, discard the back piece of filter paper (it absorbed excess liquid and radioactivity). Expose the gel to a phosphorimager cassette overnight or to X-ray film for 2–4 h.

3.4. Analysis of EMSA Results

3.4.1. Titration of L7Ae onto sR8 RNA

Representative results from the above protocol are shown in **Fig. 19.2a**. The first gel lane on the left is RNA-only. L7Ae is then titrated in subsequent lanes. Increasing concentrations of L7Ae induced a shift in the RNA band, indicating slower migration of the RNA due to protein binding. Each discrete shift represents a unique RNP. The sR8 box C/D sRNA has two K-turns, which are both recognized by the L7Ae protein. Thus L7Ae binds the RNA twice. The first shift in RNA migration quickly becomes shifted again, suggesting cooperative binding. Indeed, quantification of these bands using ImageQuant software (Molecular Dynamics) revealed a sigmoidal binding curve, indicating cooperativity (*see* **Fig. 19.2b**). Since there are two binding events, the binding constants cannot be readily distinguished. However, using RNA constructs containing only one of the L7Ae binding sites allowed us to determine the binding constant for each (5).

3.4.2. Further Applications of EMSA

A number of variations to the protocol presented here allow for analysis of multiple proteins binding to an RNA species.

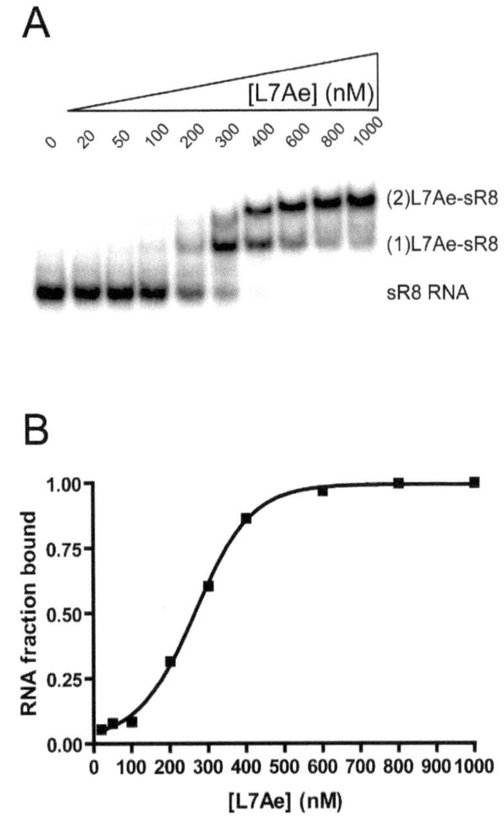

Fig. 19.2. EMSA of L7Ae binding to sR8 box C/D sRNA. **a** Titration of L7Ae with a fixed concentration of radiolabeled sR8 RNA (*see* **Table 19.2b**). L7Ae binds to the RNA twice, first forming an RNP with one L7Ae protein [(1)L7Ae-sR8] and two L7Ae proteins [(2)L7Ae-sR8]. Quantification of L7Ae binding to sR8 RNA. ImageQuant software was used to quantify the intensity of RNA and RNP bands. Band intensity was normalized and plotted using Prism 4 software (Graph Pad) as the fraction of RNA bound versus the concentration of L7Ae. The *sigmoidal curve* indicates the cooperative binding of two L7Ae proteins to sR8 RNA.

RNA-binding specificity, or demonstrating the presence of sequence-specific RNA-binding proteins in an extract (3). As an example of multicomponent RNP assembly, two other core proteins, Nop56/58 and fibrillarin, bind to form an enzymatically active box C/D sRNP in vitro. These interactions and their consequence on RNP assembly and structure have been investigated by our lab and others using EMSA (5, 11–14).

The specificity of protein binding can be determined using a variety of competitor RNAs in the binding reactions, such as specific RNA mutants. These experiments involve either radiolabeling the mutant RNA or titrating excess unlabeled mutant RNA into reactions with labeled wild-type RNA at fixed protein concentrations, often called a competition gel-shift assay. For the protocol presented here, we have included an excess of non-specific tRNA in the reactions to ensure that the shifts in RNA migration are due to specific L7Ae binding. This is an important factor since many RNA-binding proteins bind RNA non-specifically at higher concentrations.

To determine the identity of the RNA-binding protein when an extract or partially purified protein sample is used, antibodies that bind suspected proteins can be included in the binding reaction. Most often, if the antibody binds a protein that is part of the RNP, then an additional upward shift of the RNP band is observed on the gel due to formation of an antibody–RNP complex. These assays are called supershift assays. However, antibodies can also block RNA binding of the protein, resulting in a slower migrating complex or a reduction in RNA binding and band shifting.

Several experimental variables can be modified to optimize binding for a particular RNA–protein interaction. These include altering the ionic strength (salt components) or pH of the binding reaction, or adding non-ionic detergents and different carrier RNAs or proteins. The protocol shown here for L7Ae and sR8 RNA has been optimized. The buffers used are mild and a good starting point for independent investigations of other protein–RNA interactions. Electrophoresis conditions can also be modified, such as the acrylamide percentage, ratio of acrylamide to bisacrylamide, and crosslinking percentage. Likewise, different running and gel buffers can be used. Common buffers are TBE, Tris-acetate-EDTA (TAE), or Tris-glycine buffers. In this protocol, we used phosphate buffer, which is a more universal buffer and provides optimal resolution for assembly of the archaeal box C/D sRNP. Occasionally, it may be useful to include salts, like magnesium, in the running buffer, although this requires re-circulation between the top and bottom tanks during electrophoresis to prevent salt deposition on the electrodes. Variations in other physical parameters may also be useful. For instance, the archaeal box C/D sRNP requires elevated temperatures for efficient protein binding. We recently made use of low temperature binding and resolution of reactions in a cold room to determine which steps in the assembly pathway were dependent on temperature and therefore RNA or protein conformational dynamics (11).

4. Notes

1. Use caution when working with phenol. Ideally, phenol should be redistilled. Equilibrate in 0.1 M, pH 8.0, Tris-HCl buffer prior to use. Addition of 0.2%

β-mercaptoethanol will prevent oxidation and extend shelf life. Store redistilled and equilibrated phenol at –20°C and working stocks at 4°C. Discard phenol after 6 months at –20°C or if it becomes discolored (most often a pink hue).

2. The phenol and aqueous layers can become inverted, such as with solutions of very high salt or sucrose concentration. Adding a small amount of water prior to or after extraction will determine which layer is aqueous. Be careful not to collect phenol with the aqueous phase. For best results, leave some of the aqueous phase on top of the phenol to ensure that phenol is not also collected. If phenol is collected with the aqueous phase, it is removed in the chloroform extraction step. Chloroform that is carried over in the last step of phenol/chloroform extraction is removed during RNA precipitation.
3. Very dilute RNA solutions often do not precipitate completely. In the protocols presented here, this is typically not a problem. However, if a problem arises, incubation at –20°C overnight (16 h) can increase precipitation efficiency. Also, small amounts of a carrier can be added to aid precipitation, such as glycogen or a non-specific RNA that is safe to have in your reactions later (we often use tRNA from *Escherichia coli* or yeast).
4. Use Beer's Law to calculate the RNA concentration from absorbance at 260 nm (if the extinction coefficient is known) or multiply the OD value at 260 nm by the general conversion factor of 40 μg/OD/mL for RNA. This will yield a value of μg/mL RNA.
5. If your RNA is synthetic, then it does not have a 5′ phosphate and this step should be omitted. If the RNA was transcribed in vitro or purified from an extract, you will need to dephosphorylate.
6. G-25 resin should be swollen and equilibrated in TE buffer prior to use. A swinging bucket rotor will provide a cleaner separation and more efficient removal of free [γ-^{32}P]ATP.
7. Do not use scintillation fluid for Cerenkov counting. Place 1 μL of eluate in a 0.5-mL microfuge tube in a scintillation vial for counting.
8. For your particular RNA, you may choose the gel percentage based on the size of the RNA to be resolved (*see* **Table 19.1**). We keep a stock of 20% acrylamide in 1× TBE, 7 M urea in our lab and dilute it with a stock of 1× TBE, 7 M urea to give the desired final acrylamide percentage. This circumvents the task of preparing fresh every time, which is unnecessary and time consuming. Keep stocks in the dark at room temperature and discard after 1 month.

9. Pre-running the gel is necessary to heat it up. It should be very warm or hot to the touch. The heat and the urea in the gel help to keep the RNA denatured. As a general rule, electrophorese the RNA until it has migrated approximately 2/3 of the way through the gel (*see* **Table 19.1**). Monitor the gel running apparatus closely so that the gel does not overheat. The glass plates should be warm or hot to the touch during the run, but not unbearable. If the plates get too hot they will crack. If samples "smile," where the middle samples run faster than the outer samples, this indicates uneven heating of the gel. Too much smiling should be avoided by reducing the current going through the gel.
10. We do not adjust the size of the phosphorimager picture, just the contrast if necessary, then print. As an alternative, the gel can be exposed to X-ray film for 5–15 min in the dark room and the developed film can be placed under the glass gel plate.
11. Alternatively, the elution step can be performed by rocking overnight at 4°C.
12. The storage and dilution buffer for L7Ae is buffer D. Always add protein solution directly to the reaction last and not to the sides of the tube.
13. tRNA serves as a non-specific RNA in these reactions at excess molar concentrations. Its presence ensures that only specific RNA binding is observed and provides better RNP resolution on the gel.
14. Radiolabeled RNA should be mixed with non-radiolabeled RNA to make a stock with 20,000 cpm's and 0.2 pmol of RNA per μL. 20,000 cpm's per reaction is usually sufficient; however, more or less may be necessary to optimize visualization of the radioactive RNA bands.
15. *M. jannaschii* is a thermophile and efficient binding of the L7Ae protein to sR8 box C/D sRNA requires elevated temperatures. Most RNA-binding proteins bind optimally in a range from 4 to 37°C. Frequently, a carrier protein, such as bovine serum albumin (BSA) is added in excess to keep protein levels at a relatively constant concentration, prevent protein precipitation, and provide clearer RNP resolution. BSA is not included in these reactions due to aggregation and precipitation that can occur under the extreme heat conditions employed for binding. The final reaction contains 20 mM HEPES, pH 7.0, 0.15 M NaCl, 0.5 mM DTT, 0.2 mM EDTA, 10% glycerol, 1.5 mM $MgCl_2$, 1 mg/mL tRNA.
16. Glycerol is an important component of the gel. Do not omit. During drying of the gel prior to visualization,

glycerol prevents the gel from shrinking and cracking, which can make a gel unusable for publication. Glycerol is also thought to aid in RNP resolution.

17. Prior to assembling glass plates with spacers for pouring the gel, the plates should be clean and free of detergent. Glass plates can be wiped or rinsed with ethanol before assembly to remove dust. Glass plates can also be baked, although a thorough washing is often sufficient to remove contaminating RNases.

18. 1× phosphate buffer is sensitive to fungal and bacterial growth and should be prepared fresh before use from a 10× stock. Running buffer can be reused up to several times if it is re-circulated between the top and bottom tanks during or after electrophoresis. However, reuse of buffer is not recommended.

19. Use care when loading the gel so as not to mix sample with running buffer. Start by placing the tip of the pipette at the bottom of the gel and slowly filling the well. When the tip is almost empty, pull it up to the top of the well to finish. Avoid blowing bubbles out of the pipette tip, which will push the sample out of the well and dilute it. Remember, native gels do not have a "stacking layer" so compact loading is important to form sharp bands in the gel. The gel must be kept cool for optimal resolution. Cold running tap water is usually sufficient. Do not allow the temperature of the gel to rise more than a few degrees above room temperature.

20. If the gel percentage is high (>14%), the gel may not stick to the filter paper. In this case, place plastic wrap on top of the gel, flip the gel and plat over, and peel the plate away from the gel. The gel should stick to the plastic wrap. The filter paper can then be placed on top of the gel for drying. For low percentage gels (4%) the gel can very easily lose shape, making the bands in the gel wavy after drying and visualization. Use caution in transferring the gel from the plate to the filter paper. Squirting ddH_2O onto the gel will help if the gel will not adhere to one plate.

References

1. Fried, M., Crothers, D. M. (1981) Equilibria and kinetics of *lac* prepressor-operator interactions by polyacrylamide gel electrophoresis. *Nucleic Acids Res* 9, 6505–6525.
2. Garner, M. M., Revzin, A. (1981) A gel electrophoresis method for quantifying the binding of proteins to specific DNA regions: application to components of the *Escherichia coli* lactose operon regulatory system. *Nucleic Acids Res* 9, 3047–3060.
3. Buratowski, S., Chodosh, L. A. (1996) Mobility shift DNA-binding assay using gel electrophoresis, in (F. Ausubel et al., eds.), *Current Protocols in Molecular Biology*. John Wiley & Sons, Inc., New York, NY. pp. 12.2.1–12.2.8.

4. Kuhn, J. F., Tran, E. J., Maxwell, E. S. (2002) Archaeal ribosomal protein L7 is a functional homolog of the eukaryotic 15.5kD/Snu13p snoRNP core protein. *Nucleic Acids Res* 30, 931–941.
5. Tran, E. J., Zhang, X., Maxwell, E. S. (2003) Efficient RNA 2′-O-methylation requires juxtaposed and symmetrically assembled archaeal box C/D and C'/D' RNPs. *EMBO J* 22, 3930–3940.
6. Gagnon, K. T., Zhang, X., Maxwell, E. S. (2007) In vitro reconstitution and affinity purification of catalytically active archaeal box C/D sRNP complexes. *Methods Enzymol* 425, 263–282.
7. Moore, T., Zhang, Y., Fenley, M. O., Li, H. (2004) Molecular basis of box C/D RNA-protein interactions: cocrystal structure of archaeal L7Ae and a box C/D RNA. *Structure* 12, 807–818.
8. Hama, T., Ferre-D'Amare, A. R. (2004) Structure of protein L7Ae bound to a K-turn derived from an archaeal box H/ACA sRNA at 1.8 A resolution. *Structure* 12, 893–903.
9. Turner, B., Melcher, S. A., Wilson, T. J., Norman, D. G., Lilley, D. M. J. (2005) Induced fit of RNA on binding the L7Ae protein to the kink-turn motif. *RNA* 11, 1192–1200.
10. Suryadi, J., Tran, E. J., Maxwell, E. S., Brown, B. A., II (2005) The crystal structure of *Methanocaldococcus jannaschii* multifunctional L7Ae RNA-binding protein reveals an induced-fit interaction with the box C/D RNAs. *Biochemistry* 44, 9657–9672.
11. Gagnon, K. T., Zhang, X., Agris, P. F., Maxwell, E. S. (2006) Assembly of the archaeal box C/D sRNP can occur via alternative pathways and requires temperature-facilitated sRNA remodeling. *J Mol Biol* 362, 1025–1042.
12. Zhang, X., Champion, E. A., Tran, E. J., Brown, B. A., II, Baserga, S. J., Maxwell, E. S. (2006) The coiled-coil domain of the Nop56/58 core protein is dispensable for sRNP assembly but is critical for archaeal box C/D sRNP-guided nucleotide methylation. *RNA* 12, 1092–1103.
13. Omer, A. D., Ziesche, S., Ebhardt, H., Dennis, P. P. (2002) In vitro reconstitution and activity of a C/D box methylation guide ribonucleoprotein complex. *Proc Natl Acad Sci USA* 99, 5289–5294.
14. Omer, A. D., Zago, M., Chang, A., Dennis, P. P. (2006) Probing the structure and function of an archaeal C/D-box methylation guide sRNA. *RNA* 12, 1708–1720.

Chapter 20

Polysome Analysis and RNA Purification from Sucrose Gradients

Tomáš Mašek, Leoš Valášek and Martin Pospíšek

Abstract

Velocity separation of translation complexes in linear sucrose gradients is the ultimate method for both analysis of the overall fitness of protein synthesis as well as for detailed investigation of physiological roles played by individual factors of the translational machinery. Polysome profile analysis is a frequently performed task in translational control research that not only enables direct monitoring of the efficiency of translation but can easily be extended with a wide range of downstream applications such as Northern and Western blotting, genome-wide microarray analysis or qRT-PCR. This chapter provides a basic overview of the polysome profile analysis technique and the RNA isolation procedure from sucrose gradients. We also discuss possible experimental pitfalls of data normalization, describe main alternatives of the basic protocol and outline a novel application of denaturing RNA electrophoresis in several steps of polysome profile analysis.

Key words: Translational control, polysome profile, RNA isolation, sucrose gradient, RNA denaturing electrophoresis.

1. Introduction

Regulation of translation plays a very important role in the control of gene expression as it allows for a more rapid response to a variety of both intra- and extracellular stimuli than transcriptional modulation. Translational control mechanisms target mostly the initiation phase as it is the rate-limiting step of protein synthesis. For the majority of cellular transcripts, the 40S ribosome associated with several translation initiation factors (eIFs) including eIF2 (in complex with GTP and Met-tRNA$_i^{Met}$) and eIF3 interacts with mRNAs *via* the 7-methyl guanosine cap structure at

H. Nielsen (ed.), *RNA*, Methods in Molecular Biology 703,
DOI 10.1007/978-1-59745-248-9_20,

their 5′end prebound with a complex of three eIFs called eIF4F. The 48S pre-initiation complex formed in this way then undergoes specific conformational changes that enable this machinery to start scanning the 5′-untranslated region for the AUG start codon in an optimal sequence context. Upon AUG recognition, the GTP hydrolysis reaction is completed, most of the eIFs are ejected, and a large 60S ribosomal subunit joins the 40S-mRNA-Met-tRNA$_i^{Met}$ complex to produce the translation-competent 80S ribosome (monosome; for a general review on translation initiation *see* (1)).

More than one 80S monosome can be translating an mRNA at a time producing so called polysomes. The number of polysomes on an mRNA reflects the initiation, elongation and termination rates and is a measure of the translatability of the particular transcript under given conditions. Lower or higher than average association of a particular mRNA with ribosomes indicates its "strength" as well as a potential involvement of gene-specific regulatory mechanisms.

Velocity sedimentation in sucrose gradients was introduced more than 40 years ago for assessing translational fitness of the cell (2). The polysome profile analysis has been routinely used to monitor the translational status under various physiological conditions (3–5), during stress and subsequent cell recovery (6–8) (*see* **Fig. 20.1**), to reveal defects in ribosome biogenesis (9, 10), to investigate functions of proteins involved in translation (11–14), to determine the role of 5′UTR structures on translatability of corresponding mRNAs (15), and for examination of miRNA mediated translational repression (16, 17). The polysome profile analysis is especially well established in yeast translation research. However, the method can be easily modified for bacterial (18, 19), plant (20, 21) and mammalian cells (22, 23) as well as for the translation-competent cell-free systems (24). The general use of polysome analysis can be further extended by collecting

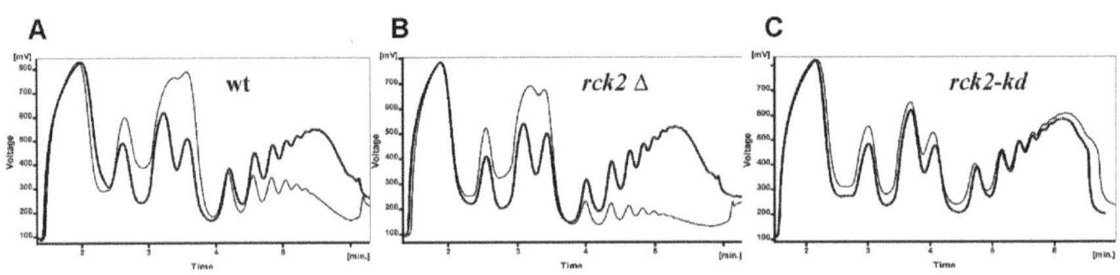

Fig. 20.1. An example of typical polysome profiles of three yeast strains subjected to an oxidative stress. One contains the wild-type *RCK2* gene (a MAPKAP kinase operating downstream of HOG signaling pathway (7)). The other two contain its mutant alleles. Cultures were harvested before (*thick lines*, no stress) or after the exposure to 0.8 mM t-butyl hydroperoxide for 30 min (*thin lines*, 30 min with tBOOH). (**a**) Wild-type; (**b**) *rck2Δ*; (**c**) *rck2-kd* (a dominant negative allele encoding catalytically inactive enzyme). In the wt and *rck2Δ* cells, the number of polysomes decrease upon stress indicating an inhibition of general translation initiation. The polysome fraction is decreased to an even higher degree in *rck2Δ*. By contrast, *rck2-kd* fails to show polysome run-off due to a block in translation elongation. The charts were generated by the Clarity software.

fractions from sucrose gradients followed by a variety of downstream applications including Western and Northern blotting, qRT-PCR (25, 26), RNase protection assay (16), and microarray analysis (27–30). High-throughput polysome fractionation using deep 96-well plates has also been reported. However, it does not seem to stand up to the high quality of resolution of the classical setup (31).

The polysome profile analysis has been described in the literature with a variety of modifications. The main concern has to do with the choice of a stabilization reagent used to prevent polysome run-off. The most widely used reagent is the antibiotic cycloheximide that binds the 60S ribosomal subunit (32) and is thought to block translation elongation by preventing release of deacylated tRNA from the ribosome E site after translocation (33, 34), thus stalling the 80S ribosomes on mRNA in a polysomal state. Usage of cycloheximide may be omitted in studies aimed at examining defects in the elongation step as these usually prevent polysome run-off that naturally occurs during the cell lysate preparation in the absence of any stabilization agent (35).

Heparin, a highly sulfated glycosaminoglycan, is routinely used to stabilize translational complexes pre-treated with cycloheximide (36, 37) and to protect them against RNase activity during preparation of cell extracts. However, inclusion of heparin in extraction buffers seems to inhibit initiation of protein synthesis (38, 39) and leads to artificial association of initiation factors with pre-initiation complexes that do not reflect their natural state in the cell at the time of lysis (25). Hence, a new strategy has recently been developed employing formaldehyde as a cross-linking reagent to fix ribosomes on mRNAs in the living yeast cells. This technique is believed to provide the best available approximation of the native 43S/48S pre-initiation complexes composition in vivo (40).

A decrease in the initiation rate results in the polysome run-off with a concomitant increase in the amount of free 80S ribosomes seen as a monosomal peak in a polysome profile. The fraction of vacant mRNA-free 80S ribosomes can be distinguished from mRNA-bound monosomes on the basis of their different sensitivity to high salt concentrations (41). The 80S couples dissociate into individual subunits at 0.8 M KCl (41) or 0.7 M NaCl (36) only if they are not associated with an mRNA.

When performing polysome analysis, a common task is calculation of ratios of particular peak areas in order to determine what proportion of the translational machinery is actively engaged in translation. Consensually, only polyribosomes are considered to be actively translating ribosomes because the monosomal peak contains an unknown proportion of mRNA-free 80S couples. Therefore the translational rate is usually expressed as the polysome-to-monosome (P/M) ratio, which, in theory, decreases with translation initiation defects but increases with defects in elongation. The P/M ratio determination may not have a true predicative value in those cases where a particular mutant causes accumulation of free ribosomal subunits as a consequence of either a defect in ribosome biogenesis or reduced ability of mRNAs to be translated or as a result of inhibition or slowing down initiation complex assembly (*see* **Fig. 20.2a**).

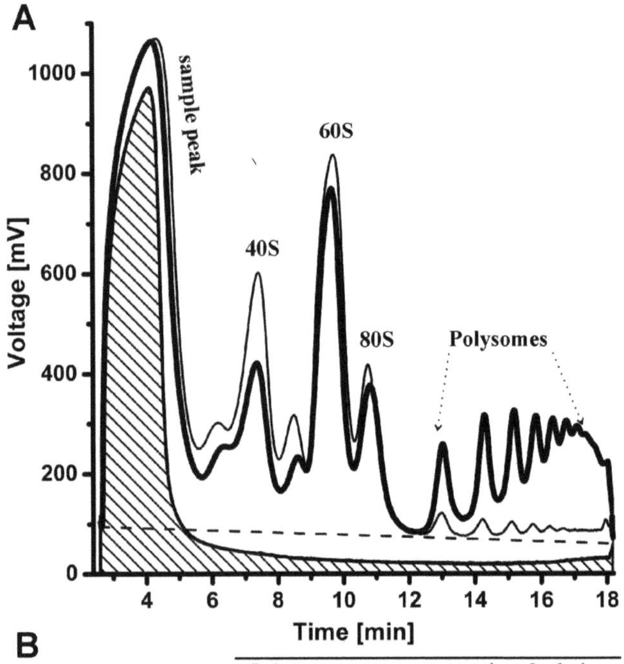

B

	Polysome to monosome ratio calculation	
	using chromatogram baseline	*via* blank subtraction
W303 *ceg1*^{ts} 24°C	3,38	3,84
W303 *ceg1*^{ts} 37°C	1,12	0,56
Change of translation efficiency	3x	6,8x

Fig. 20.2. Polysome profile normalization strategies and the P/M ratio calculations. **a** Polysome profiles of the W303 strain carrying a temperature-sensitive allele of an essential *CEG1* gene coding for a guanylyl transferase subunit of the yeast capping enzyme (45). The culture was grown at a semipermissive temperature of 24°C (*black thick line*) and then shifted to a non-permissive 37°C for additional 12 h (*dark grey thin line*). A sample peaks indicate a volume and a total absorbance of yeast cell lysates deprived of ribosomes or PEB buffer loaded on sucrose gradients (*blank*). Following peaks correspond to 40S and 60S ribosomal subunits, to the 80S monosome and to polysomes. Identities of the two unmarked peaks, which typically appear in the *ceg1*^{ts} polysome profiles, are unknown. The *dashed grey line* depicts the actual chromatogram baseline calculated by Clarity software after overlaying both chromatograms and transposing the curves to the same position. This baseline connects the lowest points of the curves. *Shaded area* corresponds to a blank tube containing only PEB. Raw data were exported into a tab-delimited-text format and displayed with the help of OriginPro® software. **b** Polysome-to-monosome area ratios were determined either using the chromatogram baselines or by subtraction of the blank area from the polysome profile areas. The inclusion of a blank tube in this experiment permitted more accurate determination of P/M ratios.

Calculation of peak areas represents another pitfall (*see* **Fig. 20.2**). In the majority of published experiments, the peak areas are subtracted either from the baseline corresponding to detector zero or from the baseline extrapolated by the application of chromatography software, which usually connects the lowest points of curves. These methods of peak area determination might not be as accurate as often believed. Discrepancies may be caused by extraction buffers containing TritonX-100 and/or other compounds exhibiting a substantial absorbance at 254 nm. The area of the first peak detected in a polysome profile (the sample peak) thus mostly reflects the amount of TritonX-100 in the sample loaded on the sucrose gradient and, indirectly, also corresponds to the sample volume. If the sample peaks do not differ substantially and if the equal sample volumes were loaded, it is recommended to overlay and compare polysome profile chromatograms based on the first sample peak. Blank tubes containing only extraction buffer can be used to circumvent many difficulties and provide us with a more realistic baseline reflecting absorbance of extraction buffer across the polysome profile and allowing for a more exact determination of the P/M ratio (*see* **Fig. 20.2**). As for the data acquisition followed by their post-analysis modifications as well as for the peak area calculations, we take an advantage of ISCO gradient analyzer connected with the data-acquisition PC card in combination with the Clarity chromatography software (DataApex Company; www.dataapex.com). This software allows not only smooth on-line data acquisition but also many logistic operations such as baseline shifting, profiles zooming in/out and peak editing, combining and dividing. The Clarity software also supports graphical editing of profile curves including their overlaying (*see* **Fig. 20.1**) as well as saving and exporting raw or edited data in various formats (*see* **Fig. 20.2**). The subtraction of blank area from polysome profiles can be carried out if the same artificial detector zero line is inserted at the beginning of all readings. Such artificial baselines ensure easy comparison and recalculation of the measured data between samples in addition to the blank sample subtraction. The calculation of profile areas can be further achieved after raw data export to the suitable spread-sheet calculator (e.g., OriginPro®) without inserting artificial detector zero line (*see* **Fig. 20.2**). Substantial differences in the net profile areas after blank subtraction in a single experiment usually indicate that unequal lysate concentrations were loaded on gradients.

Unexpected discrepancies in the polysome profile analysis may also be caused by degradation of RNA during a crude cell extract preparation and subsequent procedures. It is recommended to check the RNA quality in lysates electrophoretically prior to the analysis. We have recently introduced a simpler and less hazardous TAE/formamide agarose gel electrophoresis that is particularly suitable for RNA separation in crude cell extracts containing large amounts of proteins, DNA and other contaminating molecules. We have also demonstrated that this technique can be successfully used for analysis of unpurified or partially purified sucrose gradient fractions as well as for high quality resolution of purified polysomal RNA that is perfectly suitable for the subsequent Northern blot analysis (42) (*see* **Fig. 20.3**).

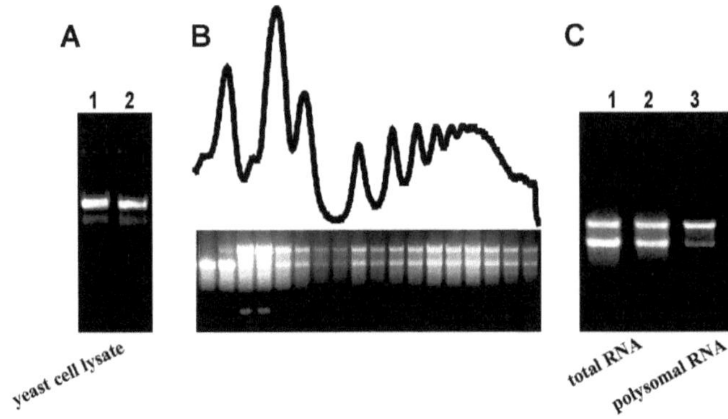

Fig. 20.3. Application of TAE-formamide agarose gel electrophoresis at various steps of the yeast polysome profile analysis. **a** Quality assessment of a crude yeast cell extract. Formamide was added to a yeast cell lysate to a final concentration of 60% (v/v). Loading dye was supplemented with 1% SDS (**Section 3.5**, **Option 1**). **b** Electrophoresis of yeast polysome profile fractions. The profile corresponds to the lysate shown in (**a**; line 1) that has been loaded onto a 7–50% sucrose gradient and centrifuged in a SW41 rotor for 3 h at 35,000 RPM at 4°C. 0.5-mL fractions were collected starting from a layer where the small ribosomal subunits sediment. RNA was coarsely purified according to the protocol described in **Section 3.5**, **Option 2. c** Comparison of whole-cell RNA (lines 1, 2) purified by the acid-phenol method directly from the yeast and RNA samples purified from polysome fractions (lines 3, 4) by the protocol presented at **Section 3.5**, **Option 3.** All samples originated from the same yeast culture.

2. Materials

2.1. Yeast Culture and Preparation of Cell Lysate

1. SC medium: 2% (w/v) glucose, 0.65% (w/v) yeast nitrogen base, 50 mg/L of each auxotrophic supplement
2. YPD medium: 2% (w/v) glucose, 1% (w/v) yeast extract, 2% (w/v) bactopeptone
3. RNAse-free deionized water (*see* **Notes 1 and 2**)
4. Cycloheximide stock 10 mg/mL
5. Polysome extraction buffer (PEB): 20 mM Tris-HCl, pH 7.4, 140 mM KCl, 5 mM $MgCl_2$, 0.5 mM DTT, 1% (v/w) TritonX-100, 0.1 mg/mL cycloheximide, 0.2 mg/mL heparin (ammonium salt)

6. Glass beads, acid washed (0.45–0.55 mm in diameter, *see* **Note 3**)

2.2. Gradient Preparation

1. Gradient solution 1: 20 mM Tris-HCl, pH 7.4, 140 mM KCl, 5 mM $MgCl_2$, 0.5 mM DTT, 0.1 mg/mL cycloheximide, 0.2 mg/mL heparin, 7% (w/v) sucrose (*see* **Note 4**)
2. Gradient solution 2: 20 mM Tris-HCl, pH 7.4, 140 mM KCl, 5 mM $MgCl_2$, 0.5 mM DTT, 0.1 mg/mL cycloheximide, 0.2 mg/mL heparin, 50% (v/w) sucrose
3. Solution 3: 20 mM Tris-HCl, pH 7.4, 140 mM KCl, 5 mM $MgCl_2$, 0.1 mg/mL cycloheximide, 60% (w/v) sucrose (*see* **Note 5**)

2.3. RNA Isolation from Polysomal Profiles

1. GuITC: 6 M guanidium thiocyanate, 0.25 M sodium acetate
2. 96 and 75% ethanol
3. RNAase-free deionized water
4. Acid phenol, pH 4.0–5.2
5. Chloroform: Chloroform:Isoamylalcohol (24:1)
6. 6 M LiCl
7. 3 M sodium acetate, pH 5.2

2.4. RNA Electrophoresis

1. 50× TAE: to prepare 1 L, add 242 g Tris, 100 mL of 0.5 M EDTA, pH 8.0, and 57.1 mL of glacial acetic acid
2. Agarose
3. Deionized formamide
4. Ethidium bromide (1 mg/mL)
5. 10× Loading Dye: 50 mM Tris-HCl, pH 7.6, 0.25% (w/v) Bromophenol Blue, 60% (v/v) glycerol

3. Methods

3.1. Yeast Culture and Preparation of Cell Lysate

1. Inoculate a yeast strain of interest into 40 mL of a suitable medium and grow it in 250-mL Ehrlenmayer flask at the desired temperature for 12–24 h to early stationary phase. Use either rich YPD medium or defined SC minimal medium depending on the type of experiment; an incubation temperature of 28°C works well for most yeast strains (*see* **Note 6** on culture conditions and **Note 7** for a procedure for making extracts from mammalian cells).
2. Inoculate approximately 40 μL of the stationary culture to 75 mL of fresh medium in a 250-mL Ehrlenmayer flask

and incubate the cells for 12–16 h with vigorous shaking until the cultures reach mid-exponential growth phase (OD_{660}=0.4–0.6).

3. At the time of harvest, add 750 μL of cycloheximide from the stock solution and chill cells by adding one spoon of crushed ice (approx. 20 g of ice, 25% of total culture volume), gently shake several times and keep on ice for 5 min. All subsequent steps have to be carried out on ice and with pre-chilled tubes and centrifuges.
4. Transfer the cells into two 50-mL Falcon tubes and centrifuge them for 5 min at 3,000×*g* at 4°C. Resuspend the pellets by adding 3 mL of ice-cold PEB and pool the aliquots in one tube (*see* **Note 8**). Centrifuge once again for 5 min at 3,000×*g*.
5. Repeat the washing step by adding 6 mL of ice-cold PEB.
6. Resuspend the cells in 700 μL of ice-cold PEB and transfer the resulting cell suspension into a pre-chilled 1.5-mL Eppendorf tube containing 450 μL of pre-chilled glass beads.
7. Break the cells by vigorous agitation with 30 oscillations/s in a bead-beater for 3 min (e.g., MM301, Retsch). Tube holders should be pre-chilled at –20°C for at least 1 h. Appropriate conditions for efficient cell lysis can vary with different equipment and should be set empirically (*see* **Note 9**).
8. Clear cell lysates by centrifugation at 8,000×*g* for 5 min.
9. Immediately proceed to loading of the cell lysates onto sucrose gradients. If necessary, lysates can be stored at –70°C for no longer than several days or shipped on dry ice at this stage.

3.2. Gradient Preparation and Centrifugation (for SW41 Beckman Rotors)

1. Measure the concentration of nucleic acids in the lysate spectrophotometrically and optionally check the RNA integrity (*see* **Section 3.5**, **Option 1**). About 10–15 OD_{260} units should be loaded on the gradient, optimally at a volume less than 400 μL, but not exceeding 800 μL (*see* **Note 10** for up-scaling).
2. Linear sucrose gradients can be prepared in several ways (*see* **Note 11**). We usually prefer making gradients with the use of a commercial gradient maker (Hoefer SG-50, **Fig. 20.4**). For preparation of one 7–50% sucrose gradient, fill chamber A with 6.3 mL of gradient solution 1 (7% sucrose) and chamber B, which is closer to the outlet, with 6.3 mL of gradient solution 2 (50% sucrose). The combined volumes make up for the maximal capacity (12 mL) of SW41 centrifugation tubes (Beckman, Ultra-ClearTM tube) and the

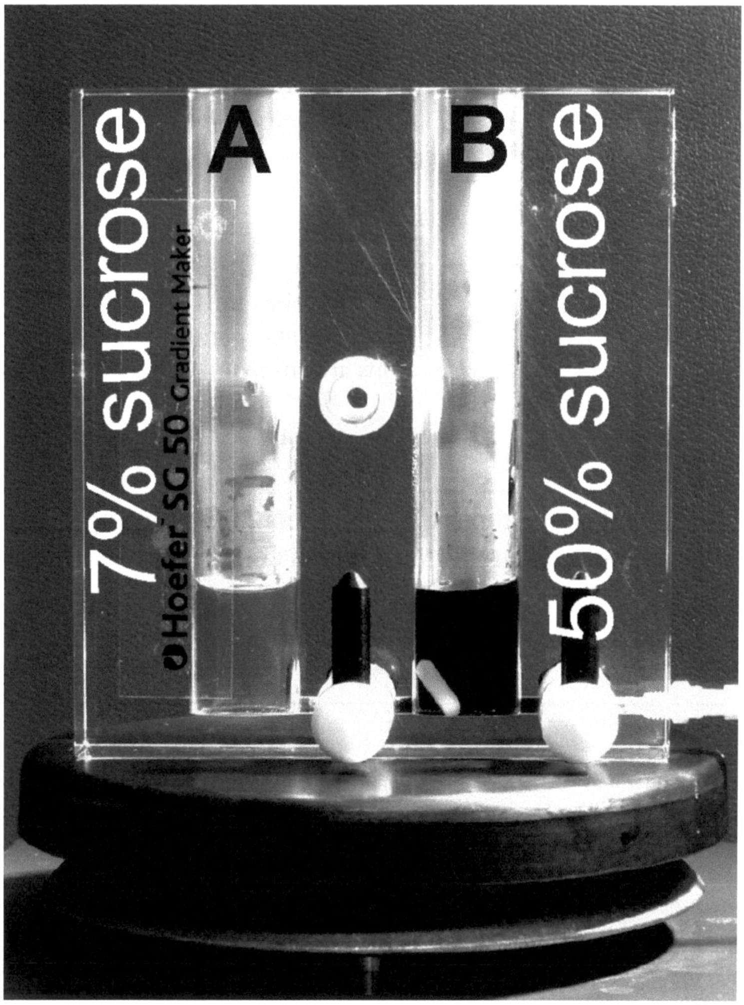

Fig. 20.4. Use of a gradient maker for preparation of linear sucrose gradients. Chamber A and B are filled with the same volume of 7 and 50% sucrose solutions, respectively. After connecting the chambers, gentle continuous mixing in chamber B generates a concentration gradient which flows into a centrifugation tube. More detailed instructions are described in **Section 3.2**, **Step 2**.

dead volume of the Hoefer SG-50 gradient maker that is around 800 μL. After filling both chambers, add a stir bar into chamber B, open the tap connecting the chambers and force out any bubbles from the connecting tube, if necessary. Open the outlet tap and suck the solution towards the end of the connected elastic tubing by a pipette. Then turn on the magnetic stirrer, adjust the position of the gradient maker, and apply an appropriate speed of swirling to provide gentle but complete mixing in chamber B. Place the end of the elastic tubing at the bottom of a centrifugation tube and start pouring the gradient by slow continuous movement of the tubing towards the top. A slow pouring is important for preparation of a high-quality undisturbed gradient and can be achieved by proper adjustment of the distance between the tube outlet and the level of the sucrose solutions in the gradient maker.

3. Carefully load the lysate on the top of the gradient. Balance pairs of tubes to be centrifuged carefully with PEB (*see* **Note 12**).
4. Put the tubes into a pre-cooled SW41 rotor according the Beckman instructions (*see* **Note 13**). Centrifugation conditions are summarized in **Table 20.1**.

Table 20.1
Gradient ranges and centrifugation conditions (using a Beckman SW41 rotor) for visualization of different translational complexes

Translational complexes to be resolved	Sucrose concentration range (%)	Time of centrifugation	Speed (rpm)	Citation
Eukaryotic 40S, 60S subunits, 80S ribosome and polysomes	4.5–45	2.5	39,000	(25)
	7–50	3	35,000	(7)
	15–50	2.5	40,000	(5)
Bacterial 30S, 50S subunits, 70S ribosome and polysomes	5–40	2.5	35,000	(19)
	10–40	2.5	35,000	(46)
40S–80S	15–40	4.5	39,000	(36)
	5–40	2	27,000	(9)
40S–60S	7.5–30	5	41,000	(25)
	5–30	8	27,000	(9)

3.3. Data Acquisition and Normalization

1. Place the centrifugation tube onto a Tube Piercer of the UA-6 UV/Vis detector (ISCO, Inc.) and carefully mount it in a holder. Start pushing up 60% sucrose (Solution 3) from the bottom of tube by switching on the peristaltic pump and adjusting flow rate to 2.4 mL/min. Monitor absorbance at 254 nm continuously.
2. Absorbance profiles can be recorded either by chart recorder which is an integral part of the ISCO instrument or by an external data-acquisition module equipped with an appropriate software. We recommend Clarity from DataApex (www.dataapex.com).

3. If various ribosomal complexes are subjected to further analysis of their RNA and/or protein content, collect fractions corresponding either to the desired peaks or to fixed volumes (*see* **Note 14**).

3.4. RNA Isolation from Polysome Profiles (for SW41 or SW28 Profiles Split into Two Fractions)

1. To prevent degradation of samples by RNases, mix them with an equal volume of GuITC and vortex well immediately upon collecting the fractions. Add an equal volume of 96% ethanol to precipitate nucleic acids from the samples and incubate them overnight at –20°C, which is usually sufficient for quantitative precipitation. Optionally, RNA can be analyzed electrophoretically at this step (*see* **Section 3.5, Option 2**).
2. Transfer samples into 28-mL centrifuge tubes and spin down precipitated nucleic acids for 20 min at 25,000×*g* at 4°C. For SW28 fractions, pool aliquots in one 28-mL centrifuge tube by repeating the centrifugation step.
3. Wash the pellets with 5 mL of 75% ethanol. Decrease the volume of 75% ethanol for easier transfer of pellets into 1.5-mL Eppendorf tubes. At this step, the isolation procedure can be interrupted and samples can be stored at –70°C.
4. Centrifuge the samples at 21,000×*g* for 15 min at room temperature. Aspirate the ethanol and apply a second short spin followed by removal of residual ethanol and air-drying of the pellet to completely get rid of ethanol (beware that over-drying can result in difficulties in resuspension of the pellet).
5. Dissolve pellets by adding 400 μL of DEPC-treated water. Add 400 μL of acidic phenol and vortex for 5 min. Incubate samples for 1 min at room temperature, then add 400 μL of chloroform and repeat vortexing for 5 min. Centrifuge samples for 20 min at 21,700×*g* at 4°C (or *see* **Note 15**).
6. Transfer the aqueous phase containing the RNA into a new 1.5-mL Eppendorf tube.
7. Adjust volumes to 750 μL by RNase-free water, then add 250 μL of 6 M LiCl (final concentration 1.5 M), vortex, and leave overnight at –20°C. Centrifuge samples for 20 min at 25,000×*g* at 4°C (*see* **Note 16**).
8. Completely remove the supernatants and wash the pellets with 1 mL of 75% ethanol. Centrifuge samples for 15 min at 21,700×*g* at 4°C.
9. Repeat Step 8.

10. Dissolve dried RNA pellets in 350 μL of RNase-free water and precipitate by adding 35 μL of sodium acetate and 1 mL of 96% ethanol. Incubate samples for at least 1 h at –20°C, followed by centrifugation for 20 min at 21,700×*g* at 4°C.
11. Wash the purified RNA by repeating the Steps 8 and 9, but decrease the volume of ethanol to 500 μL.
12. Dissolve RNA in an appropriate volume of RNase-free water (e.g., 25 μL) to a recommended concentration of 1–5 μg/μL. Measure concentration of RNA spectrophotometrically and check the integrity (*see* **Section 3.5**, **Option 3**). RNA has to be stored at –80°C.

3.5. Electrophoresis of RNA in TAE/Formamide Agarose Gels

1. Dissolve an appropriate amount of agarose to obtain 1.2–1.5% gel in 1× TAE buffer by heating in a microwave oven. Cool the agarose to approximately 50°C and cast the gel.
2. Place the gel tray with the solidified agarose gel into an electrophoresis tank and fill the tank with 1× TAE buffer.
3. Add a loading buffer to your samples (*see* **Options 1–3** below), denature RNA at 65°C for 10 min and chill on ice for 5 min.
4. Load the samples into the wells, connect power supply and run electrophoresis at a voltage of 5 V/cm.

Option 1: Analysis of the quality of cell lysates by RNA electrophoresis.

Mix 10 μL of your lysate with 17 μL of deionized formamide. Add 1 μL of ethidium bromide and 2 μL of 10× loading dye supplemented with 1% SDS.

Option 2: Rapid determination of the content of polysome profile fractions by RNA electrophoresis.

We routinely collect 0.5-mL fractions from SW41 sucrose gradients. Rough purification of RNA is carried out by adding 0.5 mL of GuITC, vortexing, and precipitation with 1 mL of 96% ethanol. After two washing steps, each with 1 mL of 75% ethanol and air-drying the pellets, RNA is dissolved in 60 μL of formamide. For RNA electrophoresis, 30-μL aliquots are mixed with 1 μL of ethidium bromide and 3 μL of a loading dye (or *see* **Note 17**).

Option 3: Electrophoresis of highly purified RNA isolated from sucrose gradients for Northern blotting, RT-PCR, and microarrays analyses.

Generally, we load 5–15 μg of RNA on a 1–1.5% agarose gel and prepare samples to meet the following criteria: formamide, 60–90% (v/v); ethidium bromide, 1–5 μg; and 1× loading dye.

4. Notes

1. Because of a high consumption of RNase-free water for polysome profile analysis, we tend to prepare it in bigger volumes; e.g., 5 L of ddH_2O in one large Pyrex bottle is stirred on magnetic stirrer with 5 mL of diethyl pyrocarbonate (DEPC) overnight, then boiled in the open Pyrex bottle for 4 h, and finally autoclaved for 40 min to completely remove residual DEPC. DEPC is a hazardous chemical; both mixing and boiling must be performed in a fume hood.
2. As a precaution against degradation of ribosomal complexes or purified RNA it is imperative to avoid contamination by RNases. We strongly recommend to heat-sterilize all glassware at 160°C for 2 h, to use all disposable plasticware (e.g., pipette tips and Eppendorf tubes) directly from dedicated bags, to submerge all non-disposable or non-sterile plasticware (for example, Beckman centrifugation tubes) into 1% (v/v) peroxide for 6 h followed by thorough rinsing in DEPC-treated water. We prepare all solutions by directly dissolving chemicals in RNase-free water without subsequent treatment with DEPC. Gloves should be worn throughout.
3. RNase-free glass beads are either commercially available (for example from Sigma-Aldrich) or can be prepared by an overnight wash with hydrochloric acid, followed by a thorough rinsing on a Büchner funnel and heat-sterilization at 180°C for 3 h.
4. We recommend preparation of gradient solutions 1, 2, and 3 a day in advance, to let them slowly cool to 4°C.
5. Solution 3 has to be equilibrated to room temperature for several hours before use. We normally evacuate the solution to prevent the formation of air bubbles when running the gradient through the ISCO UV detector cell.
6. Generally, rapidly growing cultures cultivated in rich media give very nice polysomal profiles. Translation is extremely sensitive to the physiological status of cells. To analyze effects of mutations, chemical agents or stress treatment on translation, it is essential first to add cycloheximide to the cell cultures, then to rapidly harvest and chill them on ice. All cultures to be compared should be handled in the same way during the entire procedure. The results can vary, for instance, by longer incubation on ice leading to slow dissociation of polysomes.

7. To make whole-cell extracts from mammalian cells the following protocol is recommended (43). 1×10^6 HeLa cells give enough material for one SW41 gradient. First, cells are washed twice with PBS buffer containing cycloheximide at a concentration of 100 μg/mL. After scraping cells into 1 mL of cycloheximide supplemented PBS, pellet them by centrifugation at 1,000×*g* for 2 min at 4°C. Lyse the cell pellet for 10 min on ice by adding 450 μL of lysis buffer containing 20 mM HEPES, pH 7.5, 125 mM KCl, 5 mM $MgCl_2$, 2 mM DTT, 0.5% (v/v) NP-40, and 100 μg/mL of cycloheximide; alternatively supplemented with 100 U/mL RNase inhibitor, 1× complete protein inhibitor cocktail (Roche), or 1 mM PMSF. Clear the lysate by centrifugation at 16,000×*g* for 10 min at 4°C.
8. Alternatively, PEB without TritonX-100 can be used for washing the cells. This would prevent formation of bubbles during vortexing.
9. If no breaking apparatus is available, three pulses of vigorous vortexing at the maximum speed for 40 s alternated by 2-min breaks on ice disrupt yeast cells sufficiently.
10. For analysis of the content of polysome fractions, for instance by microarrays, the whole procedure can be easily scaled-up by a factor of 3 in SW28 centrifugation tubes (38 mL). In this case 30–45 OD_{260} units should be loaded on each gradient, optimally in final volume of 1–1.5 mL but not exceeding 2.5 mL; centrifugation is set to 28,000 rpm for 5 h at 4°C; and finally a flow rate of the peristaltic pump should be adjusted to 2.5–2.8 mL/min.
11. There are several additional protocols describing how to make sucrose gradients. The very elegant, simple, and reproducible freeze-thawing method was introduced by Luthe (44). First, prepare 17.5, 25.6, 33.8, 41.9, and 50% (w/v) sucrose gradient solutions. Then pour 2 mL of each solution into 14 × 89 mm Beckman centrifugation tube, starting with the most concentrated sucrose. Each sucrose layer is frozen at –80°C for 15 min before applying the next solution. Before use, thaw the tubes at 4°C overnight to form continuous sucrose gradient. This method is suitable for preparation of multiple sucrose gradients in advance, because preformed gradients can be stored for several months.
12. If areas of different peaks between multiple profiles are going to be compared, it is advantageous to even the gradient volumes before loading the samples. To normalize the polysome profile data, it is also advantageous to run one

or two blank tubes containing the same volume of PEB as that of the samples.

13. Overnight pre-cooling of the desired rotor at 4°C is strongly recommended as it prevents its potential damage due to temperature heterogeneity of the rotor body.
14. Most commonly, two fractions are collected. The first fraction usually contains ribosomal subunits and monosomes representing a non-translated pool of transcripts (note that an inclusion of monosomes into this fraction might not be suitable for all mRNAs). The second fraction, containing polysomes, corresponds to a pool of actively translated mRNAs. For accurate collection of corresponding fractions, we recommend to determine the dead volume of the detector/collector system and a corresponding time delay between particular peak detection and its elution from the ISCO instrument. Fractionation of the polysome profile into multiple fractions of a fixed volume is particularly suitable for Northern and Western blot analyses.
15. The Steps 5 and 6 can be substituted by a Trizol reagent isolation procedure.
16. The LiCl precipitation procedure is applied in order to remove heparin from the RNA sample. However, lithium ions interfere with reverse transcriptase activity. Hence the protocol has to be extended by a second round of ethanol precipitation if the RNA is to be used as a substrate for this enzyme.
17. For the simplest and the most rapid electrophoretic analysis of RNA from polysome profile fractions, it is possible to take directly 15 μL of a collected fraction, mix it with 15 μL of formamide, 1 μL of ethidium bromide and 3 μL of 10× loading dye, and run the agarose gel. This setup is not recommended for subsequent northern analysis.

Acknowledgment

This work was supported by Czech Science Foundation grant No. 301/07/0607, Ministry of Education, Youth and Sports of the Czech Republic grant No. LC06066 (both to MP). LV was supported by The Wellcome Trusts grant No. 076456/Z/05/Z, Fellowship of Jan E. Purkyne from Academy of Sciences of the Czech Republic, and Inst. Research Concept AV0Z50200510.

References

1. Hershey, J. W. B., Merrick, W. C. (2000) Pathway and mechanism of initiation of protein synthesis, in (Sonenberg, N., Hershey, J. W. B. and Mathews, M. B., eds.), *Translational Control of Gene Expression*. Cold Spring Harbor Laboratory Press, Cold Spring Harbor, NY, pp. 33–88.
2. Warner, J. R., Knopf, P. M., Rich, A. (1963) A multiple ribosomal structure in protein synthesis. *Proc Natl Acad Sci USA* 49, 122–129.
3. Dickson, L. M., Brown, A. J. (1998) mRNA translation in yeast during entry into stationary phase. *Mol Gen Genet* 259, 282–293.
4. Kuhn, K. M., DeRisi, J. L., Brown, P. O., Sarnow, P. (2001) Global and specific translational regulation in the genomic response of Saccharomyces cerevisiae to a rapid transfer from a fermentable to a nonfermentable carbon source. *Mol Cell Biol* 21, 916–927.
5. Ashe, M. P., De Long, S. K., Sachs, A. B. (2000) Glucose depletion rapidly inhibits translation initiation in yeast. *Mol Biol Cell* 11, 833–848.
6. Uesono, Y., Toh, E. A. (2002) Transient inhibition of translation initiation by osmotic stress. *J Biol Chem* 277, 13848–13855.
7. Swaminathan, S., Masek, T., Molin, C., Pospisek, M., Sunnerhagen, P. (2006) Rck2 is required for reprogramming of ribosomes during oxidative stress. *Mol Biol Cell* 17, 1472–1482.
8. Asp, E., Nilsson, D., Sunnerhagen, P. (2008) Fission yeast mitogen-activated protein kinase Sty1 interacts with translation factors. *Eukaryotic Cell* 7, 328–338.
9. Van Ryk, D. I., Lee, Y., Nazar, R. N. (1992) Unbalanced ribosome assembly in *Saccharomyces cerevisiae* expressing mutant 5 S rRNAs. *J Biol Chem* 267, 16177–16181.
10. Martin-Marcos, P., Hinnebusch, A. G., Tamame, M. (2007) Ribosomal protein L33 is required for ribosome biogenesis, subunit joining, and repression of GCN4 translation. *Mol Cell Biol* 27, 5968–5985.
11. Valasek, L., Nielsen, K. H., Hinnebusch, A. G. (2002) Direct eIF2-eIF3 contact in the multifactor complex is important for translation initiation in vivo. *EMBO J* 21, 5886–5898.
12. Jivotovskaya, A. V., Valasek, L., Hinnebusch, A. G., Nielsen, K. H. (2006) Eukaryotic translation initiation factor 3 (eIF3) and eIF2 can promote mRNA binding to 40S subunits independently of eIF4G in yeast. *Mol Cell Biol* 26, 1355–1372.
13. Kainuma, M., Hershey, J. W. B. (2001) Depletion and deletion analyses of eucaryotic translation initiation factor 1A in *Saccharomyces cerevisiae*. *Biochimie* 83, 505–514.
14. Gross, J. D., Moerke, N. J., von der Haar, T., Lugovskoy, A. A., Sachs, A. B., McCarthy, J. E., Wagner, G. (2003) Ribosome loading onto the mRNA cap is driven by conformational coupling between eIF4G and eIF4E. *Cell* 115, 739–750.
15. Sagliocco, F. A., Vega Laso, M. R., Zhu, D., Tuite, M. F., McCarthy, J. E., Brown, A. J. (1993) The influence of 5′-secondary structures upon ribosome binding to mRNA during translation in yeast. *J Biol Chem* 268, 26522–26530.
16. Seggerson, K., Tang, L., Moss, E. G. (2002) Two genetic circuits repress the *Caenorhabditis elegans* heterochronic gene lin-28 after translation initiation. *Dev Biol* 243, 215–225.
17. Nottrott, S., Simard, M. J., Richter, J. D. (2006) Human let-7a miRNA blocks protein production on actively translating polyribosomes. *Nat Struct Mol Biol* 13, 1108–1114.
18. Irwin, C. C., Akagi, J. M., Himes, R. H. (1973) Ribosomes, polyribosomes, and deoxyribonucleic acid from thermophilic mesophilic, and psychrophilic clostridia. *J Bacteriol* 113, 252–262.
19. Xia, B., Etchegaray, J. P., Inouye, M. (2001) Nonsense mutations in cspA cause ribosome trapping leading to complete growth inhibition and cell death at low temperature in Escherichia coli. *J Biol Chem* 276, 35581–35588.
20. Breen, M. D., Whitehead, E. I., Kenefick, D. G. (1972) Requirement for Extraction of Polyribosomes from Barley Tissue. *Plant Physiol* 49, 733–739.
21. Davies, E., Larkins, B. A., Knight, R. H. (1972) Polyribosomes from Peas: an improved method for their isolation in the absence of ribonuclease inhibitors. *Plant Physiol* 50, 581–584.
22. Tscherne, J. S., Pestka, S. (1975) Inhibition of protein synthesis in intact HeLa cells. *Antimicrob Agents Chemother* 8, 479–487.
23. Wei, C. L., MacMillan, S. E., Hershey, J. W. (1995) Protein synthesis initiation factor eIF-1A is a moderately abundant RNA-binding protein. *J Biol Chem* 270, 5764–5771.
24. Tas, P. W., Martini, O. H. (1986) Effects of addition of derived 40 S subunits on translation rate and polysome profile of the reticulocyte lysate. *Biochim Biophys Acta* 866, 75–82.

25. Nielsen, K. H., Szamecz, B., Valasek, L., Jivotovskaya, A., Shin, B. S., Hinnebusch, A. G. (2004) Functions of eIF3 downstream of 48S assembly impact AUG recognition and GCN4 translational control. *EMBO J* 23, 1166–1177.
26. Nelson, P. T., Hatzigeorgiou, A. G., Mourelatos, Z. (2004) miRNP:mRNA association in polyribosomes in a human neuronal cell line. *RNA* 10, 387–394.
27. Shenton, D., Smirnova, J. B., Selley, J. N., Carroll, K., Hubbard, S. J., Pavitt, G. D., Ashe, M. P., Grant, C. M. (2006) Global translational responses to oxidative stress impact upon multiple levels of protein synthesis. *J Biol Chem* 281, 29011–29021.
28. Smirnova, J. B., Selley, J. N., Sanchez-Cabo, F., Carroll, K., Eddy, A. A., McCarthy, J. E. G., Hubbard, S. J., Pavitt, G. D., Grant, C. M., Ashe, M. P. (2005) Global gene expression profiling reveals widespread yet distinctive translational responses to different eukaryotic translation initiation factor 2B-targeting stress pathways. *Mol Cell Biol* 25, 9340–9349.
29. MacKay, V. L., Li, X., Flory, M. R., Turcott, E., Law, G. L., Serikawa, K. A., Xu, X. L., Lee, H., Goodlett, D. R., Aebersold, R., Zhao, L. P., Morris, D. R. (2004) Gene expression analyzed by high-resolution state array analysis and quantitative proteomics: response of yeast to mating pheromone. *Mol Cell Proteomics* 3, 478–489.
30. Arava, Y., Wang, Y., Storey, J. D., Liu, C. L., Brown, P. O., Herschlag, D. (2003) Genome-wide analysis of mRNA translation profiles in *Saccharomyces cerevisiae*. *Proc Natl Acad Sci USA* 100, 3889–3894.
31. Wang, Y., Ringquist, S., Cho, A. H., Rondeau, G., Welsh, J. (2004) High-throughput polyribosome fractionation. *Nucleic Acids Res* 32, e79.
32. Stocklein, W., Piepersberg, W. (1980) Binding of cycloheximide to ribosomes from wild-type and mutant strains of *Saccharomyces cerevisiae*. *Antimicrob Agents Chemother* 18, 863–867.
33. Obrig, T. G., Culp, W. J., McKeehan, W. L., Hardesty, B. (1971) The mechanism by which cycloheximide and related glutarimide antibiotics inhibit peptide synthesis on reticulocyte ribosomes. *J Biol Chem* 246, 174–181.
34. Pestova, T. V., Hellen, C. U. (2003) Translation elongation after assembly of ribosomes on the Cricket paralysis virus internal ribosomal entry site without initiation factors or initiator tRNA. *Genes Dev* 17, 181–186.
35. Ortiz, P. A., Kinzy, T. G. (2005) Dominant-negative mutant phenotypes and the regulation of translation elongation factor 2 levels in yeast. *Nucleic Acids Res* 33, 5740–5748.
36. Asano, K., Clayton, J., Shalev, A., Hinnebusch, A. G. (2000) A multifactor complex of eukaryotic initiation factors, eIF1, eIF2, eIF3, eIF5, and initiator tRNA(Met) is an important translation initiation intermediate in vivo. *Genes Dev* 14, 2534–2546.
37. Asano, K., Shalev, A., Phan, L., Nielsen, K., Clayton, J., Valasek, L., Donahue, T. F., Hinnebusch, A. G. (2001) Multiple roles for the C-terminal domain of eIF5 in translation initiation complex assembly and GTPase activation. *EMBO J* 20, 2326–2337.
38. Hradec, J., Dusek, Z. (1978) All factors required for protein synthesis are retained on heparin bound to Sepharose. *Biochem J* 172, 1–7.
39. Waldman, A. A., Marx, G., Goldstein, J. (1975) Isolation of rabbit reticulocyte initiation factors by means of heparin bound to sepharose. *Proc Natl Acad Sci USA* 72, 2352–2356.
40. Valasek, L., Szamecz, B., Hinnebusch, A. G., Nielsen, K. H. (2007) In vivo stabilization of preinitiation complexes by formaldehyde cross-linking. *Methods Enzymol* 429, 163–183.
41. Martin, T. E., Hartwell, L. H. (1970) Resistance of active yeast ribosomes to dissociation by KCl. *J Biol Chem* 245, 1504–1506.
42. Masek, T., Vopalensky, V., Suchomelova, P., Pospisek, M. (2005) Denaturing RNA electrophoresis in TAE agarose gels. *Anal Biochem* 336, 46–50.
43. Clancy, J. L., Nousch, M., Humphreys, D. T., Westman, B. J., Beilharz, T. H., Preiss, T. (2007) Methods to analyze microRNA-mediated control of mRNA translation. *Methods Enzymol* 431, 83–111.
44. Luthe, D. S. (1983) A simple technique for the preparation and storage of sucrose gradients. *Anal Biochem* 135, 230–232.
45. Schwer, B., Mao, X., Shuman, S. (1998) Accelerated mRNA decay in conditional mutants of yeast mRNA capping enzyme. *Nucleic Acids Res* 26, 2050–2057.
46. Powers, T., Noller, H. F. (1990) Dominant lethal mutations in a conserved loop in 16S rRNA. *Proc Natl Acad Sci USA* 87, 1042–1046.

Chapter 21

Prediction of Targets for MicroRNAs

Morten Lindow

Abstract

MicroRNAs (miRNAs) are small 20–22 nt long RNAs which function as post-transcriptional regulators altering the expression of genes either by blocking translation or by destabilizing mRNAs (for recent reviews *see*, e.g., Zhang et al. (J Cell Physiol, 210:279–289) and Engels and Hutvagner (Oncogene, 25:6163–6169)). A central problem in miRNA biology is to identify the mRNAs regulated by miRNAs – the miRNA targets. A large number (>10) of bioinformatics methods have been developed to address this question, but unfortunately the scarcity of experimentally validated targets makes it hard to objectively judge the performance of the methods (for an attempt *see* Sethupathy et al. (Nat Methods, 3:881–886). Nevertheless, here I will give some guidelines on how to use the existing tools to find miRNA targets.

Key words: MicroRNA, target prediction, post-transcriptional regulation, RNA-RNA interaction.

1. The Empirical Basis for microRNA Target Prediction

Since miRNA target prediction is by no means perfect yet, it is important to understand the empirical basis on which the algorithms are built. For a detailed reviews of the principles of target prediction refer to, e.g., (1–3).

1.1. Biochemical and Structural Evidence

Target finding for *plant* miRNAs was an early success in miRNA-bioinformatics: It has been shown that plant miRNA targets can be found much above noise simply by searching for sequences highly complementary to the miRNA in mRNA coding or untranslated sequences (4). The validity of predictions with high complementary to the miRNA is high and has been shown numerous times experimentally to lead to endonucleolytic cleavage of the target (5).

H. Nielsen (ed.), *RNA*, Methods in Molecular Biology 703,
DOI 10.1007/978-1-59745-248-9_21, © Springer Science+Business Media, LLC 2011

Contrary to that, *animal* miRNAs targets are generally not completely complementary to their targets over the whole sequence of the miRNA and they do not generally lead to cleavage of their targets. Instead the animal miRNA:target interaction emphasizes base pairing between the 5′ end of the miRNA and the 3′ UTR of the target (6). The critical importance of bases 2–7 from the miRNA 5′ end, often called the seed or nucleus region, has been established through comparative genomic as well as experimental studies (7–9). However, detailed in vivo studies have also demonstrated several examples of unregulated mRNAs that have perfect seed sites and conversely regulated mRNAs that lack perfect seed sites (10).

1.2. Statistical Correlation from High-Through-Put Experiments

The statistically strongest argument for the "seed-emphasis" or "seed-only" model comes from expression array measurements. Lim et al. (11) has shown that 3′ UTRs for mRNAs that significantly drop in expression upon transfection of cell lines with miRNAs are highly enriched for a motif complementary to base 2–7 of the transfected miRNA (the seed region).

Similarly, in vivo, Krützfeld (12) has shown that blocking a miRNA with an antisense molecule leads to increased concentrations of a population of mRNA matching the seed sequences of the blocked miRNA.

While the statistics are strong for the dominating role of the seed in such microarray studies of alterations in mRNA concentrations, it should be kept in mind that the effect on translation is not captured. So far only one study has attempted to address this in a high throughput fashion. Vinther et al. (13) used proteomics to measure the concentration of 504 highly expressed proteins in a cell line transfected with an miRNA and found that more complex matching models (miRanda) had a more significant overlap with the experimental observation than the Seed-only model (TargetScanS (S for *s*eed)).

Summing up from the current data we can conclude that while perfect seed matching is a good way to predict miRNA targets, it is neither universally sufficient nor necessary.

2. Guidelines for Accessing Target Predictions

The questions that bioinformatics microRNA target prediction can help answer can be grouped into four types:

1. miRNA as query: "I have a known miRNA Y – what does it target?"
2. mRNA as query: "I have an mRNA X – which known miRNA(s) target it?"

3. Both mRNA and miRNA as query: "I have a miRNA Y that I think regulate a specific mRNA X, where could the target site be?"
4. I have found a new miRNA. What does it target?

Depending on the question asked, microRNA target predictions can be accessed in two fundamentally different ways: For questions of type 1 and 2 (and to some degree type 3) precomputed data available on the web provide easy access to predictions of known miRNAs on known mRNAs. In most cases such precomputed datasets are presented in user friendly web interfaces with hyperlinks to additional information and come with additional analysis of the conservation of the target sites, which presumably pinpoints the sites most likely to be physiological important. Precomputed targets are available for many but not all model organisms.

However, in some cases the precomputed target sites cannot be used: Perhaps the miRNA is newly discovered or from an organism for which precomputed targets are not available, or the researcher has a collection of target sequences not used in the precomputed target sets. In these cases de novo prediction of target sites is necessary. If the number of sequence combinations (miRNA and possible target sequence) to search is small, web services allowing the user to upload sequence data can be used. If on the other hand there are many sequences to search it can be advantageous to run the algorithm locally on the user's own computer. Presently, only a few methods (miRanda and RNAhybrid) are available as stand-alone-programs that can be downloaded and installed. The drawbacks of running predictions locally is that results are just displayed or saved as plain text, which can be harder to interpret and summarize than results from precomputed databases. Moreover, analysis of the phylogenetic conservation of the sites is not performed by any of the current downloadable programs.

Before moving on to the list of recommended prediction services, I want to emphasize that microRNA target prediction is a field in rapid development, new methods and new useful websites are continuously appearing, and this guide cannot be exhaustive. It is advisable always to check for new developments using Google and PubMed.

3. Precomputed Predictions

Databases of precomputed predictions are typically compiled and set up as supplementary websites to publications of specific target prediction algorithms.

General advantages: Easy and fast access through a web browser, hyper linked to and from other web resources. Often integrated with phylogenetic filtering of target sites.

General disadvantages: No predictions for novel miRNAs, tied to database creator's choice of possible target sequences and organisms.

3.1. MAMI

URL: **http://mami.med.harvard.edu/**

MAMI (Meta MiR:Target Inference) is a database that has compiled predictions from five different miRNA target prediction algorithms (TargetScanS, miRanda, microT, miRtarget, and picTar). The user can query with either a known human miRNA name or an mRNA identifier, and MAMI will present a list of predicted miR:Target interactions, indicating where the algorithms agree and disagree.

Advantages: five different predictions methods in one allow fast and easy comparison. Sensitivity and specificity can be adjusted.

Disadvantages: Only available for human miRNAs.

3.2. TargetScan.org

URL: **http://www.targetscan.org**

TargetScan.org presents results obtained by running TargetScanS (8) to search for phylogenetically conserved matches between miRNAs and 3′ UTRs. No information outside the seed-match is used in this method; hence miRNAs are collapsed into families with the same seed sequence. The method is simple because it does not provide a score for miRNA:target interaction, instead it ranks the predictions by the number of sites present in each 3′ UTR.

Advantages: This method has good statistical support from microarray measurements of the targets (11, 12).

Disadvantages: Cannot find targets without perfect seed match. Available for human, mouse, rat, dog, and worm only.

3.3. miRanda

miRanda (14, 15) is based on alignment between the miRNA and putative targets, with a scoring function emphasizing matching between the seed region of the miRNA and the target. This is followed by calculation of the binding energy between target and miRNA. Finally phylogenetic conservation filters are applied.

Prediction results from the miRanda algorithm are available at two different sites.

3.4. miRbase-Targets

URL: **http://microrna.sanger.ac.uk/targets/**

Advantages: Predictions are available for all species in **www.ensembl.org**, *p*-value provided for each predicted interaction following the principles of RNAhybrid (see below).

Disadvantages: This version does not find targets without perfect seed match.

3.5. Microrna.org

URL: **http://www.microrna.org**

Advantages: Possible to detect sites without perfect seed match. The miRanda program can be downloaded and installed locally on most computers.

Disadvantages: Precomputed targets are only available for human, fruit fly, and zebra fish; no *p*-values for predictions.

3.6. PicTar

URL: **http://pictar.bio.nyu.edu**

PicTar (16–18) finds perfect seed matches, the hybridization energy between the whole miRNA and the target is calculated, and unstable duplexes discarded. Using a maximum likelihood statistic PicTar then calculates the likelihood that a transcript is regulated by two or more miRNAs in combination.

Disadvantages: Cannot find targets without perfect seed match. Only available for human, mouse, fruit fly, and worm.

4. Prediction Servers

Prediction servers are websites that allow the user to upload one or more target sequences and one or more miRNA sequences, on which target prediction algorithms are then applied.

General advantages: The user can provide his/her own sequence, making the service very flexible, but harder to use.

General disadvantages: Not feasible to search large data sets. Output can be hard to interpret.

4.1. RNAhybrid

URL: **http://bibiserv.techfak.uni-bielefeld.de/rnahybrid/**

RNAhybrid (19) is an algorithm constructed to find the lowest free energy hybridization between two RNA molecules, i.e., the most stable binding site of a miRNA on a mRNA (assume there are no proteins present). It uses extreme value statistics inspired by BLAST (20) to calculate a *p*-value for the probability of observing a binding free energy lower than the observed binding site in random sequences of the same length. RNAhybrid allows parameters to be set to enforce a perfect seed match.

Advantages: The method is algorithmically and statistically well founded and guarantees to find the binding site with the lowest free energy of hybridization. Search parameters can be modified by the user. Provides graphics showing the duplex between miRNA and predicted target.

Disadvantages: The *p*-value for a predicted site is dependent on the length of target sequence searched (a site with the same sequence in a short and a long mRNA will get a lower p-value in the short sequence, reflecting that the site is less likely to appear "by random" in the short sequence). If this is not what you want, use the binding energy as a cutoff instead.

5. Evaluating Target Predictions

Most target prediction methods provide hundreds of possible targets for a single miRNA if, for example, all human mRNAs are searched for target sites. This raises the hard but important question: Among the predictions which are the relevant targets? No general recipe can answer this question. However, often application of external biological knowledge can lead to plausible and testable hypotheses: If a phenotype from knocking down the miRNA has been observed a good starting point is of course to look for predicted targets in pathways and proteins known to be involved in that phenotype, e.g., if abolishing expression of the miRNA leads to apoptosis look for targets in genes involved in apoptosis and cell cycle regulation. Biomedical literature and pathway databases such as KEGG and BioCarta can be helpful here.

References

1. Bentwich, I. (2005) Prediction and validation of microRNAs and their targets. *FEBS Lett* 579, 5904–5910.
2. Lindow, M., Gorodkin, J. (2007) Principles and limitations of computational miRNA gene and target finding. *Cell DNA Biol.* May, 26(5):339–51.
3. Yoon, S., De Micheli, G. (2006) Computational identification of microRNAs and their targets. *Birth Defects Res C Embryo Today* 78, 118–128.
4. Rhoades, M. W., Reinhart, B. J., Lim, L. P., Burge, C. B., Bartel, B., Bartel, D. P. (2002) Prediction of plant microRNA targets. *Cell* 110, 513–520.
5. Chen, X. (2005) MicroRNA biogenesis and function in plants. *FEBS Lett* 579, 5923–5931.
6. Lai, E. C. (2002) Micro RNAs are complementary to 3′ UTR sequence motifs that mediate negative post-transcriptional regulation. *Nat Genet* 30, 363–364.
7. Brennecke, J., Stark, A., Russell, R. B., Cohen, S. M. (2005) Principles of microRNA-target recognition. *PLoS Biol* 3, e85.
8. Lewis, B. P., Burge, C. B., Bartel, D. P. (2005) Conserved seed pairing, often flanked by adenosines, indicates that thousands of human genes are microRNA targets. *Cell* 120, 15–20.
9. Stark, A., Brennecke, J., Bushati, N., Russell, R. B., Cohen, S. M. (2005) Animal MicroRNAs confer robustness to gene expression and have a significant impact on 3′ UTR evolution. *Cell* 123, 1133–1146.
10. Didiano, D., Hobert, O. (2006) Perfect seed pairing is not a generally reliable predictor for miRNA-target interactions. *Nat Struct Mol Biol* 13, 849–851.
11. Lim, L. P., Lau, N. C., Garrett-Engele, P., Grimson, A., Schelter, J. M., Castle, J., Bartel, D. P., Linsley, P. S., Johnson, J. M. (2005) Microarray analysis shows that some microRNAs downregulate large numbers of target mRNAs. *Nature* 433, 769–773.
12. Krutzfeldt, J., Rajewsky, N., Braich, R., Rajeev, K. G., Tuschl, T., Manoharan, M., Stoffel, M. (2005) Silencing of microRNAs in vivo with 'antagomirs'. *Nature* 438, 685–689.
13. Vinther, J., Hedegaard, M. M., Gardner, P. P., Andersen, J. S., Arctander, P. (2006) Identification of miRNA targets with stable isotope labeling by amino acids in cell culture. *Nucleic Acids Res* 34, e107.

14. Enright, A. J., John, B., Gaul, U., Tuschl, T., Sander, C., Marks, D. S. (2003) MicroRNA targets in Drosophila. *Genome Biol* 5, R1.
15. John, B., Enright, A. J., Aravin, A., Tuschl, T., Sander, C., Marks, D. S. (2004) Human MicroRNA targets. *PLoS Biol* 2, e363.
16. Grun, D., Wang, Y. L., Langenberger, D., Gunsalus, K. C., Rajewsky, N. (2005) microRNA target predictions across seven Drosophila species and comparison to mammalian targets. *PLoS Comput Biol* 1, e13.
17. Krek, A., Grun, D., Poy, M. N., Wolf, R., Rosenberg, L., Epstein, E. J., MacMenamin, P., da Piedade, I., Gunsalus, K. C., Stoffel, M., Rajewsky, N. (2005) Combinatorial microRNA target predictions. *Nat Genet* 37, 495–500.
18. Lall, S., Grun, D., Krek, A., Chen, K., Wang, Y. L., Dewey, C. N., Sood, P., Colombo, T., Bray, N., Macmenamin, P., Kao, H. L., Gunsalus, K. C., Pachter, L., Piano, F., Rajewsky, N. (2006) A genome-wide map of conserved microRNA targets in C. elegans. *Curr Biol* 16, 460–471.
19. Rehmsmeier, M., Steffen, P., Hochsmann, M., Giegerich, R. (2004) Fast and effective prediction of microRNA/target duplexes. *RNA* 10, 1507–1517.
20. Karlin, S., Altschul, S. F. (1990) Methods for assessing the statistical significance of molecular sequence features by using general scoring schemes. *Proc Natl Acad Sci USA* 87, 2264–2268.

Chapter 22

Outsourcing of Experimental Work

Henrik Nielsen

Abstract

With the development of new technologies for simultaneous analysis of many genes, transcripts, or proteins (the "omics" revolution), it has become common to outsource parts of the experimental work. In order to maintain the integrity of the research projects, it is important that the interphase between the researcher and the service is further developed. This involves robust protocols for sample preparation, an informed choice of analytical tool, development of standards for individual technologies, and transparent data analysis. This chapter introduces some of the problems related to analysis of RNA samples in the "omics" context and gives a few hints and key references related to sample preparation for the non-specialist.

Key words: Deep sequencing, transcriptome, miRNA profiling, mass spectrometry.

1. Introduction

One of the most significant trends in molecular biology is the shift from studies of individual to large sets of genes, transcripts, and proteins (the "omics revolution"). To some extent, this development has been driven by technological advances, in particular within hybridization array and sequencing technologies as well as by developments in the field of bioinformatics. First of all, this is a positive development that leads to new insight into important phenomena in biology and medicine. However, the use of these new technologies has important implications for the way science is conducted. More specialists, in particular within bioinformatics are needed. While this is not a problem in itself, it is becoming more frequent that individual authors are unable to fully account for the paper they have co-authored. Another

H. Nielsen (ed.), *RNA*, Methods in Molecular Biology 703,
DOI 10.1007/978-1-59745-248-9_22, © Springer Science+Business Media, LLC 2011

problem is the cost of the instruments and even of the analyses. In some research institutions the problem is solved by sharing the instruments through establishment of core facilities. In other institutions, the researchers depend on private companies to perform the analyses. Outsourcing of all or parts of an experiment makes it more difficult for the investigator to make decisions about all aspects of the analysis and have a confident understanding of the data set produced. Moreover, there is a risk of diffusion of the responsibility if parts of the experiments are not conducted by a co-author of the resulting paper. This chapter is written to provide the non-specialist with a few hints on how to navigate in this situation and to introduce a few recent and very useful references.

The key area to follow is the development in sequencing technologies referred to as "deep sequencing," "massive parallel sequencing," or "Next-Generation Sequencing (NGS)." An overview of the currently most popular platforms is given in **Table 22.1**. For a recent and comprehensive review of the different technologies and their applications as well as guidelines for selection of technology for specific purposes, *see* (1). The sequencing technologies can deliver fast, inexpensive and accurate genomic information that serves as output for many types of experiments. The present range of RNA-related applications include cataloguing the transcriptomes of cells, tissues, and organisms (RNA-seq), genome-wide profiling of epigenetic markers and chromatin structure (ChIP-seq and methyl-seq), and mapping of transcripts associated with RNA-binding proteins (RIP-seq), but many more applications are likely to follow. One of the latest additions is sequencing of individual RNA molecules without the need for library construction and amplification (2). The main pitfall is that there are several serious problems with data collection and handling as discussed in (3). First, the platforms differ in chemistries and raw data collection. Thus, they have disparate output and

Table 22.1
Major sequencing platforms

Platform	Amplification	No. of reads	Average read length	Company homepage
Roche GS FLX titanium	Yes	>1 million	>400 bp	**http://www.454.com**
IlluminaGAIIx	Yes	200 million	75–100 bp	**http://www.illumina.com**
AB SOLiD3	Yes	400 million	50 bp	**http://www.appliedbiosystems.com**
Helicos HeliScope	No	400 million	25–35 bp	**http://www.helicosbio.com**

unique error profiles that makes combination of outputs from different platforms virtually impossible. Second, short reads can be difficult to align and thus to annotate unambiguously making the results difficult to handle for non-specialists. Finally, the shear amount of data generated presents a problem in itself, both in terms of handling and presentation. Given these problems, there is a risk that the scientific literature for many years to come will be flawed with datasets that cannot be scrutinized in the usual way.

From the perspective of the individual researcher with an interest in a particular biological problem, it may be difficult to embark on the new technologies. However, there is some really good news to start with. First, there is a chance that the experiment in question has already been done by others as part of large-scale efforts to annotate genomic information. The data are most likely accessible on the internet (e.g., through genome browsers), and are not being used for the purpose in mind. An example is information on, e.g., transcriptional activity and chromatin structure of the human genome generated in the ENCyclopedia Of DNA Elements (ENCODE; **http://www.genome.gov/10005107**) project. The data are deposited to public databases and are available for all to use without restriction. Data linked to the genomic sequence are stored and visualized on the University of California, Santa Cruz browser (**http://genome.ucsc.edu/ENCODE/**) Other, non-sequence based data, like that from microarray studies, are available on public databases such as the Gene Expression Omnibus (GEO; **http://www.ncbi.nlm.nih.gov/geo/**) and ArrayExpress (**http://www.ebi.ac.uk/microarray-as/ae/**). There has probably never been a time in the history of molecular biology where it was more obvious to generate hypotheses based on data from other researchers. If such a hypothesis leads to experiments involving the construction of gene specific tools, the second good news is that there is a chance that this tool is available from a company. There is an ever-increasing list of companies providing, large scale or even genome-wide collections of expression cell lines, reporter cell lines, cell lines with tagged genes, cellular or animal knockout models, antibodies directed toward the gene product, etc.

The preparation for outsourcing experimental work involves three steps. First, an informed decision has to be made on the appropriate analysis tool. Second, a robust method that yields a representative, non-biased source of nucleic acid material has to be implemented. And third, the data handling issue should be addressed. In the following section, advice is provided for the first and second step. The data handling issue is more difficult to generalize. One source of help is the discussion forum for the sequencing community, SEQanswers (**http://seqanswers.com/**).

2. Preparation of Samples for External Analysis

There are three sources of information that should be consulted. First, the information specified by the vendor should be considered. Unfortunately, some companies have relatively little experience in working with RNA and incorrect protocols and unnecessary precautions are not uncommon. Thus, it can be time-saving to negotiate some of the steps in the protocols provided by the vendor. A very useful source of information is the homepages of core facilities at research institutions (can be found by googling "Functional Genomics Center"). This is often an open source of information that is frequently updated and contains FAQ sections. Finally, some journals provide methods papers and reviews that compare different platforms and technologies. This is of course highly recommendable, but the literature is scarce and not always updated.

2.1. Preparation of dsDNA Samples

The input material in all of the current major sequencing technologies is double-stranded DNA provided as short fragments flanked by adapters of known sequence. This is referred to as a "sequencing library." Constructing such a library is relatively simple and requires no specialized equipment. The advantage of constructing the library compared to outsourcing is reduced costs and better control of the experiment. The steps involved have been excellently reviewed in (4) that also provide a general and detailed protocol for preparation of short paired-end libraries from genomic dsDNA for sequencing on the Illumina platform. Most of the steps apply to other applications as well, including RNA-seq. In brief, the first step is fragmentation of the DNA by physical or enzymatic methods into short (100–600 bp, depending on technology) fragments. Size-fractionation may be necessary at this stage. The fragmentation leaves single-stranded overhangs on the fragments and these needs to be enzymatically blunted. Next, technology-specific primers are ligated to the fragments. The adapter-ligated material is size-selected in order to eliminate concatemers and to produce a uniform library. This is done by agarose gel electrophoresis. Alternatives to standard agarose electrophoresis are available and allow easy extraction of the material from the gel and eliminate the use of ethidium bromide staining and UV exposure that damage the DNA. If the amount of input material is small, the sequence library has to be amplified by PCR. This is a critical step that introduces bias in the library. As pointed out by in (4), a final quality control and quantification of the library sample before the sequencing step is crucial. The amount of material should be quantitated by fluorometry using a fluorescent dye such as SYBR green or Picogreen. The quality of the sequencing step is sensitive to the concentration of the input DNA and quantification based on UV absorbance at 260 nm and the OD_{260}/OD_{280} ratio to assess sample purity is not applicable in this case because significant protein contamination of the sample will only change the ratio marginally. Cloning ad sequencing by conventional Sanger sequencing of a sample of the library can serve as a final check of library integrity before embarking on the much more costly massive parallel sequencing step.

The input material in all major sequencing technologies is a few μl of sample containing dsDNA in the low nM range. For fragments of a few hundred base pairs, this corresponds to less than one ng of DNA that needs to be amplified prior to sequencing. As mentioned above, PCR-amplification inevitably introduces bias in the library. However, if μg amounts of input material are provided, amplification can be avoided.

2.2. Preparation of RNA for Transcriptome Analysis

The aim of a transcriptome analysis is to determine the types and amounts of transcripts in a sample. Originally this was done by microarray analysis but sequencing (RNA-seq) appears to have surpassed microarrays for many aspects of transcriptome analysis. The main advantages of RNA-seq are that it can detect new transcripts, discriminate very similar variants, and exceed the dynamic range of microarray analysis. Although RNA-seq is the most popular application of sequencing technology in RNA biology, it is also the most complex. Transcripts differ in their 5′ and 3′ ends; they can be alternatively spliced, edited, and expressed from alleles that only differ slightly. Transcripts can comprise elements from distant locations in the genome and derive from both strands of DNA. All of these issues should be addressed at the experimental and/or the data treatment level.

A classical protocol for RNA-seq is provided by Mortazavi et al. (5). The steps that are involved in preparation of a sample for RNA-seq are not very different from those used in classical construction of a cDNA library. First, whole cell RNA is isolated, typically using a variation of the acidic phenol/guanidinium thiocyanate method (6). Then, two passes of oligo(dT)-based chromatography are used to purify poly(A)$^+$ RNA. cDNA synthesis is usually performed using random hexamer primers rather than oligo(dT) to avoid overrepresentation of the 3′ ends. The material has to be fragmented into smaller pieces prior to sequencing. This can be done at the RNA level by a brief incubation at 94°C in a slightly alkaline buffer with high (30 mM) Mg^{2+}-concentration (5) or at the cDNA level as described in **Section 2.1**. All of the subsequent steps are similar to those concerning preparation of dsDNA described in **Section 2.1**.

The amounts of RNA required for making RNA-seq depends on the transcripts of interest. The poly(A)$^+$ RNA fraction of whole cell RNA is in the range 1–4% dependent on cell type. Thus, less than 100 ng of whole cell RNA should be sufficient to produce the 1 ng of dsDNA required as the input material for a sequencing run. Given that a typical mammalian cell contains 10–30 pg of RNA this corresponds to a few thousand cells. However, this is a naïve calculation as transcripts vary several orders of magnitude in abundance and, as a consequence, much more RNA is required to obtain enough sequence depth to quantitate low abundance transcripts. Many biological questions require that fewer cells are being analyzed and this appears to be the goal of new developments, e.g., by direct sequencing of RNA molecules (2).

It is important to prevent loss of material due to adsorption to surfaces when handling small amounts of nucleic acid. "Non-stick" (e.g., siliconized or Teflon-coated) disposable plasticware should be used and detergent (e.g., 0.02% Tween-20) included in reaction steps. In precipitation reactions, glycogen should be included to facilitate recovery and handling of the nucleic acid. A convenient way of shipping nucleic acids is to perform an ethanol precipitation, remove most of the ethanol by aspiration, and leave the material as a wet (ethanol) pellet (dry pellets will not stay at the bottom of the tube). Then, the tube is wrapped in parafilm and shipped. DNA can be shipped at ambient temperature and this is also possible with RNA for some applications.

2.3. Preparation of RNA for miRNA Profiling

Transcriptome analysis of the miRNA fraction of RNA or "miRNA profiling" is an important type of analysis in biology and medicine. The principal methods are qRT-PCR, microarray hybridization, and deep sequencing. A very useful reference for comparison of the three methods is Git et al. (7). In this paper three biological samples were analyzed on six different microarray platforms and by sequencing, and 89 miRNA were further validated by qRT-PCR. The study discloses the strengths and weaknesses of each of the methods and provides an excellent example of the difficulties in dealing with genome-wide datasets.

The steps in preparation of samples for miRNA profiling generally involves isolation of whole cell RNA using a variation of the acidic phenol/guanidinium thiocyanate method (6) followed by size-fractionation to obtain the small RNA (<200 nt) fraction. There are several methods for this, e.g., denaturing polyacrylamide gel electrophoresis or differential elution from an RNA-binding (silica disc) matrix.

miRNA profiling typically starts with 1–10 μg of whole cell RNA. The low molecular weight fraction that is isolated is typically 10–15%. Most of this is 5S rRNA, 5.8S rRNA, and tRNA and the miRNA constitutes considerably less than 1%.

The comments on handling and shipping of samples are similar to those described in **Section 2.2**.

2.4. Preparation of Protein Samples

Protein identification, e.g., of an RNA-binding protein by mass spectrometry ("mass-spec" or MS) is relatively straightforward and is typically done as a service by a core facility or commercial service. These services have been around much longer than facilities handling nucleic acids for the purposes described above and sample preparation and data handling are quite simple. General proteomic analysis and characterization of proteins with carbohydrate, lipid, or other modifications is of course a different story, but this is beyond the scope of this chapter. The amount of protein required for identification is typically 50–100 ng. In order to obtain this amount for a 50-kDa protein found in 1,000 copies per cell, 20 mg of total protein from 1.2×10^9 cells are required as starting material provided 100% efficiency of all procedures used. The protein is provided as a stained gel band or spot stained with a MS-compatible stain. Coomassie blue stain has a detection limit (100 ng) corresponding to the amount required for MS-analysis and is a good choice. Tips for sample preparation can be found at the homepages of MS-services, e.g., **http://www.albany.edu/genomics/proteomics-sample-preparation.html**.

Reference

1. Metzker, M. L. (2010) Sequencing technologies – the next generation. *Nat Rev Genet* 11, 31–46.
2. Ozsolak, F., Platt, A. R., Jones, D. R., Reifenberger, J. G., Sass, L. E., McInerney, P., Thompson, J. F., Bowers, J., Jarosz, M., Milos, P. M. (2009) Direct RNA sequencing. *Nature* 461, 814–818.
3. McPherson, J. D. (2009) Next-generation gap. *Nat Methods* 6, S2–S5.
4. Linnarsson, S. (2010) Recent advances in DNA sequencing methods – general principles of sample preparation. *Exp Cell Res* 316, 1339–1343.
5. Mortazavi, A., Williams, B. A., McCue, K., Schaeffer, L., Wold, B. (2008) Mapping and quantifying mammalian transcriptomes by RNA-Seq. *Nat Methods* 5, 621–628.
6. Chomczynski, P., Sacchi, N. (1987) Single-step method of RNA isolation by acid guanidinium thiocyanate-phenol-chloroform extraction. *Anal Biochem* 162, 156–159.
7. Git, A., Dvinge, H., Salmon-Divon, M., Osborne, M., Kutter, C., Hadfield, J., Bertone, P., Caldas, C. (2010) Systematic comparison of microarray profiling, real-time PCR, and next-generation sequencing technologies for measuring differential microRNA expression. *RNA* 16, 991–1006.

Subject Index

H. Nielsen (ed.), *RNA*, Methods in Molecular Biology 703,
DOI 10.1007/978-1-59745-248-9, © Springer Science+Business Media, LLC 2011

MIX
Papier aus verantwortungsvollen Quellen
Paper from responsible sources
FSC® C105338

If you have any concerns about our products,
you can contact us on
ProductSafety@springernature.com

In case Publisher is established outside the EU,
the EU authorized representative is:
Springer Nature Customer Service Center GmbH
Europaplatz 3, 69115 Heidelberg, Germany

Printed by Libri Plureos GmbH
in Hamburg, Germany